Information Systems Engineering and Management

Volume 9

The book series "Information Systems Engineering and Management" (ISEM) publishes innovative and original works in the various areas of planning, development, implementation, and management of information systems and technologies by enterprises, citizens, and society for the improvement of the socio-economic environment.

The series is multidisciplinary, focusing on technological, organizational, and social domains of information systems engineering and management. Manuscripts published in this book series focus on relevant problems and research in the planning, analysis, design, implementation, exploration, and management of all types of information systems and technologies. The series contains monographs, lecture notes, edited volumes, pedagogical and technical books as well as proceedings volumes.

Some topics/keywords to be considered in the ISEM book series are, but not limited to: Information Systems Planning; Information Systems Development; Exploration of Information Systems; Management of Information Systems; Blockchain Technology; Cloud Computing; Artificial Intelligence (AI) and Machine Learning; Big Data Analytics; Multimedia Systems; Computer Networks, Mobility and Pervasive Systems; IT Security, Ethics and Privacy; Cybersecurity; Digital Platforms and Services; Requirements Engineering; Software Engineering; Process and Knowledge Engineering; Security and Privacy Engineering, Autonomous Robotics; Human-Computer Interaction; Marketing and Information; Tourism and Information; Finance and Value; Decisions and Risk; Innovation and Projects; Strategy and People.

Indexed by Google Scholar. All books published in the series are submitted for consideration in the Web of Science.

For book or proceedings proposals please contact Alvaro Rocha (amrrocha@gmail.com).

Soumi Dutta · Álvaro Rocha ·
Pushan Kumar Dutta · Pronaya Bhattacharya ·
Ramanjeet Singh
Editors

Advances in Data Analytics for Influencer Marketing: An Interdisciplinary Approach

Springer

Editors
Soumi Dutta
Sister Nivedita University
Kolkata, West Bengal, India

Álvaro Rocha
University of Lisbon
Lisboa, Portugal

Pushan Kumar Dutta
Amity University
Kolkata, West Bengal, India

Pronaya Bhattacharya
Amity University
Noida, Uttar Pradesh, India

Ramanjeet Singh
Assam Down Town University
Guwahati, Assam, India

ISSN 3004-958X ISSN 3004-9598 (electronic)
Information Systems Engineering and Management
ISBN 978-3-031-65726-9 ISBN 978-3-031-65727-6 (eBook)
https://doi.org/10.1007/978-3-031-65727-6

This Springer imprint is published by the registered company Springer Nature Switzerland AG
The registered company address is: Gewerbestrasse 11, 6330 Cham, Switzerland

Contents

Introduction to Influencer Marketing and Data Analytics 1
Uttam Kaur and Prashant Kumar Siddhey

The Evolution of Influencer Marketing 17
Aun Ahmed and Tushar Rathore

From Word-of-Mouth to AI-Powered Engagement: The Dynamic Evolution of Influencer Marketing in the Digital Era 29
Sonal Purohit and Rakhi Arora

Revolutionizing Influencer Marketing: Harnessing the Power of Data Analytics and Artificial Intelligence (AI) 41
Khalid Aziz, Sakshi Dua, and Prabal Gupta

Influencer Marketing Unleashed: Leveraging Data Analytics for Success 67
Divya Bansal, Naboshree Bhattacharya, and Priyanka Shandilya

Data Analytics and AI for Influencer Marketing 89
Nurul Asmak Md. Lazim and Sharifah Hafizah Syed Ariffin

Interdisciplinary Data Analytics Transforming Influencer Marketing Strategies 103
Pawan Whig, Jhansi Bharathi Madavarapu, Nikhitha Yathiraju, and Ramya Thatikonda

Influencer Marketing Evolution: Unveiling the Power of Data Analytics and AI for Strategic Campaign Excellence 125
Rahila Khan and Shalini

Sustainability and Influencer Marketing: Measuring the Gap Between Consumer Attitudes and Purchase Behavior 149
M. L. Vasita and Vineet Chouhan

An Audience Threshold in the Millions: Exploring Influencer Marketing Activations Online ... 165
Brandon Chicotsky and Ken Corbit

Metaverse: A New Frontier for Influencer Marketing ... 181
Amandeep Dhaliwal

Influencer Marketing on New Platforms and the Metaverse for Male Grooming Brands ... 197
Nidhi Sharma, Anita Singh, and Anubha Vashisht

Impact of Influencer Marketing Through Instagram Among Youth: A Study ... 215
Pooja Verma and Monica Sainy

E-Commerce Platforms Extend Beyond Selling. A Case Study on India ... 229
Sulagna Das

Analyzing the Effectiveness of Influencer Campaigns in Driving Customer Acquisition and Retention: Perspectives from Unilever Pakistan ... 241
Arman Khan, Zahid Hussain, Muhammad Ussama Majeed, and Amarnath Padhi

Analysing the Impact of Hosting Influencer Giveaways on Purchase Intention of Luxury Items by Generation Z ... 257
Nibir Khawash, Prasenjit Chakravarty, and Sudeshna Pahari

Innovative Fusion: Attention-Augmented Support Vector Machines for Superior Text Classification for Social Marketing ... 283
Raghavendra M. Devadas, Vani Hiremani, J. Praveen Gujjar, N. Shobha Rani, and K. R. Bhavya

Fuzzy Artificial Intelligence as a Technique to Find Relative Desirability for Fake Review Detection ... 305
A. Firos and Seema Khanum

A Study on the Influence of Artificial Intelligence on Brain Plasticity with Reference to Consumer Behaviour and Workplace Dynamics ... 323
K. Umashankar and H. G. Charitra

Influential Asanas: An In-Depth Study Exploring the Dynamics of Yoga Influencers ... 341
Shilpa Sarvani Ravi, Shikha Bhagat, and Rashmi Rai

Online User Choices Data Analytics and Privacy in Digital Marketing: Legal-Ethical Landscapes in the Digital-Futuristic Arena Projecting Multiple Layers of Artificial Intelligence and Blockchain ... 359
Bhupinder Singh, Pushan Kumar Dutta, and Christian Kaunert

The Pinnacle of Influence: Unveiling the Evolution, Global Dynamics, and Multidisciplinary Tapestry of Social Media Influencer Marketing (2016–2024) ... 377
Anisha Arora and Prashant Kumar Siddhey

'Influencer Relevance Index'—A New Parameter for Selecting Influencers from Discovery Tools by Startups Based on Campaign Goals ... 397
Himadri Barman and Prakrit Saikia

Ethical Dimensions of AI Integration in Influencer Campaigns: Balancing Authenticity, Targeting Precision, and User Privacy ... 425
Ashwani Panesar and Indu Saini

Introduction to Influencer Marketing and Data Analytics

Uttam Kaur and Prashant Kumar Siddhey

Abstract The formation of the Internet came before notable alterations in the surroundings that have had a profound impact on people's conduct. Customization is becoming more and more necessary for goods and services to get a competitive edge in the fiercely competitive market of today. A new framework for marketing operations has been provided by the advancement of communication technology. Marketing has emerged as one of the strategic functions of a firm and has become essential to modern business. A whole new idea in marketing called e-marketing is emerging quickly with the development of social networking sites. People are actively creating material on networking sites, and these platforms have altered the way that we interact. According to this innovative method of marketing, "influencers" have a far bigger effect on consumers. The research investigates peoples view and whether or not that perception influences their decisions. Consequently, the aim of this article is to shows the influence of individuals' behavior and ascertains how these effects affect customers' purchasing decisions.

Keywords Influencer Marketing (IM) · Consumers · Behavior · Social media · Social networking sites · Electronic marketing

1 Introduction

Worldwide interaction between individuals has benefited greatly from the growth of ICT [1]. The Internet's beginning, which not only signifies an evolution in communication but also the handling of relationships, is unquestionably the most important addition to the field of ICT [2]. The network is currently one of the most powerful social media channels, and major force behind the process of internationalization [3]. In addition to the internationalization already discussed, ICT and the Internet have typically boosted commercial activity [4]. The industrial climate has

U. Kaur · P. K. Siddhey (✉)
Chandigarh University, University School of Business, Gharuan 140413, Punjab, India
e-mail: drpksiddhey@gmail.com

S. Dutta et al. (eds.), *Advances in Data Analytics for Influencer Marketing: An Interdisciplinary Approach*, Information Systems Engineering and Management 9,
https://doi.org/10.1007/978-3-031-65727-6_1

grown increasingly competitive, and the globe is now seen as an attractive marketplace from an international perspective [5]. Nevertheless, despite the more changing circumstances, internet access turned out to be a major potential for businesses of all kinds, making electronic marketing an essential part of marketing in the twenty-first century [6]. When associated to certain conventional strategies of marketing, the pool of consumers available through electronic marketing is incomparable [7]. Famous people have long been employed by marketers to raise awareness of and opinion of their brands; electronic marketing has built on a similar idea. Influencers are individuals who possess the information, abilities, and personality to alter an individual or team's views without being in an instance of power [8]. IM is becoming a more prevalent method of advertising in today's world. Thus, it may be inferred that IM is the skill of enlisting social media influencers to disseminate paid advertising that conveys the message of a company to their intended consumers. Thus, the objective of this investigation is to assess the degree to which influencers affect customers' conduct.

2 The Essential Element of the Twenty-First Century Advertising is Electronic Marketing

An essential component of effective commercial organizations' operations is the application of IT, particularly at the level of the worldwide economy [9]. E-commerce and the building and processing of databases of prospective customers were made possible by the advancement of IT, and these developments gave rise to new forms of marketing known as digital online advertising [10]. The amount of time the Internet has been used for marketing has increased tremendously since the first online advertising appeared in 1994.The accuracy of the procedure that is founded on digital foundation is known as digital advertising, or online marketing [11]. Additionally, online advertising is a process performed by business organizations on an online marketplace through which you may identify customer requirements, engage with market participants, and make offers using modern IT [12]. The organization's attempt to educate, promote, and distribute goods and services online is known as e-marketing [13]. While some define e-marketing as a more inclusive term including database advertising and e-client relationship management (eCRM), others define online advertising as the application of online platforms and other technological advances (such as cellphones) in conjunction with traditional techniques to achieve advertising objectives [14]. The writers may differ in their definitions of electronic marketing, but they all recognize that, in contrast with conventional media, online platforms allow for targeted communication to fewer people. As a result, advertising campaigns through electronic media tend to be more shifting, specifically divided, and build long-term connections with customers [15]. Due to the abundance of advertising methods made available by online technology, different phrases that precisely define certain electronic marketing techniques—which are mostly focused on advertising and marketing strategies—can be found in research [16]. Among the phrases that are most frequently used are online marketing, email advertising, periodicity marketing, individual marketing, social media marketing, referral advertising, and instantaneous marketing [17]. Additionally, there's blogging marketing, internet marketing, influencer advertising, content promotion, marketing for search engines, and more [18].

3 Establishment of Customer Connection in Electronic Marketing

Substantialmodifications in the digital environment are related with the dissimilar customer attitude [12]. Specifically, conventional techniques for maintaining consumer connections, increasing social networking sites to interact and, as such, view supporters as possible customers [19]. Businesses examine how to control "supporters" and turn them into customers—or, better still, devoted customers—from the aforementioned vantage point [20]. Therefore, we may draw the conclusion that although technology allows for a better considerate of customer behavior, it also generatesa dilemma from the time whenclientsuse similar method to regulate how they engage with businesses and organizations [21]. The other issue arises from the fact because electronic media is unquestionably playing a significant role in the information going viral. The explanation for this is that social networking makes it simple and quick for various groups of individuals to share material with one another, spreading viral content around the world [22]. Due to the complexity of controlling that type of marketplace environment, businesses are compelled to substitute conventional advertising campaigns with electronic marketing initiatives in order to at least partially regulate the subject matter of messages sent and then assess their effectiveness [12]. By assessing the environment in this way, businesses take into account the behaviors of different customer segments in their digital environments and attempt to reach the major customers collectively with a focus on communication; in simple terms, they are responding with particular techniques and approaches aimed at particular customers [23]. In addition to the aforementioned endeavors, businesses are focusing on keeping an eye on customer exchanges of values and reciprocal connections. The sharing of electronic content and conducting of business amongst consumers (for example, through online groups) represent a few of them [24, 25]. Entertaining and innovative material may be created and shared in a digital environment. The most tangible illustration of it is online advertising, which encompasses all actions taken to disseminate the intention of disseminating an idea via email, social networking sites, text messages, and MMS (Multimedia among other mediums [26, 27]. This creative tactic, which is based on the idea that consumers find non-marketing knowledge sources—like companions and acquaintances—more pertinent than advertising information—like company advertisements—produced various messages that have since spread within the targeted group [20, 28]. The recipient assumes the role of both transmitter and recipient and the information spreads swiftly. Studying how a digital environment functions in that way leads one to conclude that the investigation is concentrated on serving the requirements and desires of customers, which represents one of the key factors influencing the handling of client relationships [19]. It is possible to forecast upcoming customer responses to the indication of novelgoods based on understanding consumers' wants. Subsequently, due to the much strong cooperation and interaction in conventional customer relationship management, the consumer isn't a member of the customer relationship management [22].

4 The Importance of Social Media in Customer Relationship Management

Social media has revolutionized communication between people, groups, and organizations today [29]. In general, individuals find value in using social media to network and stay in touch with colleagues, share material, communicate, collaborate, and gather updates, data, and reviews [30]. Through this, information may take the shape of written words, images, audio files, or video recordings. Social media is an interactive medium that facilitates engagement with people in the community, in contrast with conventional media, which is a tool for one-way communication like radio, TV, newspapers, and journals. As a response to what they have seen, people can publicly post their feedback [5]. Thus, the media aids companies in entering novel markets, give advertisements characters, and establish a corporate identity. They also support business branding and regularly publish information that attains the organization's advertising section, raising the public's understanding about the items and enticing them to follow [31]. Even though there are numerous distinct types of social networking sites, the most widely used ones include websites and networking websites where digital content is exchanged, websites where votes and marking are done, evaluation pages, discussions, and interactive environments. Innovative use of Internet technology is made possible by social media [6]. Therefore, to draw in and engage a clearly defined target audience and promote lucrative consumer behavior, marketing and commercial operations including the creation and distribution of value and appealing content are enabled [32]. Large audiences, branding, building relationships with customers, enhancing business operations (through user feedback, hiring quality staff, finding new suppliers, etc.), boosting search engine rankings, the potential for sales via social media, alternate advertising opportunities, and other gains are all advantages that companies can gain from using these media. The content plays a crucial part in this kind of marketing and offers customers value in a variety of ways. For any material to reach its intended audience, it must be delivered through certain channels [33]. One can utilize search engines, blogs, networking sites, electronic mail, and other platforms as Internet conduits. In conclusion, as social networks are being used by individuals all over the world to interact, the best argument for moving customer relationship management to social media is the presence of their clientele [4]. Social media may be utilized not just to gather complaints and reward customers for their loyalty, but also to realize greater value for items. Professionals in advertising can now pinpoint customers with remarkable precision and establish a connection between them and the business through relevant and genuine conversations on social media sites. Real-world conversations, or more specifically, current events, trends, and topics, serve as an inspiration for the material created for immediate advertising [34]. The introduction of social media and the growth of digital platforms have significantly changed the marketing landscape in recent years. Influencer marketing is a significant trend that has arisen from this change. It is a potent tactic that uses people's reach and influence on social media to advertise goods and services. Simultaneously, optimizing and gauging the success of marketing initiatives has made the incorporation of data analytics increasingly important. This thorough investigation, explores the fields of data analytics and influencer marketing, revealing the complex interactions between these two dynamic forces that are changing the marketing landscape.

5 Influencer Marketing

5.1 The Evolution of Influencer Marketing

Influencer marketing originated as a result of social media platforms' explosive growth. With the rise in popularity of social media sites such as Instagram, YouTube, and TikTok, people with distinctive voices, skills, or specialized knowledge started gaining significant fan bases. These social media influencers have become essential in influencing the views and choices of consumers, offering companies a unique chance to engage with their target market in a more genuine and approachable way. The influence's power: Personal ties are crucial for the success of influencer marketing [35]. Influencers act as a trusted and recognizable presence between companies and customers, bridging the gap created by traditional advertising, which frequently feels impersonal. Influencers are viewed by audiences as sincere people who offer their thoughts, advice, and experiences. Brands benefit from increased engagement and conversion rates as a result of this authenticity, which converts into a potent persuasive strategy [36]. Influencers are diverse individuals who serve a certain market or group of people. Micro-influencers provide a close relationship with their audience because of their tiny but very active followings. However, because of their enormous reach, macro-influencers are a good fit for larger initiatives. The target demographic, intended degree of engagement, and brand objectives all play a role in the influencer selection process. Many firms have found amazing success by utilizing influencer marketing to its fullest [37]. There are several case studies illustrating how smart partnerships with influencers have helped companies reach new heights in a variety of industries, including fashion, beauty, technology, and lifestyle. Analyzing these success stories offers insightful knowledge about the science and art of running successful influencer marketing programmes [38].

Famous people have long been a tool utilized by sellers to upgrade product appreciation and image. The individuals are likely to belief superstars whom they desire to be like. While the concept of instant messaging remains the same, social media influencers—celebrities representing the online social media world—are utilized instead of renowned actors, actresses, athletes, or singers. Influential individuals can be hired on social apps, although the majority of them get money on YouTube and Instagram, based on the kind of company [3]. An "influencer" might be any person who possesses a certain degree of influence. Those with the capacity to mold the views of peers, and irregularly even supervisors are known as influencers based just on their expertise, skills, and personality and not their position in the organization [39]. A large following on social media and being more fascinating to a larger audience, celebrities are not always trustworthy about promoting goods or services. It is important to emphasize that influencers not only aid in developing brand awareness and drawing in new consumers, but also foster community building, action initiation, concept strengthening, and loyalty development [40]. However, combining influencers with CRM may drastically alter marketing initiatives. Because of their high level of credibility and affordability when compared to other forms of advertising, influencers are regarded by customers as more trustworthy and persuasive when they are promoting products or services [41]. This is especially true when the influencer is powerful. Influencer marketing is thought to be more reliable than celebrity marketing when it comes to social media, especially for businesses catering to younger consumers [30]. According to marketing terminology, influencers act as intermediaries between businesses and their target audiences, providing a voice those followers, potential customers, and influencers may all trust [42].

6 The Convergence of Influencer Marketing and Data Analytics

Traditional analytical tools could no longer handle the massive volume of data created by online interactions as digital marketing expanded. A new age of actionable insights from the mountains of data available to marketers was brought about with the introduction of big data and sophisticated analytics [43]. The utilization of data-driven methodology has demonstrated its invaluable nature in comprehending customer behavior, refining marketing tactics, and calculating return on investment (ROI). Data Analytics's Place in Influencer Marketing Influencer marketing relies heavily on data analytics because it provides a methodical, quantitative approach to campaign strategy, implementation, and assessment [44]. Data may be used by marketers to determine which influencers are most relevant to their target audience, evaluate campaign effectiveness in real-time, and adjust strategy based on useful information. Influencer marketing and data analytics work together to provide companies with the capacity to make wise choices and optimize the results of their marketing initiatives [45]. A multitude of metrics and key performance indicators are used to assess the effectiveness of influencer marketing efforts. These metrics include, but are not limited to, reach, engagement rate, click-through rate (CTR), conversion rate, and return on advertising spend (ROAS). Marketers may track these variables, examine patterns, and refine their approaches to get the best outcomes by using data analytics [46].

7 Challenges and Ethical Considerations

Influencer marketing has a lot of potential, but it also has drawbacks. The sector deals with problems including influencer fraud, false followers, and possible inconsistencies between influencer behavior and brand ideals [47]. Choosing wisely when choosing influencers and putting in place reliable assessment systems are crucial for overcoming these obstacles. Concerns about data privacy and consent are brought up by the rising dependence on data analytics [48]. Marketers need to put ethics first and make sure that data collection and analysis follow privacy laws. Retaining credibility and trust requires open and honest communication regarding data usage with influencers and their audiences [49].

8 Future Trends and Innovations

The field of data analytics and influencer marketing will unavoidably change as technology develops. Future marketing trends to watch include virtual influencers, augmented reality (AR) campaigns, and immersive experiences [30]. Comprehending these patterns is crucial for marketers who aim to maintain a competitive edge and fully utilize influencer marketing and data analytics. It makes sense to combine data analytics and influencer marketing with e-commerce [48]. The integration of shoppable content, affiliate marketing links, and data-driven personalized recommendations facilitates a smooth and effortless consumer experience from discovery to purchase. This connection gives advertisers a thorough grasp of consumers' purchasing paths in addition to improving the shopping experience [42].

9 Examining How Influencers Affect Customer Behaviour

The study focused on how people felt about influencers and how such changes affected their conduct. The primary inquiry that guided the investigation was whether influencers significantly affect customers' purchasing decisions and behavior. Thus, the aim of this investigation is to study the significance and function of influencers on customer behavior. Based on the above-indicated purpose the following hypothesis emerged:

H: IM positively influence consumers' behavior and buying choices.

It is assumed that influencers' impact and customers' purchasing behavior are strongly, positively, and correlatively related. The questionnaire for the survey was used to gather the data needed for this study.

10 Research Methodology

The aim of the investigation was to conclude the degree to which influencer recommendations can alter the opinions and purchasing behaviors of customers. The investigation was carried out utilizing the original investigation outcomes. The data were gathered from 468 people'sutilizing a standardized survey questionnaire. Multiple variables were utilized in the investigation, which were associated by quantitative indicators. The objective of the study was to understand customers' responses to influencer advice about how to utilize everyday consumer products. Using a descriptive statistical analysis, the survey questionnaire's primary information was processed. The degree of connection between the influencer's effect and the conduct of the consumer—which determines whether the hypothesis is accepted or rejected—was represented with the use of descriptive analysis.

11 Analysis and Discussion

The principal goal of the conducted study was to gain an understanding of customers' perceptions about influencers. A total of 468 survey forms were gathered. The study population was a convenience sample, and tables were used to display the sample's structure. Further study discarded the indicators of 51 participants because it is evident from the information contained in the obtained data that they were not influenced by any influencer. The social media platforms that participants most frequently utilize or go to seek knowledge from influencers were analyzed next. The outcomes are shown in the diagram. An analysis of the duration of time that people spend on social media platforms reveals that the majority of individuals spend between two and four hours a day on average, with photos and videos being the most popular items to view during that time (Tables 1, 2 and Figs. 1, 2).

The parameters are shown in the table below. The next series of inquiries reveals the extent to which influencers affect individuals' purchase habits. The following are the indications (Table 3).

Based on the conducted study, it can be inferred that customers' attention is positively impacted by influencer statements, even if the customers do not believe that the influencers affect their purchase behavior. On the other hand, people occasionally

Table 1 Demographic pofile (Gender)

Gender	Frequency	Percentage (%)
Female	149	31.84%
Male	319	68.16%
Total	468	

Table 2 Demographic profile (Age)

Age groups	Frequency	Percentage (%)	Follow influencers	Percentage (%)
20–29	116	24.78	116	100%
30–39	128	27.35	128	100%
40–49	94	20.09	89	94.68%
50–59	81	17.31	56	69.14%
Above 60	49	10.47	28	57.14
Total	468		417	

purchase the influencer-recommended items, actively seek them, andcomment on and suggest them to their colleagues groups. Consequently, it can be said that influencers' messaging reaches beyond their fan base to a far wider audience of interested parties through word-of-mouth marketing (Table 4).

The most important finding from this study is that, when deciding between competing products, 98% of respondents say they choose those recommended by influencers. This indicates that, indeed, IM has a substantial impact on customer buying choices and behavior, supporting the study's hypothesis.

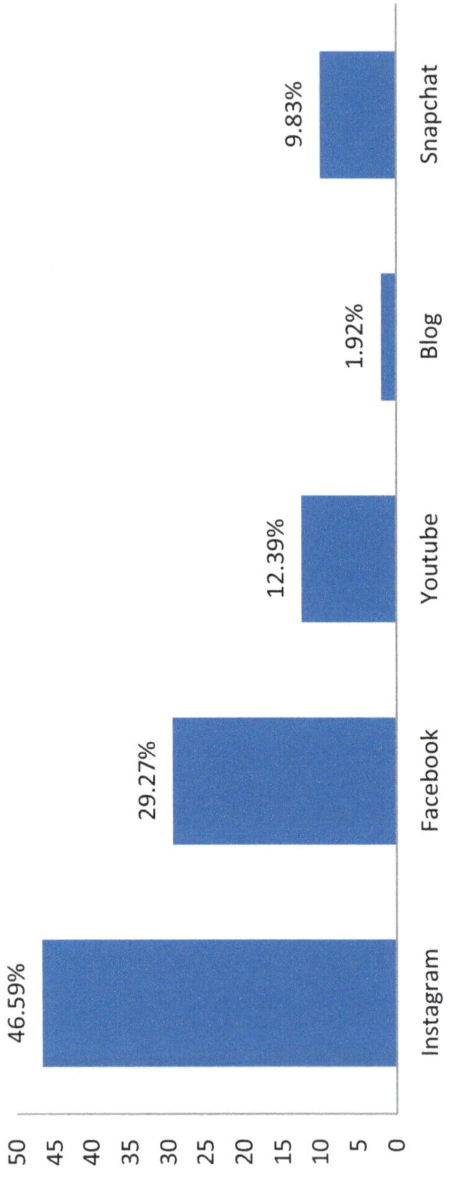

Fig. 1 Social networking sites that participants utilize most often

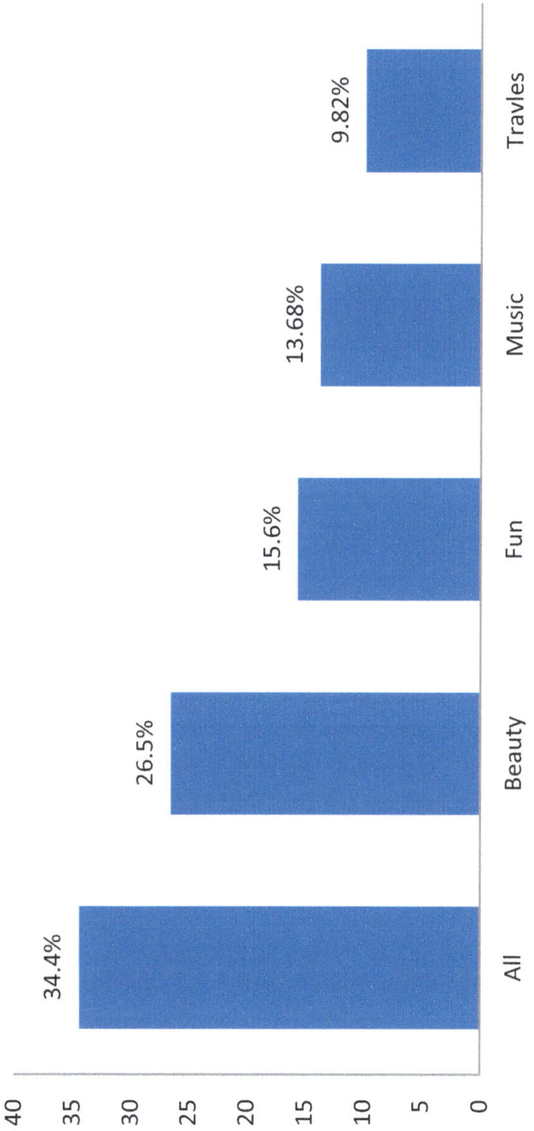

Fig. 2 Focus while selecting influences

Table 3 Parameters from the survey

	Less than 2 h	2–4 h	More than 4 h
Time spend on social media	162	239	16
	Videos	Texts	Photographs
Type of preference	126	208	83

Table 4 Survey parameters

	Yes	No
Do additional updates have an effect on the influencer receiving greater attention?	342	75
Do the public statements made by influencers affect your purchasing behavior?	329	88
Do you look for items that influencers utilize and suggest?	347	70
Do you purchase items that influencers utilize and suggest?	412	5
Do you talk about things that influencers use and suggest with the people you care about?	394	23
Do you encourage items that influencers utilize and suggest?	352	65
Do you select a product suggested by influencers above those offered by competitors?	368	49

12 Conclusion

The research question that motivated the study done for this project was: Does influencer marketing significantly affect how customers behave and make decisions about what to buy? The purpose of hypothesis H was to verify whether or not there is a positive correlation between the influencer and the customer's opinion of the product, and therefore the possible incentive for the customer to make a snap choice when making a purchase. Specifically, the statistics described above clearly show that customers react strongly to influencer referrals, support hypothesis H. Consequently, the investigation's conclusions provide fresh insight into how a business can behave strategically when rising an electronic marketing campaign that centers on influence marketing. Influencer marketing has a widespread effect on consumer behavior and buying choices. Social networking channels continue to rule the digital sphere, and influencers have significant power to shape the thoughts and tastes of their followers. Influential people's accessibility and genuineness to advertisements create a special bond with customers that build credibility and trust. Influential people's ability to provide goods and services in authentic settings, together with this relationship, is crucial in influencing people's choices to buy. Influencer marketing is dynamic, which not only mirrors the changing customer landscape but also highlights how effective it is at increasing interaction and, eventually, influencing the decisions customers make within a crowded market.

References

1. Bognar, Z.B., Puljic, N.P., Kadezabek, D.: Impact of influencer marketing on consumer behavior. Econ. Soc. Dev. Book Proc. 301–309 (2019)
2. Kadekova, Z., Holienčinova, M.: Influencer marketing as a modern phenomenon creating a new frontier of virtual opportunities. Commun. Today **9**(2) (2018)
3. Vrontis, D., Makrides, A., Christofi, M., Thrassou, A.: Social media influencer marketing: a systematic review, integrative framework and future research agenda. Int. J. Consum. Stud. **45**(4), 617–644 (2021)
4. Gamage, T.C., Ashill, N.J.: Sponsored-influencer marketing: effects of the commercial orientation of influencer-created content on followers' willingness to search for information. J. Product Brand Manag. **32**(2), 316–329 (2023)
5. Pick, M.: Psychological ownership in social media influencer marketing. Euro. Bus. Rev. **33**(1) (2021)
6. Chopra, A., Avhad, V., Jaju, A.S.: Influencer marketing: an exploratory study to identify antecedents of consumer behavior of millennial. Bus. Perspect. Res. **9**(1), 77–91 (2021)
7. Nadanyiova, M., Gajanova, L., Majerova, J., Lizbetinova, L.: Influencer marketing and its impact on consumer lifestyles. Forum Scientiae Oeconomia **8**(2), 109–120 (2020)
8. Bu, Y., Parkinson, J., Thaichon, P.: Influencer marketing: Homophily, customer value co-creation behaviour and purchase intention. J. Retail. Consum. Serv. **66**, article number 102904 (2022)
9. Lubis, A.N., Lumbanraja, P., Hasibuan, B.K.: Evaluation on e-marketing exposure practice to minimize the customers' online shopping purchase regret. Cogent Bus. Manag. **9**(1), article number 2016039 (2022)
10. Singhal, M., Gupta, A.: E-Marketing as a tool for corporate success in 21st Century. Int. J. Adv. Eng. Manag. (IJAEM) **4**(5), 380–386 (2022)
11. Priansa, D.J., Suryawardani, B.: Effects of E-marketing and social media marketing on E-commerce shopping decisions. JurnalManajemen Indonesia **20**(1), 76–82 (2020)
12. Abrar, M., Safeer, A.A., Baig, S.A., Ghafoor, A.: E-marketing and its implementation on developing social perceptions of customers through effective marketing mix of organic textile products. Am. J. Ind. Bus. Manag. **6**(12), 1194–1211 (2016)
13. Lamichhane, B.: Impact of digital marketing on consumer behavior in Pokhara. KIC Int. J. Soc. Sci. Manag. **1**(1), 13–23 (2022)
14. Al-Ababneh, H.A.: Researching global digital E-marketing trends. Eastern-Euro. J. Enterprise Technol. **1**(13), 115 (2022)
15. Herwade, M.A.R., Shendage, R.N.: Digital marketing: 21st century challenges and opportunities. Int. J. Adv. Appl. Res. **9**(3), 558–568 (2022)
16. Al-Haraizah, A., Mubarak, A., Alshidi, A.K.: The role of E-marketing tools in constructing online consumer behavior of Omani Youth in Muscat City. J. Critical Rev. **7**(13), 1270–1280 (2020)
17. Adam, M., Ibrahim, M., Idris, S., Saputra, J., Putra, T.: An investigation of e-marketing and its effect on the consumer buying decision during COVID-19 pandemic in Aceh Province, Indonesia: a mediating role of perceived risk. Int. J. Data Netw. Sci. **6**(1), 115–126 (2022)
18. Babalola, H.B., Lateef, S.A., Zekeri, A.A.: New trends of intelligent e-marketing and consumer buying behaviour: a study of selected universities in Osun State, Nigeria. JurnalAplikasiManajemen, Ekonomi Dan Bisnis **5**(1), 14–25 (2020)
19. Ipang, I., Suroso, S., Novitasari, D.: A study on the relationship of e-marketing, e-CRM, and e-loyalty: evidence from Indonesia. Int. J. Data Netw. Sci. **5**(2), 115–120 (2021)
20. Waheed, A., Jianhua, Y.: Achieving consumers' attention through emerging technologies: the linkage between e-marketing and consumers' exploratory buying behavior tendencies. Balt. J. Manag. **13**(2), 209–235 (2018)
21. Alkufahy, A., Al-Alshare, F., Qawasmeh, F., Aljawarneh, N., Almaslmani, R.: The mediating role of the perceived value on the relationships between customer satisfaction, customer loyalty and e-marketing. Int. J. Data Netw. Sci. **7**(2), 891–900 (2023)

22. Qtaishat, N.M.: The impact of E-marketing on consumer purchase decision in Jordan. WSEAS Trans. Bus. Econ. **19**, 161–168 (2022); Riyadi, S.: Effect of E-Marketing and E-CRM on E-Loyalty: an empirical study on Indonesian Manufactures. Turkish J. Physiotherapy Rehabil. **32**(3), 5290–5297 (2021)
23. Krypton, A., Naldo, H.W.S.: The effect of social media e-marketing towards consumers shopping behavior. JurnalVokasi Indonesia **6**(2), 1 (2018)
24. Coviello, N.E., Brodie, R.J., Brookes, R.W., Palmer, R.A.: Assessing the role of e-marketing in contemporary marketing practice. J. Mark. Manag. **19**(7–8), 857–881 (2003)
25. Abdallah, S., Jaleel, B.: Website appeal: development of an assessment tool and evaluation framework of e-marketing. J. Theor. Appl. Electron. Commer. Res. **10**(3), 45–62 (2015)
26. Harsono, A.: Understanding E-business and E-commerce and their relation to the conceptual framework of E-marketing: case study. Sisfotenika **1**(2), 53–68 (2011)
27. Bolos, C., Idemudia, E.C., Mai, P., Rasinghani, M., Smith, S.: Conceptual models on the effectiveness of e-marketing strategies in engaging consumers. J. Int. Technol. Inf. Manag. **25**(4), 3 (2016)
28. Hooda, S., Aggarwal, S.: Consumer behaviour towards e-marketing: a study of Jaipur consumers. Res. World **3**(2), part 2, 107 (2012)
29. Lou, C., Yuan, S.: Influencer marketing: how message value and credibility affect consumer trust of branded content on social media. J. Interact. Advert. **19**(1), 58–73 (2019)
30. Appel, G., Grewal, L., Hadi, R., Stephen, A.T.: The future of social media in marketing. J. Acad. Mark. Sci. **48**(1), 79–95 (2020)
31. Saima, Khan, M.A.: Effect of social media influencer marketing on consumers' purchase intention and the mediating role of credibility. J. Promotion Manag. **27**(4), 503–523 (2020)
32. Kumar, A., Bezawada, R., Rishika, R., Janakiraman, R., Kannan, P.K.: From social to sale: the effects of firm-generated content in social media on customer behavior. J. Mark. **80**(1), 7–25 (2016)
33. Masuda, H., Han, S.H., Lee, J.: Impacts of influencer attributes on purchase intentions in social media influencer marketing: mediating roles of characterizations. Technol. Forecasting Soc. Change **174**, article number 121246 (2022)
34. Dhun, Dangi, H.K.: Influencer marketing: role of influencer credibility and congruence on brand attitude and eWOM. J. Internet Commerce **22**(suppl 1), S28–S72 (2023)
35. Jin, S.V., Muqaddam, A., Ryu, E.: Instafamous and social media influencer marketing. Mark. Intell. Plan. **37**(5), 567–579 (2019)
36. Ye, G., Hudders, L., De Jans, S., De Veirman, M.: The value of influencer marketing for business: A bibliometric analysis and managerial implications. J. Advert. **50**(2), 160–178 (2021)
37. Childers, C.C., Lemon, L.L., Hoy, M.G.: # Sponsored# Ad: agency perspective on influencer marketing campaigns. J. Curr. Issues Res. Advert. **40**(3), 258–274 (2019)
38. Haenlein, M., Anadol, E., Farnsworth, T., Hugo, H., Hunichen, J., Welte, D.: Navigating the New Era of influencer marketing: how to be successful on Instagram, TikTok, & Co. Calif. Manag. Rev. **63**(1), 5–25 (2020)
39. Campbell, C., Farrell, J.R.: More than meets the eye: the functional components underlying influencer marketing. Bus. Horiz. **63**(4), 469–479 (2020)
40. Xiao, M., Wang, R., Chan-Olmsted, S.: Factors affecting YouTube influencer marketing credibility: a heuristic-systematic model. J. Media Bus. Stud. **15**(3), 188–213 (2018)
41. Glucksman, M.: The rise of social media influencer marketing on lifestyle branding: a case study of Lucie Fink. Elon J. Undergraduate Res. Commun. **8**(2), 77–87 (2017)
42. Eid, R., El-Gohary, H.: The impact of E-marketing use on small business enterprises' marketing success. Serv. Ind. J. **33**(1), 31–50 (2013)
43. Purba, K.R., Tan, Y.J.: Data-driven influencer marketing strategy analysis and prediction based on social media and Google analytics data. Appl. Mark. Anal. **8**(3), 314–328 (2023)
44. Huynh, T., Nguyen, H.D., Zelinka, I., Nguyen, K.V., Pham, V.T., Hoang, S.N.: Advo: A system to manage influencer marketing campaigns on social networks. Appl. Sci. **11**(14), article 6497 (2021)

45. Aswani, R., Ghrera, S.P., Chandra, S., Kar, A.K.: Outlier detection among influencer blogs based on off-site web analytics data. In: Caelli, T.J., et al. (ed.) Digital Nations–Smart Cities, Innovation, and Sustainability: 16th IFIP WG 6.11 Conference on e-Business, e-Services, and e-Society, I3E 2017, pp. 251–260. Springer International Publishing (2017)
46. Srivastava, M.: Mapping the influence of influencer marketing: a bibliometric analysis. Mark. Intell. Plan. **39**(7), 979–1003 (2021)
47. Aw, E.C.X., Agnihotri, R.: Influencer marketing research: review and future research agenda. J. Mark. Theory Practice 1–14 (2023)
48. Gabhane, D., Varalaxmi, P., Rathod, U., Hamida, A.G.B., Anand, B.: Digital marketing trends: analyzing the evolution of consumer behavior in the online space. Boletin de Literatura Oral Literary J. **10**(1), 462–473 (2023)
49. Gorea, I., Jacobson, J.: The Ethics of Using Social Media Influencers for Marketing Purposes. The SAGE Handbook of Social Media Marketing, p. 470. (2022)

The Evolution of Influencer Marketing

Aun Ahmed and Tushar Rathore

Abstract Influencer marketing (IM) is more than just a buzzword. In recent years, IM has gained widespread attention, but its roots actually extend further back than many realize. With the rise of the internet and the explosion of social media, IM has evolved into a new and influential form of marketing communication utilized by businesses of all types. This chapter delves into the historical development of IM, tracing its origins and evolution. We also explore the shift from traditional celebrity endorsements to influencer marketing, providing a comparison of these two endorsement forms. Additionally, we examine the maturation of the IM industry, addressing the challenges and opportunities it presents for marketers. The chapter concludes by addressing current issues in the IM industry, offering a comprehensive and informative look at the phenomena and concepts surrounding IM. Throughout, we strive to provide a thorough explanation of these phenomena, drawing from academic literature and theory to enhance understanding.

Keywords Influencer marketing · Influencer advertising · Marketing communications · Internet marketing · Digital marketing · Social media · Social networks

1 Historical Context

Marketing is one of the oldest practices of humankind, yet it is considered one of the youngest among the fields of business studies [6]. Similarly, the practice of influencer marketing can be traced back to ancient Roman gladiators who endorsed items for sale [63]. However, its recognition as a field of inquiry in business and marketing

A. Ahmed (✉)
Department of Humanities and Social Sciences, Indian Institute of Technology Ropar, Rupnagar, Punjab, India
e-mail: aun.iitrpr@gmail.com

T. Rathore
Balaji Institute of International Business, Pune, Maharashtra, India

S. Dutta et al. (eds.), *Advances in Data Analytics for Influencer Marketing: An Interdisciplinary Approach*, Information Systems Engineering and Management 9,
https://doi.org/10.1007/978-3-031-65727-6_2

is relatively recent. Before we explore the history of influencer marketing, it would be appropriate to understand the theoretical underpinnings of *influence.* There are various types of influence, but which one occurs in influencer marketing?

The Oxford Learner's Dictionary defines influence as "the power that somebody/something has to make somebody/something behave in a particular way" [55]. An individual who can exercise their power to shape the behavior of others is called an influential person or an influencer [52]. The phenomenon of influence is pervasive and plays a vital role in shaping people's behavior [8, 19]. An early paper by Berey and Pollay [9] studied the influence of a child on the mother's purchase behavior. The authors found that the mother's purchase behavior was related to the child's assertiveness and how child-centered the mother was. That is, the influence was related to the characteristics of both the influencer and the one being influenced. Parsons [56] conceptualized influence as "a generalized mechanism by which attitudes or opinions are determined" (p. 37). For a mechanism to be classified as an influence, it has to be both active and intentional. If an individual alters their attitudes due to observations or experiences related to a natural or social event not designed to induce a change, it would not be deemed an influence. Changes in behavior resulting from observing others is a more passive form of influence called social influence [19]. Influencer marketing relies on the former type of influence, that is, active and intentional influence. It leverages the social capital of social media personalities (the influencers) to exert influence on consumers who have formed interpersonal relationships with the influencer [5]. Online influencers' social capital comprises their follower networks, positioning, the content they create, and the follower trust they command [45]. Harnessing the social capital accumulated by everyday people with significant social media followings to endorse products and services subtly is called influencer marketing. Table 1 lists popular online influencers, the platform they are active on, their follower numbers, and their positioning.

The advent of Web 2.0 made the internet bidirectional and allowed consumers to become content creators and participate in huge online social networks [16]. Consumers became content creators who shared their content with their followers. Web 3.0 accelerated the penetration of social media platforms (e.g., Instagram and Facebook) via mobile applications on portable devices [45]. These technological developments created massive word-of-mouth (WOM) networks [16]. Consumers perceive WOM communications (for example, online product reviews) as more

Table 1 Popular online influencers

Influencer	Platform	Followers (Millions)	Positioning
Cristiano Ronaldo	Instagram	628	Sports
Zoe Sugg	Instagram	9.2	Lifestyle
Charli D'Amelio	TikTok	79.9	Fashion
Tara McKenna	Instagram	0.36	Sustainability
Taylor Swift	Instagram	283	Music, Pop culture
Elon Musk	Twitter	181.5	Business

credible and trustworthy than marketing communication [10, 70]. Influential social media users gained ardent followers by crafting a distinct persona [13]. The ardent followers developed trust and strong psychological bonds with the influencers who shared unique content around their lives, interests, and expertise [41, 64]. Therefore, marketers began incentivizing influencers to post favorably on social media such as Instagram and Facebook [12], giving rise to influencer marketing. Individuals on social media began devising ways to influence their followers and obtain compensation from marketers when they saw value in their influence [52]. Thus, three actors—influencers, followers, and marketers—working together gave rise to influencer marketing.

The influence of influencers stems from two main phenomena—opinion leadership and parasocial relationships. The concept of opinion leadership is rooted in Katz and Lazarsfeld's [39] two-step flow theory. The theory suggests that interpersonal communication is more influential in shaping people's opinions than mass media communication, which often flows to the masses via opinion leaders [15]. Opinion leadership is an individual's tendency to influence the opinions, attitudes, and behavior of others in a premeditated way [23]. Opinion leaders "exert an unequal amount of influence on the decisions of others" [58, p. 435]. Interpersonal influence transmits from opinion leaders to followers and amongst the followers connected in a network [47]. The advent of digital markets created internet opinion leaders, and social media platforms provided them with a larger audience to influence. Followers of internet opinion leaders (the influencers) perceive them as leaders in their network, having some domain-specific competence [20]. Opinion leaders are followed based on their perceived credibility and originality of content [34]. For example, Elon Musk, as a cryptocurrency opinion leader, has had a great impact on the crypto community, and his Tweets (microblogs on X.com) have influenced Bitcoin prices [46, 68].

'Parasocial relationships' [29], originally conceptualized to explain consumer-celebrity relationships, is now also applied to follower-influencer relationships. [28] define parasocial relationships as "nonreciprocal socio-emotional connections with media figures such as celebrities or influencers" (p. 1). Parasocial relationships provide social media users with companionship with influencers who are socially attractive and who have some perceived commonality with them [69]. Parasocial relationships are generally viewed as nonreciprocal or one-sided. Such relationships can, however, turn more real. For example, influencers like Lady Gaga, Paris Hilton, and Taylor Swift have directly communicated with their followers, met them in person, and even stayed with them [1].

2 From Celebrity Endorsement to Influencers

Celebrity endorsement, the use of celebrities to promote brands or products, has been around since the latter half of the nineteenth century [18]. The rise of television and cinema in the 1960s facilitated its rise [18, 33, 36] and has since been used for the promotion of entities beyond brands and products, such as political parties, non-profit organizations, and government bodies [43]. A celebrity is an "individual who enjoys public recognition" [51, p. 310] and can lend their attractiveness and credibility to the endorsed brand, product, political party, or other organizations [59]. Besides attractiveness and credibility, celebrities carry a cultural meaning. According to the meaning transfer model [51], celebrities assume a meaning from their fields of expertise (e.g., acting, sports), which they bring to the endorsement process and transfer to the brand or product being endorsed [26]. Another stream of literature put forward the match-up hypothesis to explain the effectiveness of celebrity endorsement. According to the match-up hypothesis, a celebrity's attractiveness will be effective only if there is a fit or match-up between the product and the meaning conveyed by the celebrity [37].

Celebrity endorsers generally wield public recognition because of their achievements in sports or films [26, 61]. They can transfer their attractiveness, credibility, and cultural meanings to the brands they endorse. However, celebrity endorsement can be costly and poses some risks to the brand being endorsed. Celebrities may sometimes overshadow a brand and create a negative brand image due to a controversy or an image change [18]. For example, the 2009 Tiger Woods scandal cost its sponsors—mainly Electronic Arts, Pepsi, and Nike—2% of their market value within fifteen stock trading sessions after the scandal [42]. This estimate ranges from 5 to 12 billion US dollars [54].

Because of the costs and risks associated with celebrity endorsement, companies are increasingly switching to social media influencers who have many followers on social media without any association with the sports or film industry [24]. Influencers establish a personal brand through the content they create and post on social media. They enjoy a lot of freedom in creating content for brands in alignment with their positioning and the nature of their audience (the followers). Contrarily, the content disseminated by celebrity endorsers (e.g., the message in an ad film featuring a celebrity) is fully controlled by the brands they endorse [45]. The influencers in a social network are identified mainly by their follower count [40, 70]. However, identifying influencers for a brand endorsement based solely on follower count can be inefficient and perhaps ineffective. Other factors, such as the influencer's positioning, nationality, and engagement level (e.g., likes and shares) of their posts, are essential considerations. Moreover, beyond their follower counts, influencers can vary in credibility, expertise, and perceived authenticity of the messages they post. Based on various characteristics of influencers, [12] identify five categories of influencers: celebrity-influencers, mega-influencers, macro-influencers, micro-influencers, and nano-influencers. Influencers' follower counts and cultural capital decrease, and accessibility and authenticity increase from celebrity- to nano-influencers. Celebrity-influencers are individuals who already have celebrity status and have built a large follower base on social media. For example, a celebrity sportsperson, Cristiano Ronaldo, garnered 628 million followers on Instagram and became a celebrity influencer. With some influencers, the reverse happens—they begin as influencers and become celebrities. Brooks et al. [11] have termed this phenomenon as influencer celebrification. Charli D'Amelio, starting as a teenage content creator, amassed 79.9 million TikTok and 44.4 million Instagram followers. She is a mega-influencer since she lacked celebrity status before becoming a social media influencer and has gained more than a million followers, according to Campbell and Farrell's [12] classification. Macro-, micro-, and nano-influencers have successively smaller —100 k, 10 k, and less than 10 k—followers. Since Instagram users have an average of 150 followers [66], the potential reach of an Instagram influencer with 10,000 followers will be 1.5 million users [60].

3 Maturation

Influencer marketing has evolved as a global marketing phenomenon, and companies are increasingly integrating it into their marketing strategies [45] and media mixes [12]. The market size for influencer marketing was 6.1 billion US dollars in 2016 and is expected to reach 24 billion US dollars by the end of 2024 [31].

Influencer advertising has supplemented traditional advertisements containing messages created by firms or on their behalf by advertising agencies. Influencers create their own content, disseminate it via their own online networks, and provide personal recommendations for the product or brand [5]. Brands reward influencers and, in exchange, encroach on influencers' content. This is why influencer marketing has been considered as a form of product placement[1] [4], and influencers are increasingly being seen as 'product promoters' rather than 'content creators' [38]. Consumers are aware of the commercial association between brands and influencers who are simultaneously endorsing diverse products and brands at the same time. This practice leads to influencer-product mismatch, a loss of influencer authenticity, and a low impact of their posts on purchase intention [72]. Consumers' negative reactions to product endorsement posts are more pronounced when influencers do not accompany the product in their posts [35]. The decreasing authenticity of influencers makes their posts less engaging, and followers may leave. Therefore, managing influencer authenticity is crucial for all influencer-marketing stakeholders (influencer, follower, marketer). Influencers and marketers can work together to ensure authenticity. To enhance influencer authenticity, [4] proposed two strategies: "passionate authenticity" and "transparent authenticity." Creating content based on their passion makes the process of content creation more satisfying for the influencers. Under "transparent authenticity," influencers disclose the nature of the partnership with the brand and also include their own opinion of the brand to respect their integrity.

Influencer marketing has matured and become more popular since the COVID-19 pandemic. Due to the increased time spent on the internet, consumers are actively seeking out and engaging with influencers more than ever. However, influencer marketing has become overused, perceived influencer authenticity has dwindled, and marketers have questioned its effectiveness [44]. The marketers' dissatisfaction

---

[1] PRODUCT placement (also called brand integration) is the insertion of branded products or brand identifiers in movies, TV shows, books, etc., and was developed as an alternative to traditional advertisements [71].

with influencer marketing has been attributed to three issues: (1) the presence of fake followers and likes makes the reach of influencers unverifiable and unmeasurable, (2) influencers becoming less impactful because they endorse several unrelated brands simultaneously, and (3) regulatory issues arising because of lack of transparency, for example, nondisclosure of sponsored content [65, 67]. These issues with influencer marketing make benchmarking and measurement of its effectiveness essential for influencer marketing maturation [17].

Marketing communication effectiveness is measured in terms of 'diagnostic metrics' such as brand awareness and brand loyalty and 'evaluative metrics' such as market share, sales, and profits [50]. These metrics can also be used as a measure of influencer effectiveness. Leung et al. [45] have proposed six benefits or threats that can originate from an influencer and impact the marketing communication effectiveness of the firm being endorsed by the influencer. Table 2 lists the benefits and threats that enhance and diminish marketing communication effectiveness.

Performance measurement is a prerequisite for managing influencer marketing or any form of social media marketing [57]. Once the benefits and risks of utilizing influencers are understood, performance measurement should be implemented. Marketers have limited control over influencer-generated content, which may not perfectly align with the brand identity crafted by the firm. Therefore, marketers need to add qualitative measurements to performance measures to assess the content quality of an influencer. Gräve [25] recommends measuring influencer marketing performance using post-related and sales- or brand-related quantitative and qualitative metrics, as outlined in Table 3.

Marketers can readily obtain the data required for computing the quantitative metrics listed in Table 3 from the social media platform on which they run the campaign. These data are, in fact, visible to anyone visiting the platform, but aggregating these data requires technical tools and the expertise to use them. The data must be used with caution because influencer marketing has been tainted by fake likes and followers, as mentioned previously.

Table 2 Benefits and threats of influencer marketing [45]

	Benefit/Threat	Effect on marketing communication effectiveness
1	Targeting benefits	Influencer's ability to reach niche consumer segments can complement firms' targeting efforts
2	Positioning benefits	Influencers' positioning can be leveraged to reinforce the brand positioning
3	Creativity benefits	Original and relevant content created by influencers evokes favorable consumer responses
4	Trust benefits	Influencer-authenticity perceptions and the follower-influencer bond evoke trust in marketing communications
5	Content control threat	Content deviating from the brand message can be detrimental
6	Customer retention threat	Influencers are more effective in acquiring rather than retaining customers

Table 3 Metrics for influencer marketing performance measurement

	Quantitative	Qualitative
Influencer- or post-related	No. of interactions	Post sentiment
	Reach	Influencer quality
	No. of views	Target segment fit
	No. of followers	–
Sales- or brand-related	Sales	Brand image
	Click-through rate	
	Conversion rate	
	Social media impact	
	Brand awareness	
	Cost	
	Customer acquisition	
	Return on investment	

Given that (1) an influencer is almost always active on multiple social media platforms and (2) parts of the data obtained from the platforms may be spurious, machine learning algorithms can be used instead of traditional social media analytics. [2] have computed an "influencer index" using different machine learning algorithms to measure the effectiveness of an influencer across Facebook, Twitter, and Instagram. References [7, 22] provide surveys of machine learning algorithms for social media analytics, which can also inform influencer analytics.

4 The Current Landscape

Until recently, social media influencers were real people. However, advances in artificial intelligence (AI) have led to the emergence of virtual influencers. The public has widely perceived AI as leading to the redundancy of many jobs [27]. This threat also looms over the influencer marketing industry: companies are now shifting from human influencers (HIs) to virtual influencers (VIs) [21]. VIs are artificial, computer-generated entities. They behave like real social media influencers but lend marketers and advertisers great control over their appearance and the content that they post [14]. VIs are designed by experts based on insights about their target audience gained from social listening[2] [48]. VIs are perceived as less credible and authentic than HIs. However, they are engaging and persuasive because of their technological novelty and amusement value [3]. Finding the right balance in the human-like attributes of VIs is essential for their effectiveness and acceptance. VIs' highly human-like attributes, together with human-like behavior, can evoke negative reactions and hamper their

[2] SOCIAL listening is an "active process of attending to, observing, interpreting, and responding to a variety of stimuli through mediated, electronic, and social channels" [62].

effectiveness [49]. Interested readers can visit https://www.virtualhumans.org/#influencers to learn about some popular VIs.

The future of influencer marketing will be shaped by the rising use of artificial intelligence and other information and communication technologies [32]. Instead of using influencer marketing as an ad hoc campaign, companies are expected to integrate it with their internet search ads, mobile ads, email campaigns, and traditional television and print advertisements [45]. Influencer marketing has been very effective, especially in targeting young consumers. Therefore, governments and policymakers have raised ethical concerns about influencer marketing [30]. However, the proper regulation of influencer marketing is in its infancy. So far, existing laws have been applied to regulate influencer marketing. For example, in the European Union, the extant consumer protection law also applies to influencer marketing [53].

References

1. Archer, A., Robb, C. (in press).: Ethics of parasocial relationships. In: The Ethics of Relationships: Broadening the Scope. Oxford University Press
2. Arora, A., Bansal, S., Kandpal, C., Aswani, R., Dwivedi, Y.: Measuring social media influencer index-insights from Facebook, Twitter and Instagram. J. Retail. Consum. Serv. **49**, 86–101 (2019)
3. Arsenyan, J., Mirowska, A.: Almost human? a comparative case study on the social media presence of virtual influencers. Int. J. Hum. Comput. Stud. **155**, 102694 (2021)
4. Audrezet, A., De Kerviler, G., Moulard, J.G.: Authenticity under threat: when social media influencers need to go beyond self-presentation. J. Bus. Res. **117**, 557–569 (2020)
5. Avery, J., Israeli A.: Influencer marketing. In: Harvard Business School Technical Note, p. 520, 75 Posted (2020)
6. Baker, M.J.: One more time: what is marketing? In: The Marketing Book, pp. 41–56. Routledge (2008)
7. Balaji, T.K., Annavarapu, C.S.R., Bablani, A.: Machine learning algorithms for social media analysis: a survey. Comput. Sci. Rev. **40**, 100395 (2021)
8. Bearden, W.O., Netemeyer, R.G., Teel, J.E.: Measurement of consumer susceptibility to interpersonal influence. J. Consum. Res. **15**(4), 473–481 (1989)
9. Berey, L.A., Pollay, R.W.: The influencing role of the child in family decision making. J. Mark. Res. **5**(1), 70–72 (1968)
10. Bickart, B., Schindler, R.M.: Internet forums as influential sources of consumer information. J. Interact. Mark. **15**(3), 31–40 (2001)
11. Brooks, G., Drenten, J., Piskorski, M.J.: Influencer celebrification: How social media influencers acquire celebrity capital. J. Advert. **50**(5), 528–547 (2021)
12. Campbell, C., Farrell, J.R.: More than meets the eye: the functional components underlying influencer marketing. Bus. Horiz. **63**(4), 469–479 (2020)
13. Casaló, L.V., Flavián, C., Ibáñez-Sánchez, S.: Influencers on Instagram: antecedents and consequences of opinion leadership. J. Bus. Res. **117**, 510–519 (2020)
14. Cascio Rizzo, G.L.C., Berger, J., Villarroel, F.: What drives virtual influencer's impact? Working paper, LUISS Guido Carli University (2023)
15. Cheung, M.L., Leung, W.K., Aw, E.C.X., Koay, K.Y.: I follow what you post!: the role of social media influencers' content characteristics in consumers' online brand-related activities (COBRAs). J. Retail. Consum. Serv. **66**, 102940 (2022)
16. Dellarocas, C.: The digitization of word of mouth: promise and challenges of online feedback mechanisms. Manag. Sci. **49**(10), 1407–1424 (2003)

17. Ducman, A.A.: Redefining wine brand engagement through influencer marketing and ephemeral content. Theor. Appl. Econ. **31**(1) (2024)
18. Erdogan, B.Z.: Celebrity endorsement: a literature review. J. Mark. Manag. **15**(4), 291–314 (1999)
19. Falk, E., Scholz, C.: Persuasion, influence, and value: perspectives from communication and social neuroscience. Annu. Rev. Psychol. **69**, 329–356 (2018)
20. Farivar, S., Wang, F., Yuan, Y.: Opinion leadership versus para-social relationship: Key factors in influencer marketing. J. Retail. Consum. Serv. **59**, 102371 (2021)
21. Gerrath, M.H., Olya, H., Shah, Z., Li, H.: Virtual influencers and pro-environmental causes: the roles of message warmth and trust in experts. J. Bus. Res. **175**, 114520 (2024)
22. Ghani, N.A., Hamid, S., Hashem, I.A.T., Ahmed, E.: Social media big data analytics: a survey. Comput. Hum. Behav. **101**, 417–428 (2019)
23. Gnambs, T., Batinic, B.: A personality-competence model of opinion leadership. Psychol. Mark. **29**(8), 606–621 (2012)
24. Gräve, J.F., Bartsch, F.: # Instafame: exploring the endorsement effectiveness of influencers compared to celebrities. Int. J. Advert. **41**(4), 591–622 (2022)
25. Gräve, J.F.: What KPIs are key? Evaluating performance metrics for social media influencers. Social Media + Soc. **5**(3), 2056305119865475 (2019)
26. Halonen-Knight, E., Hurmerinta, L.: Who endorses whom? meanings transfer in celebrity endorsement. J. Prod. Brand. Manag. **19**(6), 452–460 (2010)
27. Hiort, A.: Yes, virtual influencers are taking jobs. VirtualHumans.org (2022, February 15). https://www.virtualhumans.org/article/yes-virtual-influencers-are-taking-jobs
28. Hoffner, C.A., Bond, B.J.: Parasocial relationships, social media, and well-being. Curr. Opin. Psychol. **45**, 101306 (2022)
29. Horton, D., Richard Wohl, R.: Mass communication and para-social interaction: observations on intimacy at a distance. Psychiatry **19**(3), 215–229 (1956)
30. Hudders, L., Lou, C.: A new era of influencer marketing: lessons from recent inquires and thoughts on future directions. Int. J. Advert. **41**(1), 1–5 (2022)
31. Influencer Marketing Hub: The State of Influencer Marketing 2024: Benchmark Report (2024, June 24). https://influencermarketinghub.com/influencer-marketing-benchmark-report/
32. Iwashita, M.: Trend of influencer marketing and future required functions. In: 2019 20th IEEE/ACIS International Conference on Software Engineering, Artificial Intelligence, Networking and Parallel/Distributed Computing (SNPD), pp. 2–2. IEEE (2019)
33. Jain, V., Roy, S.: Understanding meaning transfer in celebrity endorsements: a qualitative exploration. J. Cetacean Res. Manag. **19**(3), 266–286 (2016)
34. Jegham, S., Bouzaabia, R.: Fashion influencers on Instagram: determinants and impact of opinion leadership on female millennial followers. J. Consum. Behav. **21**(5), 1002–1017 (2022)
35. Jin, S.V., Muqaddam, A.: Product placement 2.0:"Do brands need influencers, or do influencers need brands?" J. Brand Manag. **26**, 522–537 (2019)
36. Kaikati, J.G.: Celebrity advertising: a review and synthesis. Int. J. Advert. **6**(2), 93–105 (1987)
37. Kamins, M.A.: An investigation into the "match-up" hypothesis in celebrity advertising: when beauty may be only skin deep. J. Advert. **19**(1), 4–13 (1990)
38. Kapitan, S., van Esch, P., Soma, V., Kietzmann, J.: Influencer marketing and authenticity in content creation. Australas. Mark. J. **30**(4), 342–351 (2022)
39. Katz, E., Lazarsfeld, P.F.: Personal Influence; The Part Played by People in the Flow of Mass Communications, Glencoe, Ill. Free Press (1955)
40. Kay, S., Mulcahy, R., Parkinson, J.: When less is more: the impact of macro and micro social media influencers' disclosure. J. Mark. Manag. **36**(3–4), 248–278 (2020)
41. Kim, D.Y., Kim, H.Y.: Trust me, trust me not: a nuanced view of influencer marketing on social media. J. Bus. Res. **134**, 223–232 (2021)
42. Knittel, C.R., Stango, V.: Celebrity endorsements, firm value, and reputation risk: evidence from the Tiger Woods scandal. Manag. Sci. **60**(1), 21–37 (2014)
43. Knoll, J., Matthes, J.: The effectiveness of celebrity endorsements: a meta-analysis. J. Acad. Mark. Sci. **45**, 55–75 (2017)

44. Le, K., Aydin, G.: Impact of the pandemic on social media influencer marketing in fashion: a qualitative study. J. Cetacean Res. Manag. **26**(4), 449–469 (2023)
45. Leung, F.F., Gu, F.F., Palmatier, R.W.: Online influencer marketing. J. Acad. Market. Sci. 1–26 (2022)
46. Lichti, C., Ademi, E., Tumasjan, A.: Decentralized Opinion Leadership: A Study of Crypto Influencers in the Twitter Discourse on Bitcoin (2024)
47. Litvin, S.W., Goldsmith, R.E., Pan, B.: Electronic word-of-mouth in hospitality and tourism management. Tour. Manag. **29**(3), 458–468 (2008)
48. Looi, J., Kahlor, L.A.: Artificial intelligence in influencer marketing: a mixed-method comparison of human and virtual influencers on Instagram. J. Interact. Advertising 1–20 (2024)
49. MacDorman, K.F.: In the uncanny valley, transportation predicts narrative enjoyment more than empathy, but only for the tragic hero. Comput. Hum. Behav. **94**, 140–153 (2019)
50. McAlister, L., Srinivasan, R., Jindal, N., Cannella, A.A.: Advertising effectiveness: The moderating effect of firm strategy. J. Mark. Res. **53**(2), 207–224 (2016)
51. McCracken, G.: Who is the celebrity endorser? cultural foundations of the endorsement process. J. Consum. Res. **16**(3), 310–321 (1989)
52. McMullan, K., Laurell, C., Pitt, L.: Managing the tensions in marketer-influencer relationships. Bus. Horiz. **65**(5), 559–566 (2022)
53. Michaelsen, F., Collini, L., Jacob, C., Goanta, C., Kettner, S.E., Bishop, S., Hausemer, P., Thorun, C., Yesiloglu, S.: The impact of influencers on advertising and consumer protection in the single market (2022). https://www.europarl.europa.eu/RegData/etudes/STUD/2022/703350/IPOL_STU(2022)703350_EN.pdf
54. Morain, C.: Tiger Woods Scandal Cost Shareholders $12 Billion [University of California, Davis] (2009). Accessed from https://www.ucdavis.edu/news/tiger-woods-scandal-cost-shareholders-12-billion
55. Oxford Learner's Dictionaries.: influence noun—Definition, pictures, pronunciation and usage notes. Oxford Learner's Dictionaries (n.d.). Accessed 6 May 2024, from https://www.oxfordlearnersdictionaries.com/definition/english/influence_1?q=influence
56. Parsons, T.: On the concept of influence. Public Opin. Q. **27**(1), 37–62 (1963)
57. Peters, K., Chen, Y., Kaplan, A.M., Ognibeni, B., Pauwels, K.: Social media metrics—a framework and guidelines for managing social media. J. Interact. Mark. **27**(4), 281–298 (2013)
58. Rogers, E.M., Cartano, D.G.: Methods of measuring opinion leadership. Public Opin. Q. 435–441 (1962)
59. Schouten, A.P., Janssen, L., Verspaget, M.: Celebrity versus influencer endorsements in advertising: the role of identification, credibility, and Product-Endorser fit. In: Leveraged Marketing Communications, pp 208–231. Routledge
60. Sicilia, M., López, M.: What do we know about influencers on social media? Toward a new conceptualization and classification of influencers. In: The Palgrave Handbook of Interactive Marketing, pp. 593–622. Springer International Publishing, Cham (2023)
61. Srivastava, R.K.: Effectiveness of films, sport, celebrity or mascot to content in the advertising–a dilemma for global brands. J. Promot. Manag. **27**(5), 716–739 (2021)
62. Stewart, M.C., Arnold, C.L.: Defining social listening: recognizing an emerging dimension of listening. Int. J. Listening **32**(2), 85–100 (2018)
63. Sucui, P.: History of Influencer Marketing Predates Social Media by Centuries–But is There Enough Transparency in the 21st Century? (2020)
64. Tafesse, W., Wood, B.P.: Followers' engagement with instagram influencers: the role of influencers' content and engagement strategy. J. Retail. Consum. Serv. **58**, 102303 (2021)
65. Taylor, C.R.: The urgent need for more research on influencer marketing. Int. J. Advert. **39**(7), 889–891 (2020)
66. Thompson, A.: How Many Instagram Followers Does the Average Person Have in 2023? [Blog post] (2023). Accessed from https://www.hashtagsforlikes.co/blog/instagram-followers-how-many-does-the-average-person-have/

67. Tobin, J.: Three influencer marketing problems that brands themselves can be created (2019). https://www.forbes.com/sites/forbesagencycouncil/2019/10/14/three-influencer-marketing-problems-that-brands-themselves-have-created/#284463bb6a3e. Accessed 11 May 2024
68. Treku, D.N., Wang, B., Escobari, D., Oraby, T.: Cryptocurrency Market: The Interplay Between Microblog Dynamic Opinion Leader Sentiments and Bitcoin Pricing (2023)
69. Tsiotsou, R.H.: The role of social and parasocial relationships on social networking sites loyalty. Comput. Hum. Behav. **48**, 401–414 (2015)
70. De Veirman, M., Cauberghe, V., Hudders, L.: Marketing through Instagram influencers: the impact of number of followers and product divergence on brand attitude. Int. J. Advert. **36**(5), 798–828 (2017)
71. Wiles, M.A., Danielova, A.: The worth of product placement in successful films: an event study analysis. J. Mark. **73**(4), 44–63 (2009)
72. Zniva, R., Weitzl, W.J., Lindmoser, C.: Be constantly different! How to manage influencer authenticity. Electron. Commer. Res. **23**(3), 1485–1514 (2023)

From Word-of-Mouth to AI-Powered Engagement: The Dynamic Evolution of Influencer Marketing in the Digital Era

Sonal Purohit and Rakhi Arora

Abstract Influencer marketing has emerged as an important component of the brand marketing campaigns for enhanced customer relationships. This chapter discusses the evolution of influencer marketing from passionate enthusiasts to the contemporary artificial intelligence (AI) powered virtual agents shedding light on the social media era to the integration of technology and data analytics. We explore the integrative dynamics of technology, consumer dynamics and industry trends that have impacted the influencer marketing for brand-customer relationship management. Shedding light on the transformative impact of technology particularly artificial intelligence, we show how the field has been refined through the data driven approach. The chapter further delves into the ethical considerations related to the influencer marketing campaigns. It is crucial for the brands to integrate transparency, authenticity and fairness in the campaigns showcasing responsible practices and maintain a balance between innovation and responsibility. The future of influencer marketing would be driven by data for highly personalized and targeted influencer marketing campaigns to forge deeper connections and drive success in the ever-changing marketing landscape.

Keywords Influencer marketing · Artificial intelligence · Virtual agents · Technology

S. Purohit
Symbiosis Institute of Business Management, Symbiosis International University, Nagpur, India
e-mail: sonal.purohit@sibmnag.edu.in

R. Arora (✉)
University School of Business-MBA, Chandigarh University, Ajitgarh, Punjab, India
e-mail: rakhi.e7437@cumail.in

S. Dutta et al. (eds.), *Advances in Data Analytics for Influencer Marketing: An Interdisciplinary Approach*, Information Systems Engineering and Management 9,
https://doi.org/10.1007/978-3-031-65727-6_3

1 Introduction

The marketing landscape has evolved over time with titanic shifts in the way promotions are done. One of the areas that have created a huge impact on the way brands connect with the customers is influencer marketing. Influencer marketing not only reflects a shift in the promotional approaches but also the changing dynamics of consumer behavior and communication channels [1]. One of the reasons for the emergence of influencer marketing is the advent of digital platforms. This growth is fueled by the consumers turning towards the digital platforms for information and entertainment. Since then, the influencer marketing trend has broadened the scope of promotions to include individuals with huge number of followers promoting and endorsing brands on social media [2]. These individuals unknowingly laid the foundation for a powerful marketing paradigm. However, while the term influencer marketing seems recent the roots can be traced back to a long term. It has evolved from the community leaders promoting to the local merchants and reached to the celebrities endorsing the brands. The early stages of influencer marketing can be traced back to the end of 1990s and the beginning of 2000s. In the early days, brands partnered with celebrities such as actors, musicians and sportspeople to reach and connect with masses. With the rise of bloggers, the brands connected with the bloggers in niche areas covering their target audience. The word-of-mouth marketing began encouraging loyal customers to recommend products through personal networks. These approaches majorly relied on personal networks with limited potential for targeted promotions. The advent of social media platforms such as Facebook, Instagram and Twitter in the beginning of 21st century marked the beginning of a new era in the brand marketing and promotions [3, 4]. The early influencers on social media were enthusiasts or experts in the niche areas. The audience with similar interests began following these influencers due to their authenticity and relatability with the masses. Brands recognized the potential of partnering with these influencers to reach their followers for a targeted campaign. The influencers realized their professional value as brand advocates and started collaborating with the brands strategically. Gradually, this turned into business with a focus on content creation, engagement with the audience and collaborations with the brands as their income sources [5]. Brands were benefitted by the huge follower base with specific interests enabling more targeted and personalized marketing efforts [6]. The celebrities also started to connect with the fans directly on social media converging the traditional celebrity endorsement with influencer marketing [7]. While influencer marketing has gained momentum due to presence of huge number of customers on social media, the onset of artificial intelligence (AI) and its integration to the marketing and related activities has further reshaped this phenomenon. Looking at the complexity of the influencer marketing phenomenon it is pertinent to comprehensively elaborate on the topic. Therefore, in this chapter we delve on the influencer marketing phenomenon and shed light on its evolution, current landscape and future with a special emphasis on the technology and its impact on influencer marketing.

2 The Rise of Social Media and a New Breed of Influencers

The rise of social media opened up opportunities for varied categories of influencers. These categories were based on the number of followers. Hence there were nano, micro, macro and mega influencers [8]. Brands got highly interested in the macro and mega influencers due to their followers in millions [9]. Hence they collaborated with these influencers to reach targeted audience in the niche segments. Hence, anyone with a social media presence could become an influencer. Not necessarily traditional celebrities, these influencers yielded significant influence in niche categories that the brand could use for customer engagement. By collaborating with these influencers the brands could promote the products to diverse consumer segments [10]. The increased screen times contributed to the rise of influencer marketing campaigns.

At the same time, celebrities in diverse industries recognized the potential of social media to connect and communicate with fans directly [11]. The platforms such as Instagram and Facebook provided an opportunity to personalize the connection by showing more authentic and personal angles of their lives. This was a major shift from the traditional celebrity endorsements where the endorsement was done in a more formal manner in the form of indirect communication through advertisements. Through social media celebrities became influencers and engaged with the audience in a more authentic way. This integration of the celebrity status with the influencer status was key revolution in the influencer marketing landscape [12]. The content went beyond the traditional boundaries and included videos, pictures, texts and stories. The campaign effectiveness was measured through the likes, comments and shares of the content.

2.1 *Challenges and Criticism*

While the celebrity and influencer marketing opened up huge opportunities to the brands it also introduced challenges. One of the major issues pertained to the authenticity of the influencers [13]. Concerns were raised over the influencers sharing false information for brand promotion rather than being true brand advocates. Issues started to rise over the transparency of communication and authenticity of the content [14]. Moreover, brands were troubled by not only fake influencers but also fake followers. Due to these reasons the followers started to question the authenticity of the brand creating a negative impact on the brand image [15]. A few campaigns faced backlash due to questionable authenticity and this lead to a call for more transparency, clear guidelines and ethical practices and standards within the influencer marketing industry. Additionally, the challenges related to selection of influencers, campaign management and return on investment remained prevalent [16–18]. This was because of the poor content produced by the influencers that was irrelevant to the audience thus affecting the quality of engagement [19]. Other reasons were related to the wrong choice of influencers by the brand that was merely based on the number of followers affecting the quality of audience that they reached [20, 21].

3 Integration of Technology and Data Analytics

The emergence of artificial intelligence (AI) and data analytics techniques further transformed the influencer marketing. Brands started to make use of AI and data analytics to improve their influencer marketing campaigns. This not only allowed them to address the concerns but also refined their targeted marketing approach.

3.1 Data Analytics in Influencer Marketing

The increasing usage of digital platforms by numerous people for wide range of activities has brought in a wealth of data that can be used by the brands. Such data can be used by brands to shape their influencer marketing strategies. Data analytics has laid a foundation for dynamic activities such as identification of key performance indicators, measurement of the campaign success and gain insights into the behavior of consumers [22]. Pertaining to the influencer marketing, brands began using data analytics to assess the rates of engagement by evaluating likes, shares, reach and conversations gained through influencer collaborations [23]. Hence, the impact of influencer marketing campaigns could be measured offering actionable data to the brands that they could use to refine their strategies opening up opportunities for more targeted and effective approach [24].

3.2 The Integration of AI in Influencer Marketing for Influencer Selection

The integration of artificial intelligence (AI) served as a significant milestone towards development of the influencer marketing. Through AI algorithms brands were deployed to identify and select the influencers leveraging machine learning techniques for evaluation of huge datasets [25]. The use of AI driven tools for influencer selection proved instrumental in identifying the influencers whose audiences were closely aligned to the target audiences of the brands [26]. Machine learning algorithms could assess factors such as audience engagement, content relevance, and sentiment analysis, facilitating more data-driven and objective influencer collaborations [27]. Additionally, the sophisticated tools could evaluate vast datasets encompassing audience demographics (Age, gender, location, income, and other relevant characteristics), psychographics (Interests, values, lifestyle, and online behavior patterns) and emotional preferences (Understanding the emotional triggers and sentiments within the target audience) to process information for selection of right influencer [28]. Thus, AI could evaluate not only the audience aligned to brands target markets but also the ones who resonated to the brand on a deeper subconscious level [29]. For instance, through AI brands could not only identify fashion bloggers but also the ones whose content evoked emotions like excitement that aligned with the brand's message. This laser-sharp targeting ensures campaigns hit the mark, fostering deeper connections and driving genuine, engaged participation.

3.3 Personalization Through AI Driven Analytics

With the growth in the availability of data, brands started to increase the personalization in their influencer marketing campaigns. The analysis of data driven by AI techniques facilitates a deeper understanding of the consumer behavior and preferences allowing for highly customized and personalized influencer partnerships [30]. The AI algorithms could give in depth insights into the consumer behavior, their interactions with the influencer content in the past and hence create a more refined consumer profile for personalized marketing [31, 32]. This shift towards personalization not only made the influencer content more relevant but also shaped a more authentic and resonant consumer-brand relationship.

3.4 Content Optimization

The integration of data driven AI analytics not only allowed personalization in connecting the consumer with influencer but also assisted in crafting messages that resonated to the consumers on a profoundly personal level. This was possible through techniques such as sentiment analysis, natural language processing and predictive modeling. Sentiment analysis could evaluate the past data to understand the consumer sentiment and gain understanding of the content that resonated to the customer [33]. Natural language processing assessed the data to determine the effective keywords, phrases and writing styles that provoked certain emotions and actions [34, 35]. For instance, AI could evaluate millions of beauty tutorials to identify accurate elements that inspired purchase among viewers. This allowed creation of content that elicited emotions of the customers driving an emotional connect, engagement and purchase intention.

3.5 Real Time Monitoring and Optimization

Traditionally, the measurement of campaign effectiveness was not accurate and depended on a number of factors. The advent of technology has facilitated accurate measurement of campaign effectiveness. Particularly, AI functionality allows real time monitoring and optimization. The brands can leverage AI for assessment of the consumer sentiment or reaction to the influencer campaign in real time. They can also monitor the engagement of consumers through likes, comments and shares to understand the performance of the influencer campaign [36]. Based on real-time insights, brands and influencers can adjust content, influencer selection, or even platform focus to maximize campaign impact [37]. The integration of AI serves as a continuous feedback loop that allows constant improvements for enhancing the effectiveness of the campaign [38]. For instance, if a clothing brand partners with a fashion influencer, AI can monitor the response of audience to outfit selection allowing the influencer to adjust their styling and messaging creating room for maximum engagement and brand interaction. Through this approach brand can ensure relevancy and impact thus maximizing the marketing ROI.

3.6 AI Enabled Virtual Influencers

While the importance of human influencers cannot be denied, AI has ushered a new era of AI enabled virtual influencers. A virtual influencer is a digital, computer-generated character that is given a personality to behave like humans [39]. These influencers are in the form of avatars and are enabled by AI. There are various types of these virtual influencers. These AI Influencers, designed to resemble humans, are deliberately created as computer-generated imagery or animated digital avatars [40]. The human like influencers have their own personalities and look like real. A few popular human-like virtual influencers are Shudu, Lil Maquila, Bermuda having their own page on Instagram with millions of followers. Some of them work with top fashion brands such as Prada, Dior and Calvin Klein. The virtual influencers have opened up new opportunities to the brands. Hence, the avatars or VI-influencers can be created that perfectly match with the brand image and values [41]. Through these influencers the brands can experiment with innovative story telling. Their interaction with the audience is round the clock facilitating consistent brand messaging. The global reach of these influencers help the brand in overcoming the language and cultural barriers to tap into diverse audiences. Such a technology can play a role in creating an intersection between content creation and influencer selection.

3.7 Ethical Considerations

While the integration of technology with the influencer marketing seems to bring in promising advantages for both brands and customers, it also posits ethical challenges. The issues are related to privacy of data, algorithmic biases and authenticity of content [42–44]. Data privacy is a significant concern among the consumers that has arisen from the algorithmic tapping into their behavior due to brands interest. It is essential for the brands to ensure transparency and follow responsible data collection processes during influencer marketing. Biased responses can arise from the improper handling of the training datasets. It is important for brands to recognize and mitigate these biases to ensure fairness and inclusivity. Exposing the audience to misleading or unauthentic information is also crucial for maintaining genuine human-centric connections. Brands and influencers must address these concerns to ensure responsible use of AI to ensure a future where innovation shaping a thriving landscape for data-driven, authentic influencer marketing emphasizing transparency and ethical practices.

4 The Future of Influencer Marketing

The future of influencer marketing unfolds with a huge potential and opportunities through an intersection of technological innovations, predictive intelligence, shift in consumer behavior, human centric experiences and changing market dynamics triggered by the power of AI and its applications. This section explores the emerging trends in influencer marketing in consideration of these dynamics.

4.1 Predictive Targeting

The power of AI can be harnessed in future to identify most relevant influencers before the emergence of trends. Advance predictive modeling techniques can analyze vase datasets to assess the social media trends for identification of emerging trends, hashtags and communities before they expand in popularity. With the help of real time data analysis, such modeling can analyze the consumer behavior to predict their intentions, preferences and emotional triggers [37]. Moreover, the analysis of past data on campaigns and content can assess the influencers that have the potential to be the best fit for brands in terms of their resonance with the target audience. Such a predictive power facilitates a positive engagement of brands with the rising influencers who are resonating to the emerging trends [45]. It can also be used to craft messages and content that can induce the sense of excitement and anticipation among audience. Through these the brands can remain ahead of the curve in the rapidly evolving digital landscape.

4.2 Micro-Moment Marketing

The future of marketing delves into the micro-moments where customers make the choices and decisions. Brands can be empowered by the AI to float hyper-personalized influencer content in such a micro-moment when consumer is about to take a purchase related decision [46]. Taking roots from the digital marketing where the content floats based on the location of the customer, the micro-moment marketing can enable the floating of influencer content based on location based targeting [47]. For instance, an influencer promoting a local restaurant appears with a personalized offer when the customer passes by the restaurant. Further, the micro-moment marketing can be used for real-time product recommendations when the customer is shopping online. Moreover, tailored content based on the browsing history, interests and preferences and even contemporary emotional state of the customer can create a hyper-personalized experience. The micro-moment marketing can assist the brands for increased conversion rates by delivering content at precise moments [48]. Such a personalized experience can foster the customer relationship with the brands fostering trust and loyalty. By using a targeted approach brands can maximize the effectiveness of influencer campaigns and optimize the marketing spend.

4.3 Integration with Marketing Automation

AI can facilitate the integration of influencer marketing with the broader marketing strategies. Brands can track and compare the influencer marketing content across channels gaining a holistic view of the campaign effectiveness [49]. AI can identify ideal influencers, negotiate deals, and manage logistics, freeing up valuable human resources. AI can also make adjustments to the budget allocations across the various influencer campaigns based on real time insights maximizing the returns on investments [50]. Hence, with automation, larger and more complex campaigns can be managed efficiently.

4.4 Influencer Marketing in Ephemeral Content

Ephemeral content is becoming common on the social media platforms [51, 52] and has revised the way in which audience engage with the content and influencers. Influencers are adapting to the temporary nature of ephemeral content by extending the scope of their posts to real-time, exclusive and authentic content [53]. Ephemeral content is suitable for real time and unfiltered interaction as desired by the customers.

4.5 Livestreaming

Majority of the social media platforms such as Facebook, Instagram, YouTube provide the influencers with the facility of hosting live events for a direct interaction with the customer [54]. Livestreaming has offered a dynamic platform to the influencers for active participation and engagement with the customers [55] that has not only enhanced the sense of community but also fostered authenticity into the influencer marketing strategies. To summarize, livestreaming opens new dynamics of influencer brand collaborations in the future.

4.6 Ethical Considerations and Industry Standards

The evolution of influencer marketing has also brought in the need for ethical considerations and heightened industry standards [56]. The influencer marketing industry is not immune to the ethical issues of privacy, transparency, authenticity and responsible content generation. While there are no clear guidelines on the ethical practices, in future the development of more ethical frameworks may provide directions for addressing of the concerns. The influencers must follow these guidelines to navigate the future of influencer marketing.

5 Conclusion

The influencer marketing journey has evolved over a period of time witnessing varied milestones adapting to the industry trends and dynamics, consumer behaviors and technology innovations. Each of the era has brought in its own unique set of challenges. Influencer marketing emphasizes the prominence of authentic connections in the digital world. From the roots of influencer marketing that originated with the passion of enthusiasts to the journey that encompassed individuals as influencers and celebrities as influencers to the advent of virtual influencers, the spirit of influencer marketing lies in genuine authentic connections between the audiences and the influencers. With the technology advancements leading to a future that is driven by data, the influencer marketing is heading towards a collaborative approach where the human expertise is combined with the technological capabilities to deliver successful and impactful results. Brands must harness the power of technology to increase the effectiveness of influencer marketing campaigns through personalization, optimization and real time campaign management. While the industry moves towards a revolution ethical considerations to maintain trust and credibility remain a cornerstone for the future success of influencer marketing campaigns. The future of influencer marketing campaigns will be shaped by key trends that shape the industry. While the campaigns rely on the brand-client relationships, the future must strike a balance between technology developments, creativity and ethical considerations to ensure the presence and sustainability of influencer marketing campaigns as impactful components of the digital marketing ecosystem.

References

1. Bakker, D.: Conceptualising influencer marketing. J. Emerg. Trends Market. Manag. **1**(1), 79–87 (2018)
2. Dimitrieska, S., Efremova, T.: The effectiveness of the influencer marketing. Econ. Manag. **18**(1), 109–118 (2021)
3. Aw, E.C.X., Agnihotri, R.: Influencer marketing research: review and future research agenda. J. Market. Theory Practice 1–14 (2023)
4. Kadekova, Z., Holienčinova, M.: Influencer marketing as a modern phenomenon creating a new frontier of virtual opportunities. Commun. Today **9**(2), 90–104 (2018)
5. Trivedi, J., Sama, R.: The effect of influencer marketing on consumers' brand admiration and online purchase intentions: an emerging market perspective. J. Internet Commer. **19**(1), 103–124 (2020)
6. Trivedi, J.P.: Measuring the comparative efficacy of an attractive celebrity influencer vis-à-vis an expert influencer-a fashion industry perspective. Int. J. Electron. Customer Relat. Manag. **11**(3), 256–271 (2018)
7. Guruge, M.C.: Comparison between attributes related to celebrity endorsement and social media influencer marketing: a conceptual review. Sri Lanka J. Market. **4**(1), 17–37 (2018)
8. Brewster, M.L., Lyu, J.: Exploring the parasocial impact of nano, micro and macro influencers. In: International Textile and Apparel Association Annual Conference Proceedings, pp. 1–3. Iowa State University Digital Press, USA (2020)
9. Harshitha, S., Shetty, R., Sairam, P.S.: Social media marketing: B2B marketing via nano influencers. J. Univ. Shanghai Sci. Technol. **23**(7), 1377–1387 (2021)
10. Borges-Tiago, M.T., Santiago, J., Tiago, F.: Mega or macro social media influencers: who endorses brands better? J. Bus. Res. **157**, 113606 (2023)
11. Appel, G., Grewal, L., Hadi, R., Stephen, A.T.: The future of social media in marketing. J. Acad. Mark. Sci. **48**(1), 79–95 (2020)
12. Huynh, T., Nguyen, H.D., Zelinka, I., Pham, X.H., Pham, V.T., Selamat, A., Krejcar, O.: A method to detect influencers in social networks based on the combination of amplification factors and content creation. PLoS ONE **17**(10), e0274596 (2022)
13. Le, K., Aydin, G.: Impact of the pandemic on social media influencer marketing in fashion: a qualitative study. J. Cetacean Res. Manag. **26**(4), 449–469 (2023)
14. Ahuja, Y., Loura, I.: An investigative study of influencer marketing: nuances, challenges and impact. J. Global Bus. Adv. **13**(5), 561–583 (2020)
15. Wellman, M.L., Stoldt, R., Tully, M., Ekdale, B.: Ethics of authenticity: social media influencers and the production of sponsored content. J. Media Ethics **35**(2), 68–82 (2020)
16. Syed, T.A., Mehmood, F., Qaiser, T.: Brand–SMI collaboration in influencer marketing campaigns: a transaction cost economics perspective. Technol. Forecast. Soc. Chang. **192**, 122580 (2023)
17. Doshi, R., Ramesh, A., Rao, S.: Modeling influencer marketing campaigns in social networks. IEEE Trans. Comput. Soc. Syst. **10**(1), 322–334 (2022)
18. Santiago, J.K., Castelo, I.M.: Digital influencers: An exploratory study of influencer marketing campaign process on instagram. Online J. Appl. Knowl. Manag. (OJAKM) **8**(2), 31–52 (2020)
19. Ebert, I., Sindermann, D.: An ethical view on influencer marketing–dynamic interaction between individual and economy or a simple data-driven advertising model? In: The Regulation of Social Media Influencers, pp. 74–97. Edward Elgar Publishing, Cheltenham (2020)

20. Chopra, A., Avhad, V., Jaju, A.S.: Influencer marketing: an exploratory study to identify antecedents of consumer behavior of millennial. Bus. Perspect. Res. **9**(1), 77–91 (2021)
21. Giles, D.C., Edwards, L.: Instagram and the rise of the social media 'influencer'. In: Twenty-First Century Celebrity: Fame in Digital Culture, pp. 155–173. Emerald Publishing Limited, United Kingdom (2018)
22. Attri, R., Bhagwat, J.: Influencer marketing in data-driven world. Manag. Dyn. **22**(2), 8–29 (2022)
23. Purba, K.R., Tan, Y.J.: Data-driven influencer marketing strategy analysis and prediction based on social media and Google Analytics data. Appl. Market. Anal. **8**(3), 314–328 (2023)
24. Huynh, T., Nguyen, H.D., Zelinka, I., Nguyen, K.V., Pham, V.T., Hoang, S.N.: Advo: a system to manage influencer marketing campaigns on social networks. Appl. Sci. **11**(14), 64–97 (2021)
25. Kamal, M., Bablu, T.A.: Machine learning models for predicting click-through rates on social media: factors and performance analysis. Int. J. Appl. Mach. Learn. Comput. Intell. **12**(4), 1–14 (2022)
26. Rutter, R.N., Barnes, S.J., Roper, S., Nadeau, J., Lettice, F.: Social media influencers, product placement and network engagement: using AI image analysis to empirically test relationships. Ind. Manag. Data Syst. **121**(12), 2387–2410 (2021)
27. Chaitanya, K., Saha, G.C., Saha, H., Acharya, S., Singla, M.: The impact of artificial intelligence and machine learning in digital marketing strategies. Euro. Econ. Lett. (EEL) **13**(3), 982–992 (2023)
28. Hassan, A.: The usage of artificial intelligence in digital marketing: A review. In: Applications of Artificial Intelligence in Business, Education and Healthcare, pp. 357–383 (2021)
29. Sharma, S., Bumb, A.: Femluencing: integration of femvertising and influencer marketing on social media. J. Interact. Advert. **22**(1), 95–111 (2022)
30. Alkurd, R., Abualhaol, I., Yanikomeroglu, H.: Big-data-driven and AI-based framework to enable personalization in wireless networks. IEEE Commun. Mag. **58**(3), 18–24 (2020)
31. Vashishth, T.K., Sharma, V., Sharma, K.K., Kumar, B., Panwar, R., Chaudhary, S.: AI-driven learning analytics for personalized feedback and assessment in higher education. In: Using Traditional Design Methods to Enhance AI-Driven Decision Making, pp. 206–230. IGI Global Publishing, USA (2024)
32. Melnykova, N., Shakhovska, N., Gregus, M., Melnykov, V., Zakharchuk, M., Vovk, O.: Data-driven analytics for personalized medical decision making. Mathematics **8**(8), 1211–1237 (2020)
33. Barbosa, B., Saura, J.R., Zekan, S.B., Ribeiro-Soriano, D.: Defining content marketing and its influence on online user behavior: a data-driven prescriptive analytics method. Ann. Oper. Res. 1–26 (2023)
34. Dash, G., Sharma, C., Sharma, S.: Sustainable marketing and the role of social media: an experimental study using natural language processing (NLP). Sustainability **15**(6), 5443–5467 (2023)
35. Said, N., Farouk, A.: Pet health monitoring through micro-influencer-boosted CTR: a machine learning and NLP approach. Adv. Intell. Inf. Syst. **8**(2), 34–47 (2023)
36. Borchers, N.S., Enke, N.: Managing strategic influencer communication: a systematic overview on emerging planning, organization, and controlling routines. Public Relat. Rev. **47**(3), 102041 (2021)
37. Perakakis, E., Mastorakis, G., Kopanakis, I.: Social media monitoring: an innovative intelligent approach. Designs **3**(2), 24–39 (2019)
38. Musiyiwa, R., Jacobson, J.: Sponsorship disclosure in social media influencer marketing: the algorithmic and non-algorithmic barriers. Social Media+ Soc. **9**(3), 20563051231196870 (2023)
39. Gerlich, M.: The power of virtual influencers: impact on consumer behaviour and attitudes in the age of AI. Adm. Sci. **13**(8), 178–193 (2023)
40. Yu, J., Dickinger, A., So, K.K.F., Egger, R.: Artificial intelligence-generated virtual influencer: examining the effects of emotional display on user engagement. J. Retail. Consum. Serv. **76**(3), 103560 (2024)

41. Lou, C., Kiew, S.T.J., Chen, T., Lee, T.Y.M., Ong, J.E.C., Phua, Z.: Authentically fake? how consumers respond to the influence of virtual influencers. J. Advert. **52**(4), 540–557 (2023)
42. Chimbga, B.: Exploring the ethical and societal concerns of generative AI in Internet of Things (IoT) environments. In: Southern African Conference for Artificial Intelligence Research, pp. 44–56. Springer Nature, Switzerland, Cham (2023)
43. Ye, G., Hudders, L., De Jans, S., De Veirman, M.: The value of influencer marketing for business: a bibliometric analysis and managerial implications. J. Advert. **50**(2), 160–178 (2021)
44. Pavlidou, V., Otterbacher, J., Kleanthous, S.: Fairness issues in algorithmic digital marketing: marketers' perceptions. In: European, Mediterranean, and Middle Eastern Conference on Information Systems, pp. 319–338. Springer Nature, Switzerland, Cham (2022)
45. Nguyen, C., Nguyen, T., Luu, V.: Predicting the relationship between influencer marketing and purchase intention: focusing on Gen Z consumers. In: International Econometric Conference of Vietna, pp. 467–481. Springer International Publishing, United Kingdom, Cham (2022)
46. Kilimas, F.R., Chandra, A., Rustagi, N.: Marketing 4.0: emerging technologies that are refining digital marketing. Stakeholder Wellbeing Value Creation **12**(3), 83–106 (2022)
47. Kahn, B.E.: Using visual design to improve customer perceptions of online assortments. J. Retail. **93**(1), 29–42 (2017)
48. Lee, L., Inman, J.J., Argo, J.J., Böttger, T., Dholakia, U., Gilbride, T., Tsai, C.I.: From browsing to buying and beyond: the needs-adaptive shopper journey model. J. Assoc. Consum. Res. **3**(3), 277–293 (2018)
49. Mero, J., Vanninen, H., Keränen, J.: B2B influencer marketing: Conceptualization and four managerial strategies. Ind. Mark. Manag. **108**(8), 79–93 (2023)
50. Stone, M., Aravopoulou, E., Ekinci, Y., Evans, G., Hobbs, M., Labib, A., Machtynger, L.: Artificial intelligence (AI) in strategic marketing decision-making: a research agenda. Bottom Line **33**(2), 183–200 (2020)
51. Ortová, N., Hejlová, D., Weiss, D.: Creation of a code of ethics for influencer marketing: the case of the Czech Republic. J. Media Ethics **38**(2), 65–79 (2023)
52. Cardell, K., Douglas, K., Maguire, E.: 'Stories': social media and ephemeral narratives as memoir. In: Mediating Memory, pp. 157–172. Routledge (2017)
53. Ducman, A.A.: Redefining wine brand engagement through influencer marketing and ephemeral content. Theor. Appl. Econ. **31**(1) (2024)
54. Hilvert-Bruce, Z., Neill, J.T., Sjöblom, M., Hamari, J.: Social motivations of live-streaming viewer engagement on Twitch. Comput. Hum. Behav. **84**, 58–67 (2018)
55. Fietkiewicz, K.J., Dorsch, I., Scheibe, K., Zimmer, F., Stock, W.G.: Dreaming of stardom and money: Micro-celebrities and influencers on live streaming services. In: Social Computing and Social Media. User Experience and Behavior: 10th International Conference, SCSM 2018, Held as Part of HCI International 2018, Las Vegas, NV, USA, 2018, Proceedings, Part I 10, pp. 240–253. Springer International Publishing (2018)
56. Borchers, N.S., Enke, N.: I've never seen a client say:'Tell the influencer not to label this as sponsored: an exploration into influencer industry ethics. Public Relat. Rev. **48**(5), 102235 (2022)

Revolutionizing Influencer Marketing: Harnessing the Power of Data Analytics and Artificial Intelligence (AI)

Khalid Aziz, Sakshi Dua, and Prabal Gupta

Abstract In the dynamically evolving landscape of digital marketing, the emergence of influencer marketing has marked a paradigm shift in how brands connect with their audiences. This chapter delves into the transformative role of data analytics and AI in reshaping influencer marketing strategies. We begin by exploring the fundamental concepts of influencer marketing, highlighting its significance in the contemporary digital era. This chapter then transitions into an in-depth analysis of how data analytics has become an indispensable tool in identifying the right influencers, understanding audience demographics, and measuring campaign effectiveness. We further examine the innovative applications of AI in influencer marketing, from predictive analytics for trend forecasting to the utilization of machine learning algorithms for optimizing content engagement and authenticity. A significant focus is placed on the ethical considerations and challenges posed by the integration of AI, including issues of data privacy and the authenticity of influencer content. Real-world case studies are interwoven throughout the chapter, providing practical insights into successful strategies and common pitfalls. This chapter aims to equip marketers, businesses, and academics with a thorough comprehension of the synergies between influencer marketing, data analytics, and AI, and how these can be leveraged to create more impactful, data-driven marketing campaigns in the digital age.

Keywords Influencer marketing · Digital marketing · Data analytics · Artificial Intelligence (AI) · Predictive analytics · Machine Learning (ML) · Content engagement

---

K. Aziz · S. Dua
SCA, Lovely Professional University (LPU), Phagwara, Punjab, India
e-mail: khalid.12105643@lpu.in

S. Dua
e-mail: sakshi.26557@lpu.co.in

P. Gupta (✉)
ECE, SCSE, Lovely Professional University (LPU), Phagwara, Punjab, India
e-mail: prabal.gupta9@gmail.com; prabal.20263@lpu.co.in

S. Dutta et al. (eds.), *Advances in Data Analytics for Influencer Marketing: An Interdisciplinary Approach*, Information Systems Engineering and Management 9,
https://doi.org/10.1007/978-3-031-65727-6_4

1 Introduction

The incorporation of Artificial Intelligence (AI) and Data Analytics into marketing has opened new frontiers, transforming traditional marketing landscapes into dynamic, data-driven environments. These technologies have not only revolutionized the approach brands cooperate with clients but have also paved the way for more personalized, efficient, and impactful marketing strategies. As we delve into the realm of influencer marketing, the implications of AI and Data Analytics become particularly pronounced, offering a wealth of opportunities for brands to engage with their audiences on a more profound level. AI's vast potential in marketing is well-documented, aiding in the proliferation of data and facts bases, refining software's information supervision proficiencies, and scheming progressive algorithms [1]. This technological evolution has fundamentally altered the interaction between brands and users, making marketing efforts more customer-centric and responsive to real-time needs. AI tools enable marketers to determine optimal content and channel strategies swiftly, thanks to the rich insights gleaned from data analytics, thereby enhancing user engagement and conversion rates. Maximize Market Research Pvt. Ltd. [2] highlights the dynamic and expanding nature of the influencer marketing market, underscoring its significant growth from a valuation of US$ 13.04 billion in 2022 to an anticipated US$ 82.74 billion by 2029, with a Compound Annual Growth Rate (CAGR) of 30.2%. The widespread access to high-speed, affordable internet has revolutionized advertising strategies, enabling marketers to efficiently connect with and captivate their desired demographics via influencer marketing. Platforms like Facebook, Instagram, YouTube, and Pinterest have become essential for influencer marketing, providing a robust user base for marketers to enhance customer engagement and interaction. These platforms facilitate content-driven publicity, allowing for the creation of a digital identity that fosters deeper connections with audiences. AI and Facts Analytics: The addition of artificial intelligence (AI) and information analytics in influencer promotion is evident in the emphasis on operation supervision, hunt and detection, analytics and reportage, and influencer management. AI and analytics play crucial roles in identifying the right influencers and optimizing campaigns for better performance and engagement, thereby ensuring more targeted and effective marketing strategies [2].

The pandemic has accelerated the adoption of influencer marketing as traditional advertising shoots were halted, and businesses sought cost-effective promotional methods. The increased reliance on social media during this time further boosted the appeal of influencer marketing. Despite its growth, the influencer marketing sector faces challenges such as measuring impact and dealing with fake followers. However, the advancements in AI and data analytics offer solutions by enhancing the ability to track performance and authenticity, thus improving overall marketing effectiveness.

Figure 1 compares the market size of the influencer marketing industry in two different years, 2022 and 2029. The heights of the bars represent the market size in US dollars, with the market projected to grow significantly from 2022 to 2029. The label "CAGR 30.2%" denotes the compound annual growth rate, indicating an average annual growth rate of 30.2% over the period. This high growth rate suggests a rapidly expanding industry, signifying increasing investment and interest in influencer marketing over the forecast period [2]. Figure 2. breaks down the influencer marketing market by region for the year 2022. Each slice of the pie corresponds to a different region, such as North America, Europe, Asia Pacific, Middle East and Africa, and South America. The size of each slice represents the percentage of the total market attributed to each region, with North America having the largest share in this example. This distribution helps in understanding which regions are the most significant players in the influencer marketing industry.

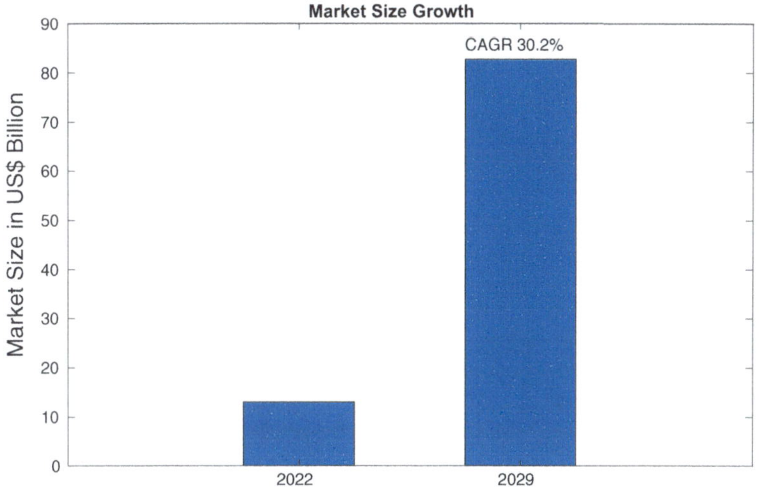

Fig. 1 Influencer marketing industry

Fig. 2 Regional analysis in influencer marketing industry

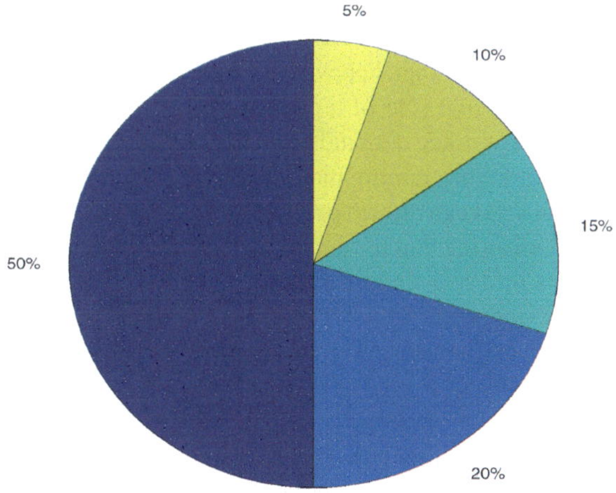

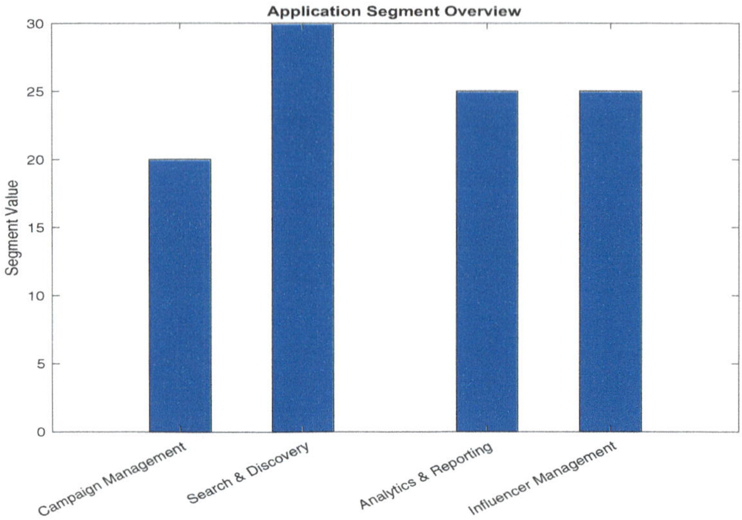

Fig. 3 Application segment influencer marketing

Figure 3 provides an overview of different application segments within the influencer marketing industry, spread across various years. It showcases the value or importance of application segments such as Campaign Organization, Hunt and Detection, Analytics and Reportage, and Influencer Management. Each bar's height indicates the segment's value or focus on a specific year, illustrating the evolution and projected emphasis on different influencer marketing tools and strategies over time. This chart can be useful for identifying trends and predicting future industry shifts in application preferences [2].

AI can assist in determining the optimal pricing approaches involving the examination of market trends, competitor pricing, and consumer demand. AI can help in identifying the right customer segments and positioning products effectively by analyzing consumer behavior and preferences (Fig. 4). This involves using AI to manage the lifecycle of a product, from development to sales, ensuring that products meet market demands. AI can optimize promotional activities by personalizing marketing communications and predicting the most effective channels and messages. AI can inform product design by analyzing customer feedback and market trends, ensuring that products meet the evolving needs of consumers. AI can provide insights into real-time market conditions, allowing for dynamic pricing and inventory management based on current demand fluctuations. AI can play a strategic role in planning marketing campaigns, resource allocation, and long-term marketing strategies by forecasting future market developments [1].

As we delve into the transformative role of data analytics and AI in influencer marketing, it's imperative to establish the fundamental metrics that guide strategic decision-making in this domain. Two pivotal metrics, Return on Investment (ROI)

Fig. 4 AI assistance in marketing

and Cost-Per-Engagement (CPE), serve as the bedrock for assessing the effectiveness and efficiency of influencer campaigns.

Return on Investment (ROI) is a universally recognized metric used across various marketing channels to assess the success of an speculation. Within the realm of influencer marketing, the ROI equation is given by:

$$ROI = \frac{Revenue\,from\,Campaign - \text{Cos}\,t\,of\,Compaign}{Cost\,of\,Compaign} \times 100 \quad (1)$$

This equation quantifies the return generated from influencer marketing efforts relative to the cost incurred. A higher ROI signifies a more profitable campaign, indicating that the influencer's content resonated well with the target audience, driving substantial revenue in comparison to the campaign's expenses [3].

Cost-Per-Engagement (CPE), on the other hand, offers a granular view of the campaign's engagement efficiency. It is calculated as:

$$CPE = \frac{Total\,Cost\,of\,Compaign}{Total\,Engagement} \quad (2)$$

Engagements—likes, comments, shares, and clicks—are vital indicators of audience interaction and interest. The CPE metric helps marketers understand the cost associated with each act of engagement, providing insights into the value derived from the campaign. Lower CPE values are indicative of higher engagement rates achieved at a lower cost, highlighting the campaign's efficiency in captivating the audience's attention [3].

Both ROI and CPE are instrumental in navigating the influencer marketing landscape. They allow brands to measure the tangible impact of their campaigns, guiding resource allocation and strategy optimization. As we progress into the era of data-informed marketing, the integration of data analytics and AI tools further empowers

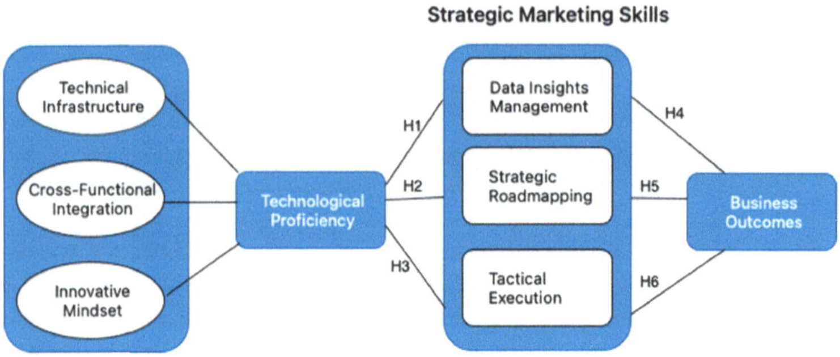

Fig. 5 Organization's skills and abilities in the realm of Artificial Intelligence (AI)

marketers to enhance these metrics, tailoring influencer selection, content creation, and campaign execution to achieve unparalleled levels of campaign performance and audience engagement. The figure presents a conceptual model outlining the relationship between Artificial Intelligence (AI) capabilities and structural presentation, specifically within the context of Business-to-Business (B2B) marketing. The model posits that AI competencies are a composite construct formed by three components, which in turn influence B2B marketing capabilities and ultimately organizational performance (Fig. 5).

In the study conducted by Mikalef et al. [4], Technological Proficiency stands out as the central construct in their model, representing an organization's skills and abilities in the realm of Artificial Intelligence (AI). This proficiency is determined by three key factors: Technical Infrastructure, denoting the technical and technological framework necessary to support AI applications; Business, assessing the extent to which AI is integrated across various business domains; and Innovative Mindset, reflecting the organization's forward-thinking and innovative approach towards adopting AI. Strategic Marketing Skills, identified as a mediating construct between AI Competencies and Business Outcome, is further disaggregated into three components: Data Insights Management, gauging the ability to efficiently handle large volumes of data using AI for actionable marketing insights; Strategic Roadmap, emphasizing the strategic influence of AI on the development of marketing strategies; and Tactical Execution, highlighting the role of AI in the operationalization and application of marketing plans.

The research model posits six hypotheses (H1 to H6) suggesting causal relationships between these constructs: H1 proposes a optimistic effect of AI Capabilities on Information Organization competences; H2 suggests a positive effect of AI Competencies on scheduling competences; H3 postulates an optimistic effect of AI Capabilities on Operation capabilities; H4 anticipates a optimistic impact of Information Management capabilities on Organizational Performance; H5 predicts a positive impact of Planning capabilities on Organizational Performance; and H6 hypothesizes a positive impact of Implementation capabilities on Organizational Performance.

The extensive incorporation of AI across various sectors, particularly within B2B marketing, highlights its capacity to offer valuable insights into customer behaviors and enhance operational efficiencies. However, refining AI competencies to optimize organizational performance demands thoughtful consideration. Research suggests that the proficiency in AI plays a direct role in shaping B2B marketing capabilities and, consequently, the overall effectiveness of the organization, underscoring the strategic importance of AI in augmenting marketing outcomes [4].

In the domain of B2B operations, AI's influence extends beyond operational efficiency, enabling businesses to navigate dynamic and uncertain market landscapes. Examining this through the lens of dynamic capabilities reveals specific microfoundations linked to AI, showcasing how organizations can efficiently deploy AI for B2B marketing operations. This perspective not only underscores the significance of AI in adapting to market shifts but also emphasizes its role in generating business value through strategic marketing operations [5]. The intersection of AI capabilities with circular business models (CBMs) provides a glimpse into the future of sustainable business practices. AI functions as an efficiency accelerator, assisting firms in overcoming challenges tied to CBM implementation. By integrating intelligence, automating processes, and orchestrating ecosystems, AI facilitates the transition toward more sustainable and efficient business operations. This alignment between AI and CBMs elucidates the broader implications of AI in influencing environmental and market performance [6]. Despite the apparent advantages, the adoption of AI among small and medium-sized enterprises (SMEs) encounters obstacles, including competitive pressure and the imperative for organizational readiness. Research underscores the pivotal role of top management commitment and organizational readiness in facilitating successful AI adoption. Addressing these challenges becomes imperative for SMEs to effectively harness AI and enhance their competitive positioning in the market [7].

2 Literature Review

Artificial Intelligence (AI) and Machine Learning (ML) are reshaping the marketing landscape by augmenting data management capabilities and facilitating the development of advanced algorithms. The incorporation of AI in marketing brings about significant changes in how brands engage with consumers, allowing for real-time personalization and targeted content delivery. AI's role extends to analyzing competitors' campaigns and understanding customer expectations, with ML algorithms improving in performance as they process more data [1]. This transformative potential is particularly evident in Business-to-Business (B2B) marketing, where AI provides insights into customer behaviors and enhances operational efficiencies [4].

The ever-changing landscape of influencer marketing demands continuous vigilance to understand the latest challenges and risks confronting both brands and influencers. Alipour et al. [8] conducted a comprehensive review of existing literature to identify and categorize the threats associated with influencer marketing. Their study revealed 11 notable challenges, which encompassed issues such as counterfeit followers and comments, inadequate disclosure practices, improper influencer selection, excessive brand oversight, insufficient control over content by brands, inappropriate platform choices, adverse effects on brand perception, detrimental impacts on consumer behavior (especially among younger demographics), absence of formal agreements, rising market saturation, and concurrent advertising campaigns. Of particular significance were concerns surrounding the selection of incompatible influencers, platform mismatches, and the detrimental influence of influencers on brand image. As influencer marketing strategies grow in complexity, there's an increasing demand for data-centric approaches to fine-tune tactics and optimize returns on investment. Doshi et al. [9] crafted an agent-based model (ABM) to simulate the intricate dynamics of influencer advertising campaigns across diverse scenarios, aiding brands in uncovering the most effective influencer marketing strategies. Their model, grounded in probabilistic graph theory, factors in real-world elements such as consumer interests, behaviors, willingness to spend, brand investment limits, influencer engagement in diffusion, and product characteristics (luxury vs. non-luxury). Through the evaluation of customer acquisition costs and conversion rates, the study showcased the contextual nature of influencer marketing, offering insights into the most suitable influencer types for various contexts. For instance, it highlighted that as products transition from luxury to non-luxury, the effectiveness of celebrities diminishes, while that of nano-influencers thrives. Conversely, when consumer interest wanes, nano-influencers' performance declines while celebrities' effectiveness rises.

Acknowledging the rising significance of online influencer marketing (OIM), [10] delivered an in-depth exploration of this phenomenon, elucidating OIM's definition and delineating its distinctive characteristics. Drawing from interviews with industry experts and consumers and grounded in social capital theory, the authors posited that OIM harnesses influencer assets (follower networks, personal branding, communication content, and follower trust) to bolster a firm's marketing communication efficacy. They articulated six innovative propositions that illuminate the advantages and potential pitfalls of OIM, laying the groundwork for an OIM theory. Moreover, the authors outlined essential strategies for proficiently managing OIM and identified critical gaps between academic literature and practical application, suggesting avenues for further scholarly inquiry. The integration of artificial intelligence (AI) within organizations, particularly in B2B contexts, has demonstrated promising enhancements in marketing capabilities and overall organizational performance. A study based on survey responses from European firms illustrates how AI competencies directly impact B2B marketing capabilities, thereby influencing the overall success of the organization [4]. This underscores the strategic importance of effectively organizing AI competencies to fully leverage their potential in steering business outcomes. Furthermore, AI serves as a facilitator of dynamic capabilities in B2B marketing operations, highlighting its value in managing marketing activities in unpredictable environments and contributing to the maintenance of a competitive edge [6]. The synergy between AI capabilities and circular business models (CBMs) positions AI as a catalyst for efficiency, assisting companies in overcoming challenges associated with the implementation of AI-enabled CBMs. Integrated intelligence, process automation, and ecosystem orchestration emerge as vital factors for the successful adoption of AI-enabled CBMs, offering insight into the future of sustainable business practices [6].

Research on AI adoption among small and medium-sized enterprises (SMEs) in Malaysia emphasizes the significance of top management commitment and organizational readiness. Contrary to competitive pressure, employee adaptability, and external support, these factors play a pivotal role in successfully adopting AI, suggesting a focus on enhancing management commitment and readiness to improve AI outcomes [7]. A global perspective on AI technologies in marketing highlights economic disparities, the need for glocalization, and consumer privacy concerns. This analysis sheds light on the heterogeneity of AI adoption across different levels—country, company, and consumer—and introduces research opportunities focusing on human–machine interaction and automated content analysis [11].

Applying the dynamic capabilities perspective to investigate the influence of AI on supply chain analytics during the COVID-19 pandemic reveals how alliance management capabilities, enhanced by AI-driven analytics, enhance organizational performance. This research underscores the significance of AI in navigating operational challenges amid environmental dynamism [12]. The progression of data-driven marketing, characterized by the integration of emerging technologies, confronts issues such as privacy concerns and the necessity for organizational change. A systematic review identifies the advantages and obstacles of utilizing consumer data, advocating for a framework to steer businesses in implementing data-driven marketing strategies [13]. The digitization of B2B customer journeys, facilitated by AI-empowered tools, requires a more profound comprehension of managing these digital interactions. An empirical study pinpoints four crucial management activities—analyze, design, engage, and guide—supported by AI tools, contributing to the literature on digital marketing and customer journey management in the B2B context [14].

Exploration into cognitive trust in AI among frontline employees in Pakistan indicates that the positive impact of AI features on trust is counteracted by disruptions in work routines. This study makes a valuable contribution to the literature on technology trust, particularly in developing economies, emphasizing the significance of data governance in establishing cognitive trust in AI [15]. The involvement of AI in the field of education necessitates an assessment of the feasibility of AI platforms for pedagogical methods. An examination of challenges and approaches to AI adoption in education proposes the utilization of tools such as "ChatGPT" as complementary aids, addressing challenges across user experience, operations, environment, technology, and ethics [16]. A review of literature on AI and the Internet of Things (IoT) within smart home ecosystems (SHE) delineates the advantages and challenges associated with these technologies in strategic decision-making. This analysis establishes a framework for comprehending value generation through AI and IoT, providing insights into novel business models and corporate advancement within the realm of SHE [17].

Research on the role of AI in accelerating the business environment, especially in the fashion industry, shows that AI-driven innovation has the potential to impact the environment and market in weather services. This study provides a framework that demonstrates the potential of intelligence in improving mitigation, adaptation, and resilience for climate action [17]. Although influencer marketing is often explored in the B2C context, its application in B2B marketing presents unique opportunities and challenges. A study based on interviews with senior business leaders across multiple B2B industries shows the business impact in the B2B space. It identifies four different management strategies used by B2B organizations, expanding the impact of marketing beyond customer experience, and highlighting the importance of professional and business products in service delivery [18]. Collaboration between brands and social media influencers (SMIs) is the foundation of business marketing. Using business value analysis, the research examines the business's relationship with SMIs, identifying key management strategies that reduce the risk of business waste and increase sustainability outside of business. This exploration of collaborative management provides a foundation for effective collaborative marketing and highlights the importance of management in the complexities of influencer marketing [19].

The evolution of social networks within the business mix underscores the importance of influencers and thought leaders in shaping users' understanding and decision-making. A qualitative literature review develops a research vocabulary that supports the understanding of thought leadership in the commercial business process, transferring knowledge from B2C influencer marketing to B2B thought leadership. This connection fosters a unified approach to business impact in B2B and B2C environments [20]. The ethics of business values have come under strict scrutiny, especially in terms of promoting events on platforms like TikTok. A study investigating the impact of promotional advertising on youth persuasion and product outcomes highlights the need for clarity in marketing interventions. It shows that advertising moral support can improve advertising awareness and understanding of emotional support, suggesting ethical behavior that promotes consumer trust and informed decision-making [21].

The impact of AI on customer experience in digital marketing, particularly conversational AI models like ChatGPT, demonstrates the technology's ability to transform user engagement strategies into products. This study investigates the impact of ChatGPT on consumers, business type, technology awareness, and demographic factors, providing insights into the effective integration of AI in enhancing digital marketing efforts and customer satisfaction [22, 23]. Another study explores the evolving landscape of consumer behavior and decision-making, focusing on the intersections of social media and artificial intelligence (AI). In the context of social media, the study delves into the five stages of consumer behavior, from problem detection to post-purchase evaluation. It highlights the impact of social media marketing on customer behavior, noting a trend where consumers increasingly discover and purchase products on social media platforms, especially in categories like clothing, cosmetics, and home goods.

The study underscores the influential role of influencers and branded accounts in shaping consumer purchase decisions, surpassing the impact of traditional celebrities. Emphasizing the importance of a strong online presence and reputation management, also discusses how AI revolutionizes understanding and influencing customer behavior. It explores the significance of AI in processing large datasets (big data) for customized marketing campaigns, identifying trends, and patterns that traditional analytical approaches may miss. The report further delves into the implications of AI for influencer marketing [24]. Barnes and Rutter [25] using Big data and analyze posts from the top 75 social media influencers over a period of 12 months. They used a convolutional neural network (Inception V3 CNN) to identify the types of products featured in 226,801 images. The study aimed to understand the efficiency of engagement based on the type of product and influencer. The results revealed that general influencers tended to perform the best in terms of overall engagement. However, travel influencers stood out for achieving higher engagement regardless of the type of product. Moreover, the research highlighted that certain types of influencers were more effective for specific product categories. The findings emphasize the importance of matching the right influencer with the appropriate product to achieve maximum impact through influencer marketing (Table 1).

3 The Impact of Data Analytics in Influencer Marketing

Data analytics has transformed influencer marketing by enabling brands to measure, understand, and optimize their campaigns more effectively than ever before. Two pivotal analytical methods, the calculation of engagement rates and audience segmentation through K-Means clustering, exemplify this transformation.

3.1 Engagement Rate Calculation

One of the most critical metrics in influencer marketing is the Engagement Rate, which offers a direct measure of how actively involved the audience is with the influencer's content. The engagement rate can be mathematically represented as:

$$Engagement\ Rate = \frac{(Total\ Engagement\ on\ a\ Post)}{Total\ Followes} \times 100 \tag{3}$$

This equation calculates the percentage of the influencer's followers who have interacted with a specific post through likes, comments, shares, or other forms of engagement. A higher engagement rate signifies a more engaged and potentially more influential audience. It not only helps in assessing the performance of individual posts but also in evaluating the influencer's overall ability to resonate with and activate their audience [26].

Table 1 Different approaches for influencer marketing

Work	Findings	Key Points	Citation
Examined the role of AI and ML in reshaping marketing landscape	AI augments data management and facilitates advanced algorithms for personalization, targeted content delivery, and analyzing competitors	AI's transformative potential in marketing	Haleem et al. [1]
Explored the influence of artificial intelligence (AI) skills on business-to-business (B2B) marketing capacities and company effectiveness	AI competencies directly impact B2B marketing capabilities, influencing overall organizational success. AI serves as a facilitator of dynamic capabilities in B2B marketing operations	Strategic importance of organizing AI competencies for business outcomes	Mikalef et al. [4], Madanaguli et al. [6]
Conducted a systematic literature review to identify influencer marketing threats	Identified 11 threats, including fake followers, lack of disclosure, inappropriate influencer selection, brand control issues, unsuitable platforms, negative brand influence, impact on consumption habits, lack of contracts, saturation, and parallel advertising	Prominent threats: inappropriate influencer selection, unsuitable platforms, negative brand influence	Alipour et al. [8]
Created an agent-driven simulation framework (ASIM) to emulate influencer marketing campaigns	The circumstance-dependent nature of influencer marketing suggests that as products vary from luxury to non-luxury, celebrity performance declines while nano-influencers improve. As customer interest decreases, nano-influencer performance declines, and celebrities improve	Optimal influencer type varies based on product nature and customer interest	Doshi et al. [9]
Provided an overview of online influencer marketing (OIM) and its unique features	Online influencer marketing (OIM) harnesses the resources of influencers (follower networks, personal branding, communication content, trust) to improve the effectiveness of marketing communication. Identified advantages, risks, and tactics for overseeing OIM	Defining OIM and its unique features, proposing a theory of OIM	Leung et al. [10]
Examined AI adoption among SMEs in Malaysia	Top management commitment and organizational readiness play a pivotal role in successful AI adoption, more than competitive pressure, employee adaptability, and external support	Importance of management commitment and readiness for AI adoption in SMEs	Lada et al. [7]

(continued)

Table 1 (continued)

Work	Findings	Key Points	Citation
Provided a global perspective on AI technologies in marketing	Highlighted economic disparities, the need for glocalization, and consumer privacy concerns. Introduced research opportunities on human–machine interaction and automated content analysis	Global challenges and research opportunities in AI marketing technologies	Kopalle et al. [11]
Investigated the influence of AI on supply chain analytics during COVID-19	Alliance management capabilities, enhanced by AI-driven analytics, improve organizational performance and help navigate operational challenges amid environmental dynamism	AI's role in supply chain analytics and navigating dynamic environments	Dubey et al. [12]

3.2 *Audience Segmentation Model (K-Means Clustering)*

To further refine influencer marketing strategies, audience segmentation becomes indispensable. The K-Means clustering algorithm is a powerful tool for segmenting an influencer's audience into distinct groups based on shared characteristics. The objective function of K-Means clustering is given by Punj and Stewart [27]

$$arg_s \min \sum_{i=1}^{k} \sum_{x \in s_i} ||x - \mu||^2 \tag{4}$$

This mathematical model seeks to partition the audience into k clusters, where each point x belongs to the cluster with the nearest mean μi, minimizing the variance within each cluster. This segmentation allows marketers to tailor their strategies to different audience segments, enhancing the relevance and effectiveness of their campaigns [27].

The engagement rate calculation provides a quick and straightforward metric for gauging content effectiveness, enabling marketers to identify high-performing content types and topics. Meanwhile, the application of K-Means clustering for audience segmentation reveals deeper insights into the diverse interests and preferences within the influencer's audience, facilitating more personalized and targeted marketing efforts. By harnessing these data analytics tools, marketers can significantly enhance their influencer marketing strategies, driving higher engagement, and fostering more meaningful connections with their target audience. In the realm of influencer marketing, data analytics serves as a cornerstone for strategic decision-making, enabling brands to optimize their influencer collaborations for maximum impact. Haleem et al. [1] provides a foundation by discussing how AI and machine learning, integral components of data analytics, transform marketing strategies through advanced data management and algorithm design.

The utilization of AI for analyzing consumer behavior and competitor campaigns directly pertains to the strategic deployment of influencer marketing campaigns. Rodriguez-Garcia et al. [17] focused on smart home ecosystems, this study offer insights into how AI and data analytics, including IoT, contribute to strategic decision-making. The principles discussed could be applicable to influencer marketing by highlighting the importance of data-driven strategies in targeting and engagement. Kopalle et al. [11] discusses the role of AI in marketing on a global scale, focusing on economic disparities, localization needs, and consumer privacy concerns Insights from this can illuminate how data analytics shapes influencer marketing across different markets and cultures, emphasizing the need for personalized and ethically grounded approaches.

4 Ethical Considerations and Challenges

4.1 Data Privacy and Consent

In the domain of data-centric influencer marketing, the procedure relies on the gathering and examination of substantial volumes of personal data, encompassing demographic specifics and interaction patterns. This circumstance brings forth noteworthy privacy apprehensions, particularly regarding the consent of individuals whose data is being amassed. Stringent regulations, like the European Union General Data Protection Regulation (GDPR) and the US California Consumer Privacy Act (CCPA), have been implemented, imposing stringent directives on practices related to data collection. These regulations highlight the imperative need for clear and explicit consent when utilizing gathered data [28]. To address the substantial challenge of data privacy in business dealings, the concept of a "privacy gap" emerges as a robust mathematical methodology. This approach ensures the preservation of confidentiality for vital information, even when aggregated to form a comprehensive perspective. The essence of this principle is encapsulated in the equation:

Differential Privacy for Data Anonymization:

$$\Pr[\kappa(D) \in S] \leq e^{\epsilon} \Pr[\kappa(D\prime) \in S] + \delta \tag{5}$$

- $\Pr[\kappa(D) \in S]$ represents the probability of a specific outcome S occurring when a query K is applied to dataset D.
- D, D' are datasets that differ by at most one element, highlighting the impact of an individual's data on the outcome.
- ϵ(epsilon) is a small positive value that quantifies the degree of privacy protection, with lower values indicating stronger privacy.
- e^{ϵ} represents the exponential function of ϵ, setting a bound on how much the inclusion or exclusion of a single data point can change the outcome probabilities.

- δ is a parameter that allows for a small probability of privacy breach, ideally as close to zero as possible.

  This equation underscores the ethical imperative of implementing data anonymization techniques that respect user privacy while enabling meaningful data analysis for influencer marketing strategies [29].

Authenticity Score for Influencer Content:

$$A = \frac{1}{N}\sum_{i=1}^{N} aunthencity(ci) \quad (6)$$

- A represents the Authenticity Score, reflecting the overall authenticity of the influencer's content.
- N is the total number of content pieces analysed.
- $aunthencity(ci)$ evaluates the authenticity of each content piece C_i potentially based on criteria such as transparency, disclosure of sponsorships, and alignment with the influencer's genuine opinions and lifestyle. Calculating the Authenticity Score offers a quantitative approach to ensuring that influencer campaigns promote genuine and transparent content, fostering trust and credibility among audiences [30].

4.2 Authenticity and Disclosure

The authenticity of influencer content is another ethical cornerstone. Influencers are trusted by their followers for their genuine insights and recommendations. However, the commercial nature of influencer-brand collaborations can sometimes blur the lines between authentic opinions and paid endorsements. The Federal Trade Commission (FTC) guidelines mandate clear disclosures when influencers have a financial or personal relationship with the brands they promote [31].

4.3 Algorithmic Bias and Fairness

AI-driven tools used in influencer marketing, from predictive analytics to recommendation systems, are not immune to biases inherent in their training data. These biases can manifest in skewed influencer recommendations, favouring certain demographics over others, thus perpetuating inequality and limiting diversity within influencer campaigns [32].

4.4 *The Impact on Mental Health*

The relentless pursuit of perfection, often portrayed in influencer marketing campaigns, can have detrimental effects on the mental health of both influencers and their followers. The pressure to conform to unrealistic standards of beauty, lifestyle, and success can exacerbate issues of self-esteem, anxiety, and depression [33]. Ethical influencer marketing practices should prioritize authenticity and inclusivity, promoting content that reflects a wide range of experiences and body types.

5 Case Studies and Success Stories

Mikalef et al. [4] examines a set of hypotheses using a survey-based approach. In this study, the authors developed a questionnaire based on previously published constructs or adapted them to fit their research context. They used a 7-point Likert scale for responses, which is standard for measuring subjective concepts. Before the main study, they conducted a pre-test with 22 organizations to ensure reliability and validity.

For the main study, the authors partnered with a panel service company to identify and collect data from respondents who were high-level IT managers in organizations operating in the Nordic region. The study aimed to assess organizations' use of AI in B2B marketing. Data collection took place between March and April 2022, with 155 complete responses received.

The authors use appropriate questions to ensure participants are familiar with AI applications for B2B marketing and have the knowledge to answer the questions correctly. They also examined sample bias by comparing early and late respondents and found no significant differences; This suggests that the sample is biased by company size, business, AI experience or feedback.

The pie charts represent significant aspects of a study on influencer marketing, with a focus on industry distribution, firm size, and respondents' positions within their organizations. Here's a general explanation of each pie chart and its implications for the study:

Respondents' Position Distribution Pie Chart This chart displays the roles of survey respondents within their organizations, including CEO/President, CIO, Head of Digital Strategy, Senior Vice President, Director, and Manager. The variety in respondents' positions indicates the study's comprehensive approach to gathering insights from multiple levels of organizational hierarchy. This diversity ensures a rich understanding of how influencer marketing is perceived and implemented across different leadership and management roles, offering a holistic view of strategic decision-making processes in influencer marketing campaigns.

The combined insights from these pie charts contribute to a nuanced understanding of influencer marketing's role and effectiveness across various industries, company sizes, and management levels. The industry distribution reveals the universal applicability of influencer marketing, the firm size distribution highlights strategic differences in its application, and the respondents' position distribution underscores the strategic importance of influencer marketing at all levels of decision-making within organizations.

Together, these charts underscore the multifaceted nature of influencer marketing, revealing its significance as a versatile and impactful marketing strategy in the digital age (Fig. 6).

Firm Size Distribution Pie Chart The firm size distribution chart illustrates the breakdown of participating organizations based on the number of employees, categorized into 1–9, 10–49, 50–249, and 250 + employees. This distribution is crucial for understanding how influencer marketing strategies may vary between small and medium enterprises (SMEs) and larger corporations.

Smaller firms might leverage influencer marketing for its cost-effectiveness and direct audience engagement, while larger firms might focus on scaling their influencer campaigns to broader audiences, reflecting different objectives and resource allocations in influencer marketing initiatives.

Industry Distribution Pie Chart This chart showcases the percentage of participants from various industries, including Bank and Financial data, Consumer Goods,

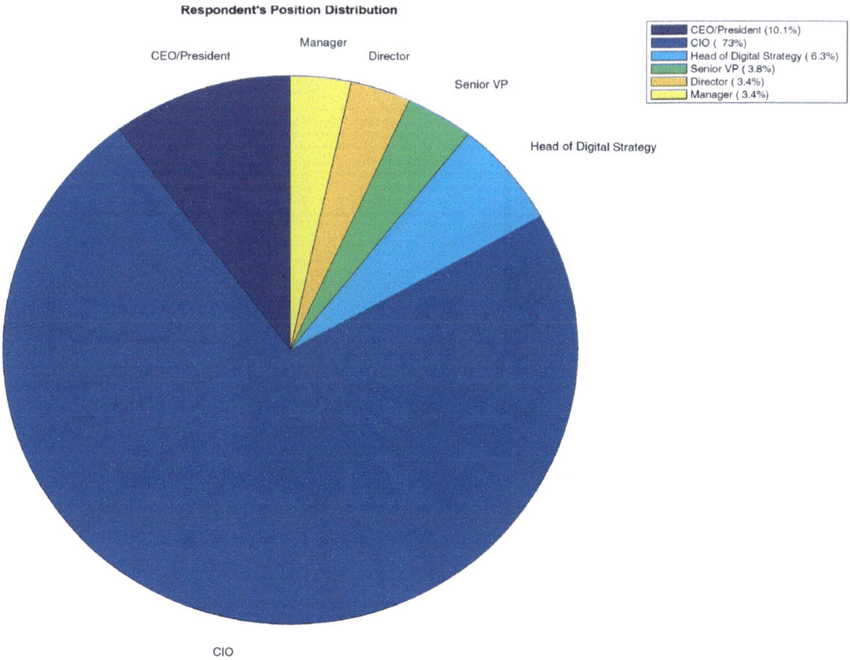

Fig. 6 **a** Respondent's Position Distribution. **b** Firm Size Distribution. **c** Industry Distribution

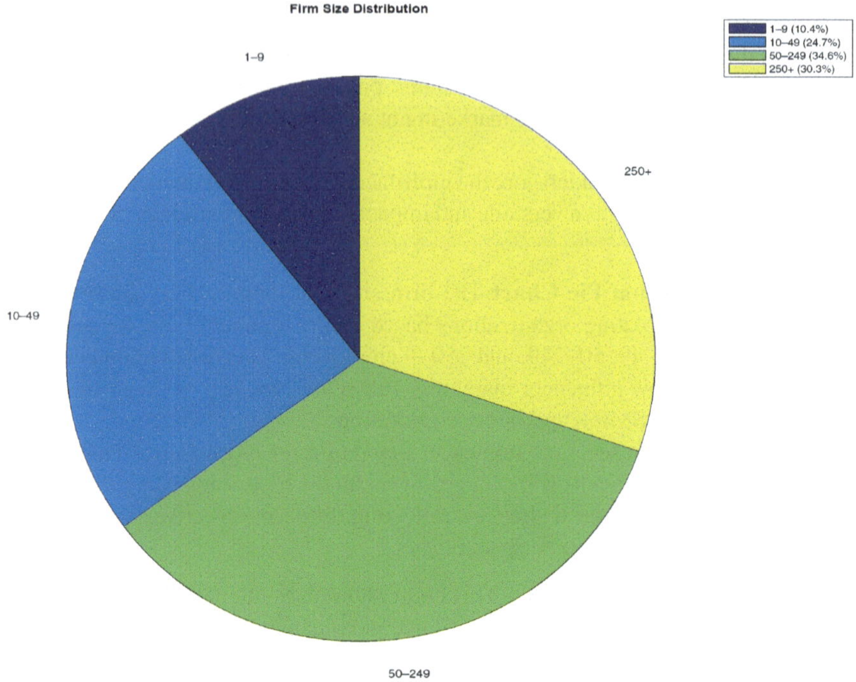

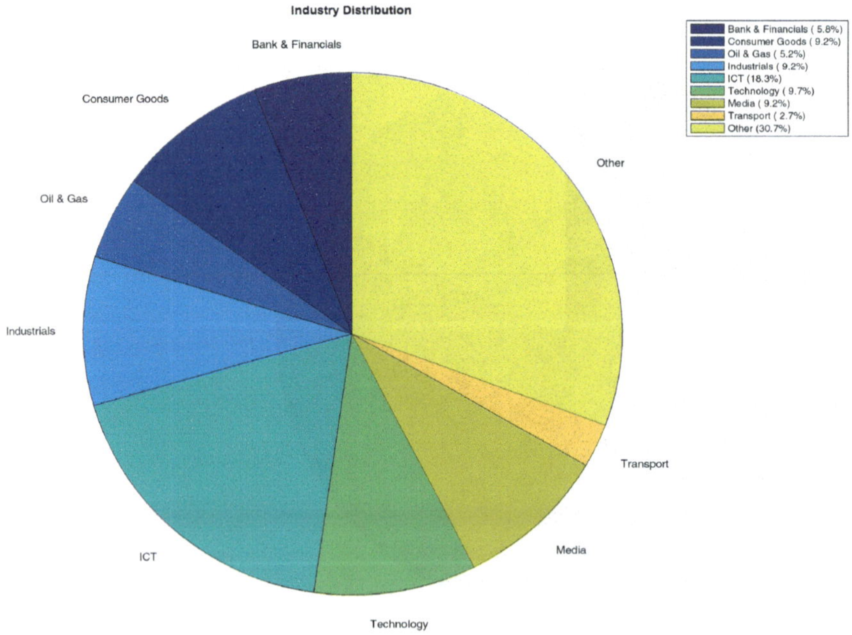

Fig. 6 (continued)

Oil and Gas, Industrials, ICT and Telecommunications, Technology, Media, Transport, and Others. The diversity in industry representation highlights the widespread relevance of influencer marketing across different sectors. It suggests that the impact of influencer marketing is not confined to traditionally consumer-focused industries but extends to a broad spectrum of fields, each with unique audience engagement strategies and marketing goals.

In another study [26], participants were recruited from Amazon's Mechanical Turk (MTurk) and targeted US residents to fit the study's social networking goals. Final data included 538 participants with a mean age of 33 years and a gender distribution of 62%. Most were white (76%), followed by African Americans (11.2%) and Asians (10.2%). Educational backgrounds vary, with nearly half completing a bachelor's degree. When it comes to social media use, the majority have Facebook (94%), YouTube (85%) and Instagram (70%) accounts. Participants follow influencers across multiple platforms, primarily in lifestyle, food and fashion.

6 Procedure

The study employed a screening process to identify regular social media users who followed at least one influencer. Participants completed an online survey on Qualtrics, answering questions about their social media habits, thoughts on influencers they follow, and demographic information. The survey completion was compensated financially.

7 Measurement

The value of influencer-created content was evaluated by participants' responses to the message on a 7-point semantic differential scale that captures both informational and entertainment value. Influencers' trustworthiness is measured by a four-item scale that measures people's trustworthiness (adapted from Munnukka et al. 2016). Different indicators and indicators are used to measure the trust and awareness of the products that produce them. Purchase intentions and engagement with following influencers are also evaluated (Fig. 7).

A Partial Least Squares (PLS) route molding tactic was utilized for data analysis. This method is suitable for theory development and prediction, especially in complex models with many latent variables. The study aimed to explore new theoretical frameworks rather than confirm established theories, making PLS-SEM a suitable choice.

This methodology provides a comprehensive framework for understanding the influence of social media marketing, combining quantitative measures of content value, influencer credibility, and consumer trust.

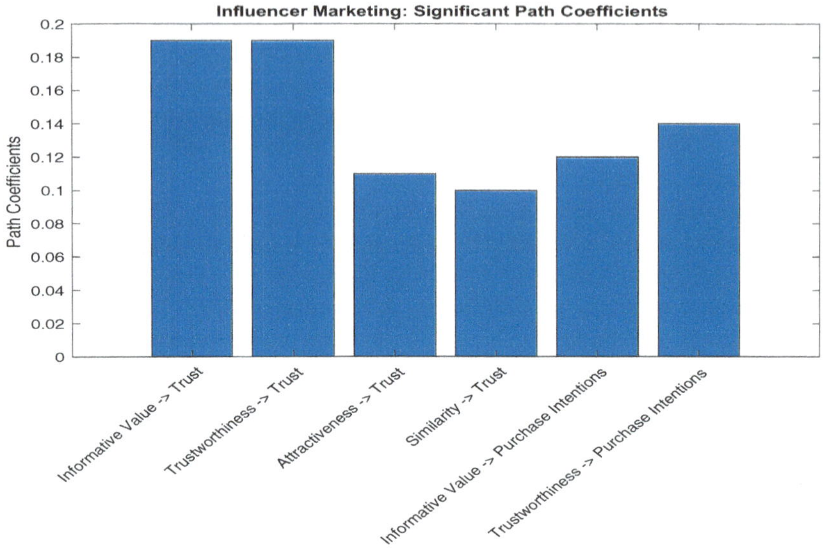

Fig. 7 Influencer marketing: significant path coefficients

To mathematically represent the methodology and results of a study focused on influencer marketing, we can consider a simplified approach based on the constructs and their relationships as described. For instance, we could model the relationship between influencer credibility (IC), informativeness and entertainment value of content (IE), and consumer trust in branded posts (CT), leading to purchase intention (PI) as follows.

8 Model Representation

Let's denote:

- IC as Influencer Credibility
- IE as Informativeness and Entertainment Value
- CT as Consumer Trust
- PI as Purchase Intention

A Simple Linear Model could be:

$$CT = \beta_0 + \beta_1 \times IC + \beta_2 \times IE + \epsilon \tag{7}$$

$$PI = \alpha_0 + \alpha_1 \times CT + \epsilon' \tag{8}$$

- Where β_0, α_0 *are intercepts.*
- β_1, β_2, α_0 are coefficients representing the effect of each predictor on the outcome.
- ϵ, ϵ' *are error terms.*

9 Tools and Technologies in Revolutionizing Influencer Marketing

The integration of advanced tools and technologies is paramount in revolutionizing influencer marketing through the strategic application of data analytics and artificial intelligence (AI). These technologies not only streamline the influencer selection process but also enhance content personalization, engagement analysis, and the overall effectiveness of influencer marketing campaigns.

9.1 Artificial Intelligence (AI) and Machine Learning (ML)

AI and ML are at the forefront of analyzing vast amounts of data to identify patterns and insights that humans may overlook. In influencer marketing, AI algorithms can predict campaign success, suggest optimal influencer-brand matches based on audience demographics, and even forecast future trends in consumer behavior. Tools like IBM Watson offer cognitive insights into unstructured social media data, enabling brands to discover influencers who resonate most with their target audience [34].

9.2 Natural Language Processing (NLP)

NLP technologies interpret and understand human language in a meaningful way. For influencer marketing, NLP is used in sentiment analysis to gauge public opinion and emotional responses to influencer content. This technology helps brands to measure the impact of influencer campaigns on brand sentiment and to tailor their messaging for maximum engagement [35].

9.3 Social Media Analytics Tools

Social media analytics tools, such as BuzzSumo and Hootsuite Insights, provide detailed metrics on influencer performance, including reach, engagement, and content effectiveness. These platforms leverage data analytics to offer real-time insights, enabling marketers to adjust their strategies promptly to optimize campaign results.

9.4 Influencer Marketing Platforms

Platforms like Traackr and AspireIQ streamline the influencer marketing workflow by offering end-to-end solutions, from influencer discovery and campaign management to analytics and reporting. These tools use AI to match brands with influencers whose followers mirror their target market, ensuring campaign alignment with audience preferences.

9.5 Predictive Analytics

Predictive analytics tools use historical data to make forecasts about future events. In influencer marketing, these tools can predict the potential ROI of influencer partnerships, helping brands to allocate their budgets more effectively. Google Analytics and Adobe Analytics are prominent examples that offer predictive insights based on user interaction data.

10 Future Directions in Influencer Marketing: Harnessing the Power of Data Analytics and Artificial Intelligence (AI)

As the digital landscape continues to evolve, the future of influencer marketing lies in the increasingly sophisticated use of data analytics and artificial intelligence (AI). These technologies promise to refine how brands engage with their audiences, offering more personalized, efficient, and impactful marketing strategies. Future advancements in AI will enable even more granular personalization of marketing content.

AI algorithms will be capable of analyzing individual consumer behaviour, preferences, and engagement history to match them with influencers who speak directly to their interests and needs. This hyper-personalization will not only improve audience targeting but also foster deeper connections between consumers and brands. Predictive analytics will play a crucial role in forecasting the outcomes of influencer marketing campaigns, allowing brands to make data-driven decisions with greater accuracy. By analyzing past campaign data, market trends, and consumer behaviour, predictive models can identify potential challenges and opportunities, guiding strategy development and resource allocation.

The integration of AR and VR technologies with influencer marketing will offer immersive experiences that captivate audiences in unprecedented ways. Influencers will be able to use these technologies to create more engaging and interactive content, from virtual try-ons for fashion and beauty products to immersive tours for travel destinations. This evolution will redefine audience engagement, making influencer marketing campaigns more memorable and impactful. Future developments in natural language processing (NLP) and sentiment analysis will provide deeper insights into consumer emotions and attitudes towards brands and influencers. Enhanced sentiment analysis tools will be able to detect nuances in consumer feedback across social media platforms, enabling brands to fine-tune their messaging and improve audience sentiment more effectively.

Blockchain technology holds the potential to introduce unprecedented transparency and trust in influencer marketing. By recording contracts and transactions on a secure, immutable ledger, blockchain can help verify the authenticity of influencers' followers, engagements, and content. This transparency will address concerns about influencer fraud and enhance the credibility of influencer marketing campaigns. As data analytics and AI become more ingrained in influencer marketing, ethical considerations and consumer privacy will come to the forefront. Brands and marketers will need to navigate the delicate balance between personalization and privacy, ensuring that consumer data is used responsibly. Adhering to ethical standards and regulations will be crucial in maintaining consumer trust and loyalty.

11 Conclusion

As a result, the combination of influencer marketing with data analytics and artificial intelligence has completely transformed digital marketing, providing influential brands with unprecedented opportunities to engage with their target audience. This chapter highlights the importance of business value in today's digital age and shows that data analysis can play a key role in adjusting strategies for selecting influencers, understanding audience dynamics, and measuring the success of marketing campaigns. Additionally, the use of a new concept of intelligence, from predictive analytics to machine learning algorithms, will increase the efficiency and accuracy of business interventions.

However, this chapter also highlights the ethical and decision-making issues involved in linking intelligence to intelligence. Data analysis in Influencer Marketing. Issues such as privacy and authenticity of influencer content are still important issues that require careful attention and responsible management.

Based on research data from around the world, this section gives an idea about the successes and failures of this rapid change. It is useful for business people, business and education by providing a good understanding of how to use business methods through data analysis and intelligence to create value in the digital age. effective, data-driven marketing. Finally, this chapter underscores the need for a balanced approach that allows for advances in technology while maintaining ethical standards and building relationships with audiences.

References

1. Haleem, A., Javaid, M., Asim Qadri, M., Pratap Singh, R., Suman, R.: Artificial intelligence (AI) applications for marketing: A literature-based study. Int. J. Intell. Netw. **3**, 119–132. KeAi Communications Co. (2022). https://doi.org/10.1016/j.ijin.2022.08.005
2. Maximize Market Research Pvt. Ltd.: Influencer Marketing Market: Global Industry Analysis and Forecast (2023–2029) (Report ID 63136). Published Date: June 2023. Accessed from MAXIMIZE MARKET RESEARCH PVT. LTD., Website: www.maximizemarketresearch.com
3. De Veirman, M., Cauberghe, V., Hudders, L.: Marketing through Instagram influencers: the impact of number of followers and product divergence on brand attitude. Int. J. Advert. **36**(5), 798–828 (2017). https://doi.org/10.1080/02650487.2017.1348035
4. Mikalef, P., Islam, N., Parida, V., Singh, H., Altwaijry, N.: Artificial intelligence (AI) competencies for organizational performance: A B2B marketing capabilities perspective. J. Bus. Res. **164** (2023). https://doi.org/10.1016/j.jbusres.2023.113998
5. Mikalef, P., Conboy, K., Krogstie, J.: Artificial intelligence as an enabler of B2B marketing: a dynamic capabilities micro-foundations approach. Ind. Mark. Manag. **98**, 80–92 (2021). https://doi.org/10.1016/j.indmarman.2021.08.003
6. Madanaguli, A., Sjödin, D., Parida, V., Mikalef, P.: Artificial intelligence capabilities for circular business models: research synthesis and future agenda. Technol. Forecasting Soc. Change **200** (2024). https://doi.org/10.1016/j.techfore.2023.123189
7. Lada, S., Chekima, B., Karim, M.R.A., Fabeil, N.F., Ayub, M.S., Amirul, S.M., Ansar, R., Bouteraa, M., Fook, L.M., & Zaki, H.O.: Determining factors related to artificial intelligence (AI) adoption among Malaysia's small and medium-sized businesses. J. Open Innovat. Technol. Market Complex. **9**(4) (2023). https://doi.org/10.1016/j.joitmc.2023.100144
8. Alipour, S.M., Ghaffari, M., Zare, H.: Influencer marketing research: a systematic literature review to identify influencer marketing threats. Manag. Rev. Q. (2024). https://doi.org/10.1007/s11301-024-00412-5
9. Doshi, R., Ramesh, A., Rao, S.: Modeling influencer marketing campaigns in social networks. IEEE Trans.Comput. Soc. Syst. **10**(1), 322–334 (2023). https://doi.org/10.1109/TCSS.2022.3140779
10. Leung, F.F., Gu, F.F., Palmatier, R.W.: Online influencer marketing. J. Acad. Mark. Sci. **50**, 226–251 (2022). https://doi.org/10.1007/s11747-021-00829-4
11. Kopalle, P.K., Gangwar, M., Kaplan, A., Ramachandran, D., Reinartz, W., Rindfleisch, A.: Examining artificial intelligence (AI) technologies in marketing via a global lens: Current trends and future research opportunities. Int. J. Res. Market. **39**(2), 522–540 (2022). https://doi.org/10.1016/j.ijresmar.2021.11.002
12. Dubey, R., Bryde, D.J., Blome, C., Roubaud, D., Giannakis, M.: Facilitating artificial intelligence powered supply chain analytics through alliance management during the pandemic crises in the B2B context. Ind. Market. Manag. **96**, 135–146 (2021). https://doi.org/10.1016/j.indmarman.2021.05.003
13. Rosário, A.T., Dias, J.C.: How has data-driven marketing evolved: Challenges and opportunities with emerging technologies. Int. J. Inf. Manag. Data Insights **3**(2) (2023). https://doi.org/10.1016/j.jjimei.2023.100203
14. Rusthollkarhu, S., Toukola, S., Aarikka-Stenroos, L., Mahlamäki, T.: Managing B2B customer journeys in digital era: four management activities with artificial intelligence-empowered tools. Ind. Mark. Manag. **104**, 241–257 (2022). https://doi.org/10.1016/j.indmarman.2022.04.014
15. Shamim, S., Yang, Y., Ul Zia, N., Khan, Z., Shariq, S.M.: Mechanisms of cognitive trust development in artificial intelligence among front line employees: An empirical examination from a developing economy. J. Bus. Res. **167** (2023). https://doi.org/10.1016/j.jbusres.2023.114168
16. Ali, O., Murray, P.A., Momin, M., Dwivedi, Y.K., Malik, T.: The effects of artificial intelligence applications in educational settings: challenges and strategies. Technol. Forecasting Soc. Change **199** (2024).https://doi.org/10.1016/j.techfore.2023.123076

17. Rodriguez-Garcia, P., Li, Y., Lopez-Lopez, D., Juan, A.A.: Strategic decision making in smart home ecosystems: A review on the use of artificial intelligence and Internet of things. In: Internet of Things (Netherlands), vol. 22. Elsevier B.V (2023). https://doi.org/10.1016/j.iot.2023.100772
18. Mero, J., Vanninen, H., Keränen, J.: B2B influencer marketing: conceptualization and four managerial strategies. Ind. Mark. Manag. **108**, 79–93 (2023). https://doi.org/10.1016/j.indmarman.2022.10.017
19. Syed, T.A., Mehmood, F., Qaiser, T.: Brand–SMI collaboration in influencer marketing campaigns: a transaction cost economics perspective. Technol. Forecasting Soc. Change **192** (2023). https://doi.org/10.1016/j.techfore.2023.122580
20. Neuhaus, T., Millemann, J.A., Nijssen, E.: Bridging the gap between B2B and B2C: thought leadership in industrial marketing–a systematic literature review and propositions. Ind. Mark. Manag. **106**, 99–111 (2022). https://doi.org/10.1016/j.indmarman.2022.08.006
21. van der Bend, D.L.M., Gijsman, N., Bucher, T., Shrewsbury, V.A., van Trijp, H., van Kleef, E.: Can I @handle it? The effects of sponsorship disclosure in TikTok influencer marketing videos with different product integration levels on adolescents' persuasion knowledge and brand outcomes. Comput. Hum. Behav. **144** (2023). https://doi.org/10.1016/j.chb.2023.107723
22. Abdelkader, O.A.: ChatGPT's influence on customer experience in digital marketing: investigating the moderating roles. Heliyon **9**(8) (2023). https://doi.org/10.1016/j.heliyon.2023.e18770
23. Javaid, M., Haleem, A., Singh, R.P., Khan, S., Khan, I.H.: Unlocking the opportunities through ChatGPT Tool towards ameliorating the education system. BenchCouncil Trans. Benchmarks, Standards Eval. **3**(2), 100115 (2023). https://doi.org/10.1016/j.tbench.2023.100115
24. Panda, V., Mishra, A., Sharma, M.: Turning data into insights: leveraging artificial intelligence for better understanding of social media consumer behaviour. In: 2023 International Conference on Sustainable Emerging Innovations in Engineering and Technology (ICSEIET), pp. 271–275. Ghaziabad, India (2023). https://doi.org/10.1109/ICSEIET58677.2023.10303309
25. Barnes, S., Rutter, R.: Using big data and AI to examine product engagement in social media influencer posts. In: 2019 4th International Conference on Information Technology, Information Systems and Electrical Engineering (ICITISEE), pp. 35–39. Yogyakarta, Indonesia (2019). https://doi.org/10.1109/ICITISEE48480.2019.9003991
26. Lou, C., Yuan, S.: Influencer marketing: how message value and credibility affect consumer trust of branded content on social media. J. Interact. Advert. **19**(1), 58–73 (2019). https://doi.org/10.1080/15252019.2018.1533501
27. Punj, G., Stewart, D.: Cluster analysis in marketing research: review and suggestions for application. J. Mark. Res. **20**(2), 134–148 (1983). https://doi.org/10.2307/3151680
28. Voigt, P., Von dem Bussche, A.: The EU General Data Protection Regulation (GDPR). Springer International Publishing (2017)
29. Dwork, C. (2008). Differential privacy: a survey of results. In: TAMC, vol. 4978, pp. 1–19. https://doi.org/10.1007/978-3-540-79228-4_1
30. Boerman, S.C.: The effects of the standardized Instagram disclosure for micro-and meso-influencers. Comput. Hum. Behav. **103**, 199–207 (2020). https://doi.org/10.1016/j.chb.2019.09.015
31. Federal Trade Commission.: Disclosures 101 for Social Media Influencers (2019)
32. Barocas, S., Hardt, M., Narayanan, A.: Fairness and abstraction in sociotechnical systems. In: ACM Conference on Fairness, Accountability, and Transparency (FAT) (2019)
33. Fardouly, J., Diedrichs, P.C., Vartanian, L.R., Halliwell, E.: Social comparisons on social media: the impact of Facebook on young women's body image concerns and mood. Body Image **13**, 38–45 (2015). ISSN 1740-1445
34. IBM: (2021). https://afluencer.com/artificial-intelligence-influencer-marketing/
35. Hutto, C., Gilbert, E.: VADER: a parsimonious rule-based model for sentiment analysis of social media text. In: Proceedings of the Eighth International AAAI Conference on Weblogs and Social Media (ICWSM 2014), vol. 8, no. 1, Article 21 (2014). https://doi.org/10.1609/icwsm.v8i1.14550

36. Belanche, D., Casaló, L.V., Flavián, M., Ibáñez-Sánchez, S.: Understanding influencer marketing: the role of congruence between influencers, products and consumers. J. Bus. Res. **132**, 186–195 (2021). https://doi.org/10.1016/j.jbusres.2021.03.067
37. Lou, C., Yuan, S.: Influencer marketing: how message value and credibility affect consumer trust of branded content on social media. J.. Interact. Advertising **19**(1), 58–73 (2019). https://doi.org/10.1080/15252019.2018.1533501
38. Barocas, S., Hardt, M., Narayanan, A.: Fairness and abstraction in sociotechnical systems. In: ACM Conference on Fairness, Accountability, and Transparency (FAT*) (2019)

Influencer Marketing Unleashed: Leveraging Data Analytics for Success

Divya Bansal, Naboshree Bhattacharya, and Priyanka Shandilya

Abstract As a result of the ongoing digital transformation, consumers are gaining increased knowledge as media users and are growing more discerning when it comes to commercial messages. Standing out in the midst of a multitude of advertisers has become exceptionally challenging for brands, prompting many to embrace Influencer marketing. For those at the forefront of purchasing decision-making, Influencer marketing is emerging as the most crucial and innovative approach to marketing in the past decade. This is primarily due to the inherent level of trust established between influencers and their audience, a trust that is extremely challenging for a brand to cultivate independently with its customers. The objective of this research chapter is to explore the reasons behind companies adopting influencer marketing as a promotional strategy and to gain insights into the steps involved in crafting an influencer marketing campaign specifically on Instagram. The aim is to delve deeper into the subject, focusing on themes such as influencer marketing as a whole and the intricacies of executing influencer marketing campaigns.

Keywords Consumer decision-making · Data-driven · Influencer marketing platforms

D. Bansal (✉) · P. Shandilya
Amity University, Noida, Uttar Pradesh, India
e-mail: divyadogra26@gmail.com

N. Bhattacharya
Education, Amity University, Ranchi, India

S. Dutta et al. (eds.), *Advances in Data Analytics for Influencer Marketing: An Interdisciplinary Approach*, Information Systems Engineering and Management 9,
https://doi.org/10.1007/978-3-031-65727-6_5

1 Introduction

1.1 Influencer Marketing

In today's landscape of online marketing, influencer marketing has become widely accepted as a conventional approach. It has gained significant attention over time and is frequently mentioned in mainstream media. The Influencer Marketing Hub has evolved into a well-established website featuring numerous articles that elucidate the complexities of influencer marketing, as well as various other forms of online marketing. Influencer marketing represents a fusion of traditional and contemporary marketing tools, merging the concept of celebrity endorsement with modern, content-driven marketing campaigns. What sets influencer marketing apart is that the outcomes of the campaigns result from collaborations between brands and influencers [1].

However, influencer marketing extends beyond celebrities and is centered around individuals who may not perceive themselves as famous in offline contexts.

The term influencer is defined as someone who possesses:

The ability to influence others' purchasing decisions because of their authority, experience, position, or relationship with their target audience. A dedicated following within a specific niche, with whom they actively interact. The size of their following is contingent on the scale of their niche topic. A significant misconception made by traditional media is the inability to distinguish between celebrities and online influencers [2].

It's crucial to understand that most influencers have methodically cultivated a dedicated and passionate audience. The fact that these individuals follow influencers rather than a brand is not coincidental. The audience is primarily interested in the influencers' opinions and is generally indifferent to the brand itself. It's advisable not to impose rules and business directives onto your influencers. Their audience belongs to them, and they have the ability to disengage, potentially taking their followers with them.

An early illustration of influencer marketing featured YouTube sensation PewDiePie, who collaborated with the creators of a horror film set in the catacombs beneath Paris. Together, they produced a series of videos showcasing challenges in the catacombs, aligning perfectly with PewDiePie's 111 million subscribers. The content garnered nearly double the views of the movie's trailer, resulting in a win–win situation for all parties involved.

Influencers, in contrast to celebrities, can emerge from any background or location. Their influence stems from having substantial followings on the internet and social media platforms. An influencer might be a prominent fashion photographer on Instagram, a well-regarded cybersecurity blogger active on Twitter, or a respected marketing executive with a strong presence on LinkedIn. In every industry, influential individuals exist; the key is to identify them. While some influencers may boast hundreds of thousands or even millions of followers, others may appear more like everyday individuals with perhaps only 10,000 followers, or even fewer. Nevertheless, they have established themselves as experts in their respective fields, becoming the go-to sources for answering people's questions. Depending on their areas of expertise, they are the ones who curate engaging social media posts, share high-quality images, create entertaining videos, and facilitate informative online discussions.

The integration of data analytics in marketing holds significant importance in today's digital landscape. The abundance of digital tools has transformed the customer experience, making it essential for businesses to adapt and stay relevant. With customers using various avenues to discover products or services, determining effective marketing strategies can be challenging. Data analytics offers a solution by providing valuable insights that enhance business relevance and positioning in saturated markets.

The primary goal is to make a brand stand out and attract customers. Instead of adopting a broad outreach approach, data analytics enables businesses to use precise information to target strategies and enhance the user experience. Effective data analytics relies on appropriate tools, playing a crucial role in testing strategies, resolving issues, generating innovative concepts, monitoring performance, and aligning every aspect to enhance the bottom line and increase profitability.

In the realm of influencer marketing, the integration of data analytics is crucial for success. Data analysis involves a methodical examination and interpretation of data to unveil valuable insights, facilitating data-driven decision-making. This strategic use of data analysis is instrumental in achieving success in influencer marketing. It empowers businesses to make well-informed decisions, from pinpointing suitable influencers to optimizing campaigns, surpassing objectives, and maximizing return on investment (ROI).

In recent times, influencer marketing has become a potent tool for brands looking to connect with consumers. Collaborations with influencers can efficiently tap into new audiences, fostering engagement and driving sales. However, as influencer marketing gains prominence, it becomes increasingly imperative for brands to gauge the effectiveness of their campaigns and comprehend their ROI. This article explores the significance of influencer marketing analytics and provides a guide on assessing the success of campaigns [4].

1. Reach: Reach denotes the number of individuals exposed to your campaign. It can be quantified by considering the influencer's follower count, the reach of their posts, and the overall exposure achieved by your campaign.
2. Engagement: Engagement encompasses the interactions generated by your campaign, including likes, comments, and shares. Elevated engagement levels signify that your campaign is striking a chord with audiences and fostering meaningful interactions.
3. Conversion rate: The conversion rate gauges the number of individuals who take a specific action, such as making a purchase or subscribing to a newsletter, due to your campaign. This metric is pivotal for evaluating the return on investment (ROI) of your campaign and comprehending the influence of influencer marketing on your overarching business objectives.

4. Audience demographics: Understanding the demographics of the audiences reached by your campaign aids in gaining insights into your target market. Information such as age, gender, location, and interests can be instrumental in tailoring your marketing efforts accordingly.
5. Cost per engagement (CPE): CPE is a metric that calculates the expense associated with each engagement generated by your campaign. By monitoring CPE, you can assess the cost-effectiveness of your campaigns and make informed, data-driven decisions to refine your strategy.
6. Influencer sentiment: Influencer sentiment gauges the overall sentiment of the content produced by an influencer. This assessment can provide insights into the impact of an influencer's brand image and messaging on your campaign, aiding in decisions about potential future collaborations.

2 Understanding Influencer Marketing

2.1 Definition and Types of Influencers

Influencers have become ubiquitous in the marketing landscape due to the notable return on investment (ROI) associated with influencer marketing compared to conventional digital marketing channels. However, to optimize marketing campaigns, it is crucial to understand the distinctions among various types of influencers [5–7].

1. **Nano influencers (1 K–10 K followers)**: Nano influencers typically have 1,000 to 10,000 followers on their social media networks. Known for their highly engaged following and commendable engagement rates, nano influencers passionately endorse brands and products. Their authenticity resonates with followers, fostering close relationships as they actively engage with their audience to nurture connections.
2. **Micro influencers (10 K–100 K followers)**: With follower counts ranging from 10,000 to 100,000, micro influencers have a significant but relatable following. Specializing in specific niches, they maintain high engagement rates and a targeted audience, facilitating the creation of tailored sponsorships with brands. Their influence is marked by authenticity and niche expertise.
3. **Macro influencers (100 K–1 M followers)**: Macro influencers, with follower counts ranging from 100,000 to 1 million, exhibit a broader appeal compared to micro influencers. Often internet celebrities, they may be social media stars, bloggers, vloggers, or podcasters. While boasting a larger audience, macro influencers may have a lower engagement rate, having cultivated their followers over an extended period.
4. **Mega or celebrity influencers (1 M + followers)**: Mega influencers and celebrities surpass the 1 million follower mark, attracting audiences of considerable breadth. As often recognized celebrities, collaborating with mega influencers requires a substantial marketing budget. It is crucial to acknowledge that the

audience associated with mega influencers tends to be broad, reflecting their widespread popularity.

3 The Rise of Influencer Marketing

3.1 Evolution of Social Media and Its Role

In recent times, brands seeking to broaden their reach, enhance brand visibility, and drive conversions have increasingly turned to influencer marketing as a prominent strategy. A substantial 93% of marketers have embraced influencer marketing, recognizing its effectiveness. In 2020, 69.8% of marketers expressed their intention to amplify their influencer marketing budget. Influencer marketing has historical roots predating the social media era. Before social media, people relied on print ads, radio, and television for product suggestions. Early marketers identified the influence of prominent figures, understanding their ability to shape consumer purchasing decisions.

3.2 The Path to Modern Influencer Marketing

Phase 1: The Pioneers

Beginning in the 18th century, marketers began to leverage the power of prominent personalities. A noteworthy example is Josiah Wedgwood, a British potter whose cream-colored artwork won Queen Charlotte's favour in 1765, earning him the formal title "Her Majesty's Potter." Recognizing Queen Charlotte's significant influence, Wedgwood strategically utilized his newfound status to promote his pottery as "Queensware," establishing it as the world's inaugural luxury brand. The royal endorsement acted as a catalyst, drawing people to his business and generating high demand for his work shortly thereafter.

Phase 2: Fictional Characters

The next era saw a rise in the use of fictional figures as influencers. A well-known example is Coca-Cola's marketing of Santa Claus in 1932. During the Great Depression, Coca-Cola used the joyful image of Santa Claus to increase beverage sales. By conveying joy during a challenging period, the company redirected their target audience's focus and facilitated consumers in recalling the joyful attributes associated with the brand [8].

Source: https://www.coca-cola.com/gb/en?utm_source=aspire.io&utm_medium=site

Phase 3: Celebrity Endorsements

Subsequently, the era of celebrity endorsements emerged. In contrast to fictional figures such as Santa and Mikey, celebrities are genuine individuals with actual preferences and extensive followings stemming from their widespread fame [9, 10].

Phase 4: Reality TV

This phase introduced reality television, where the emphasis was on depicting "reality." To a certain extent, the stars on these shows are viewed as more relatable and honest than traditional celebrities.

Phase 5: Influencer Marketing

When compared to superstars and even reality TV personalities, influencers are more like the average populace. The majority of them do not live in large houses or fly in private aircraft. Instead, they share relatable content that covers the highs, lows, and obstacles in their life. Their consistent sincerity has earned them a great level of trust and authority among their followers [11].

3.3 Impact on Consumer Behavior

In today's saturated markets, brands face the challenge of standing out and offering something distinctive to consumers in order to generate brand awareness. To succeed, companies must incorporate at least one of Cialdini's Principles of Persuasion, convincing consumers that their brand provides added value. This is precisely the role played by influencers.

A well-executed influencer marketing campaign has the potential to develop trust, loyalty, and authenticity, convincing customers that making a purchase from your business is an opportunity they cannot pass up. Influencers may help to create brand awareness, nurture brand loyalty, and drive increased conversions by leveraging the trust and loyalty they've built with their followers and transferring their established social proof. Finally, this is the key to influencing consumer behaviour and purchasing decisions [12].

3.4 Industry Trends and Statistics

In recent times, influencer marketing has emerged as one of the most popular and impactful forms of online promotion. Given the vast number of internet users regularly exploring social media platforms for entertainment, inspiration, and product suggestions, it's hardly surprising that marketers are leveraging the influence of prominent figures on these platforms for promotional activities. The global value of the influencer marketing market reached 21.1 billion U.S. dollars by 2023, marking a more than threefold increase since 2019. As the realm of influencer endorsements evolves into a mature industry, the scale and worth of influencer marketing platforms continue to expand annually, rendering collaborations between brands and creators more lucrative than ever before [13, 14, 18].

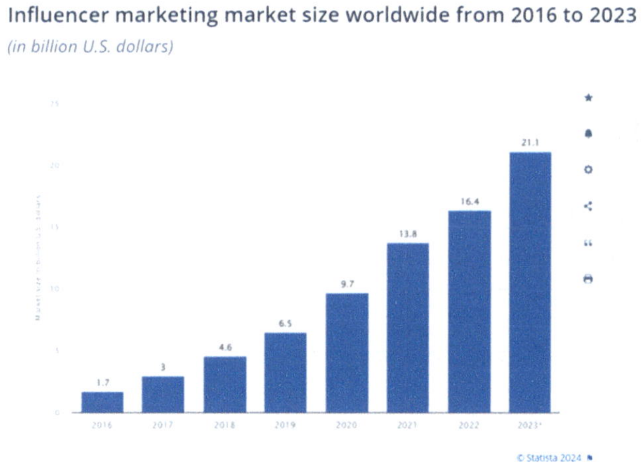

Source: https://www.statista.com/statistics/report-content/statistic/1294803

4 Benefits and Challenges of Influencer Marketing

4.1 Advantages of Influencer Marketing

- Facilitates reaching a broad and relevant audience, enhancing engagement and conversions.
- Establishes trust and credibility through collaborations with influencers.
- Expands brand visibility by tapping into the influencer's follower base and amplification capabilities.
- Offers cost-effectiveness, especially for niche markets, in comparison to traditional marketing approaches.
- Accelerates the sales cycle by leveraging recommendations from influencers.
- Streamlines the process by delegating content creation and distribution responsibilities to influencers.

Disadvantages of Influencer Marketing:

- Inappropriate influencer partnerships may negatively impact brand alignment and audience targeting.
- Mistakes in campaigns, such as subpar content or controversial behavior, can harm the brand's reputation.
- Precise impact measurement and ROI assessment are challenging due to the absence of standardized metrics and attribution.
- Inherent risks exist, as influencers may become involved in scandals or controversies, or changing trends may affect brand associations [15, 16].

4.2 Influencer Marketing Case Studies: Real-World Examples of Success

Case Study 1: Daniel Wellington's Instagram Triumph

A notable instance of successful influencer marketing is exemplified by the Swedish watch company, Daniel Wellington. The global recognition of their watches can be attributed significantly to their strategic utilization of Instagram influencers. Partnering with diverse fashion and lifestyle influencers, the brand orchestrated the sharing of captivating images featuring influencers adorned with Daniel Wellington watches. This collaboration cultivated an aura of aspiration and a particular lifestyle associated with the product. The watches' uncomplicated and refined design seamlessly aligned with Instagram's aesthetics, propelling the brand's follower count to unprecedented heights. Through the effective use of influencers, Daniel Wellington underwent a remarkable transformation from a modest business to a global powerhouse in the watch industry [17].

Case Study 2: Airbnb's #WeAccept Initiative

Facing a challenge related to combating discrimination within its community, Airbnb launched the #WeAccept campaign, a response to this issue. To amplify the reach of this initiative, Airbnb partnered with civil rights activist and influencer DeRay McKesson. McKesson utilized his influential platform to propagate the message of inclusivity and acceptance. His involvement provided Airbnb's campaign with an authentic and influential voice, leaving a positive impact on both the brand's image and the broader societal discourse. The campaign not only raised awareness but also translated into increased bookings for Airbnb [17].

5 Introduction to Data Analytics

Data analytics is a multifaceted process that involves the collection, organization, and analysis of extensive datasets to unveil patterns, trends, and valuable insights. These insights, derived through techniques such as statistical analysis, data mining, predictive modeling, and machine learning, empower informed decision-making and strategic optimization for businesses [18]. The application of data analytics to influencer marketing proves instrumental in extracting insights into audience demographics, engagement rates, and content preferences. This aids in identifying pertinent influencers, monitoring real-time campaign performance, optimizing partnerships, and understanding the impact of influencers on consumer decision-making [19].

In essence, data analytics in influencer marketing facilitates data-backed decision-making, refining audience targeting and segmentation, gauging campaign effectiveness, and augmenting credibility and authenticity. By leveraging these insights, brands gain a competitive edge, optimizing strategies and achieving better business outcomes. The integration of data analytics in influencer marketing empowers brands to comprehend consumer behavior, identify trends, and make well-informed decisions.

In conclusion, the role of data analytics in influencer marketing is pivotal. It enables brands to make data-driven decisions, understand their target audience, assess the impact of influencer marketing campaigns, optimize partnerships, and refine overall marketing strategies [20]. Through this approach, brands can stay ahead in the market by utilizing valuable insights to enhance strategies and achieve more effective outcomes.

6 Data-Driven Decision Making

Data analytics enables data-driven decision making in influencer marketing. Brands can use data analytics to analyze key metrics such as audience demographics, engagement rates, and content preferences. This data-driven approach helps brands understand their target audience better and make informed decisions when selecting influencers to collaborate with [20].

Optimizing Strategy: Data analytics allows brands to optimize their influencer marketing strategies. By analyzing data on audience engagement, conversion rates, and other key metrics, brands can identify which influencers are driving the most effective results and make adjustments to their strategies accordingly.

Measuring Campaign Effectiveness: Data analytics plays a crucial role in measuring the impact and effectiveness of influencer marketing campaigns. Brands can track the reach, engagement, and conversion rates generated by influencer partnerships, providing valuable insights into the ROI of these campaigns. Analytics creates a common language between marketing and finance for the first time by allowing the two functions to clearly see marketing's impact on financial performance. In today's rapidly changing world, the significance of accurate weather forecasts cannot be overstated. Therefore, the use of data analytics in influencer marketing allows brands to enhance their decision-making processes, optimize their strategies, and measure the effectiveness of their influencer marketing campaigns. In today's rapidly changing world, the use of data analytics in influencer marketing allows brands to enhance their decision-making processes, optimize their strategies, and measure the effectiveness of their influencer marketing campaigns [18]. In conclusion, data analytics plays a vital role in influencer marketing by enabling data-driven decision making, optimizing strategies, and measuring campaign effectiveness. In conclusion, the use of data analytics in influencer marketing allows brands to make informed decisions, optimize their strategies, and measure the effectiveness of their campaigns. In conclusion, data analytics plays a crucial role in influencer marketing by providing brands with valuable insights that enhance decision-making processes, optimize strategies, and measure campaign effectiveness. In conclusion, data analytics plays a crucial role in influencer marketing by providing brands with valuable insights that enhance decision-making processes, optimize strategies, and measure the effectiveness of their campaigns. In conclusion, data analytics plays a crucial role in influencer marketing by providing brands with valuable insights that enhance decision-making processes, optimize strategies, and measure the effectiveness of their influencer marketing campaigns.

Data analytics serves as a catalyst for data-driven decision-making within influencer marketing. Brands harness the power of data analytics to scrutinize pivotal metrics, including audience demographics, engagement rates, and content preferences. This approach empowers brands to comprehend their target audience more effectively, facilitating informed decisions when selecting influencers for collaborations [20].

- **Optimizing Strategy**:

The utilization of data analytics empowers brands to optimize their influencer marketing strategies. Through an in-depth analysis of data related to audience engagement, conversion rates, and other critical metrics, brands can discern which influencers yield the most impactful results. Consequently, brands can make necessary adjustments to enhance the overall effectiveness of their strategies.

- **Measuring Campaign Effectiveness**:

Data analytics plays a pivotal role in quantifying the impact and effectiveness of influencer marketing campaigns. By tracking metrics such as reach, engagement, and conversion rates resulting from influencer partnerships, brands gain valuable insights into the return on investment (ROI) of these campaigns. This measurement provides a comprehensive understanding of the campaigns' success and informs future decision-making.

In conclusion, the integration of data analytics in influencer marketing empowers brands to enhance decision-making processes, optimize strategies, and measure the effectiveness of their campaigns. This approach ensures a more informed and strategic approach to influencer marketing, aligning brands with the dynamic landscape of consumer behavior. The Synergy of Influencer Marketing and Data Analytics:

The integration of influencer marketing and data analytics is a powerful combination that facilitates effective decision-making and optimized strategies for brands. This synergy provides valuable insights into the performance and impact of influencer marketing campaigns, empowering brands to make informed, data-driven decisions. By analyzing key metrics such as audience demographics, engagement rates, and content preferences, brands can identify the most effective influencers tailored to their target audience.

- **Optimizing Strategies**:

Data analytics allows brands to refine and optimize their influencer marketing strategies. Through a meticulous analysis of data on audience engagement, conversion rates, and other vital metrics, brands can discern which influencers yield the most impactful results. This insight enables brands to make strategic adjustments, allocating resources more efficiently for maximum impact.

- **Measuring Campaign Effectiveness**:

A pivotal role of data analytics is in measuring the impact and effectiveness of influencer marketing campaigns. Brands can track metrics like reach, engagement, and conversion rates resulting from influencer partnerships. This detailed measurement provides a clear understanding of the return on investment (ROI) of these campaigns, guiding brands in refining and enhancing their future strategies.

In essence, the amalgamation of influencer marketing and data analytics empowers brands to identify effective influencers, target audiences more accurately, and optimize content strategies for heightened engagement and ROI. This collaborative approach enables brands to make data-driven decisions, allocate resources efficiently, and continually refine strategies for optimal effectiveness. The insights derived from this combination not only enhance decision-making processes but also contribute to the overall success of influencer marketing campaigns in the dynamic landscape of consumer behavior.

The integration of data analytics in influencer marketing allows brands to gain a comprehensive understanding of their target audience, identify the most effective influencers to collaborate with Data analytics in influencer marketing enables brands to gain a better understanding of their target audience and their preferences, allowing them to identify the most suitable influencers to By analyzing data on consumer behavior, engagement metrics, and campaign performance, brands can identify the most influential and effective influencers for their target audience. The use of data analytics in influencer marketing allows brands to gain a deeper understanding of their target audience, identify the most relevant influencers, and measure By leveraging data analytics, brands can identify the most relevant influencers for their target audience, track the success of influencer collaborations, and measure the return on analyzing data from various sources, such as social media metrics, audience engagement, and purchase behavior, brands can identify the most effective influencers to collaborate with using data analytics in influencer marketing allows brands to accurately measure the effectiveness of their influencer campaigns and make data-driven decisions to optimize their strategies.

7 Tools and Technologies

7.1 Analytics Tools for Marketing

Contemporary brands are not merely in pursuit of influencers; rather, they are in search of the valuable data-driven insights that these influencers can provide. This curated article reveals the top-tier influencer analytics tools, each boasting distinctive features designed to enhance your influencer marketing strategy. From segmenting audiences to monitoring real-time engagement, these tools act as clandestine assets for brands aiming to fully capitalize on influencer collaborations. An effective influencer analytics tool is comparable to a dependable navigation system in turbulent waters. Undertaking an influencer campaign is akin to navigating a stormy sea, where numerous activities, individuals, and budgets need coordination. Relying solely on intuition is insufficient. Even the most popular and precise influencer analytics tools in the market may fall short unless the right one is selected based on specific needs.

In the current digital era, influencer marketing has risen to prominence in advertising, with influencers serving as influential advocates for brands. However, to unlock the complete potential of influencer partnerships, data analytics has emerged as an indispensable tool [21].

7.2 *Navigating the Maze of Influencer Marketing Platforms*

In 2024, over 80% of brands are allocating substantial budgets to influencer marketing. Influencer platforms serve as valuable aids, streamlining processes for both agencies and brands engaged in managing and collaborating with influencers. An influencer marketing platform is specialized software crafted to support brands in executing their influencer marketing campaigns.

With extensive hands-on experience and direct partnerships with various influencer platforms, a profound understanding of the continually evolving intricacies of influencer software has been cultivated. When selecting optimal influencer platforms, the approach delves beyond superficial features. It involves a deep exploration of the essential functionalities that truly make a difference, ranging from the precision of influencer targeting tools to the authenticity measures applied to influencers. Evidence suggests that the right influencer platform can enhance a brand's influencer marketing strategy, while conversely, a misaligned choice can impede marketing efforts significantly.

- Grin
- Brandwatch Influencer Marketing Software
- Upfluence
- Creator.co
- 5. LTK (formerly rewardStyle)
- Influencity
- indaHash
- #paid

7.3 *Integrating Data Analytics Into Influencer Campaigns*

In the realm of influencer marketing, data analytics involves the systematic collection and examination of data to extract insights into the performance and efficacy of influencer campaigns. This methodical, data-driven approach empowers influencers and brands to make well-informed decisions, optimize strategies, and attain their marketing objectives.

Audience Insights:

Utilizing data analytics allows influencers and brands to delve deeper into the demographics, interests, and behaviors of their target audience. By scrutinizing audience data, it becomes possible to pinpoint influencers whose characteristics align most closely with the preferences of the target audience, ensuring a more impactful partnership.

Influencer Selection:

Analytics becomes a guiding force for brands in identifying the most suitable influencers for their campaigns. Through the assessment of an influencer's engagement rates, follower demographics, and past campaign performance, brands gain the insights needed to make informed decisions when selecting their partners.

Campaign Performance Tracking:

Real-time tracking of campaign performance metrics such as reach, engagement, click-through rates, and conversion rates is made possible by data analytics. This information empowers influencers and brands to gauge the success of their campaigns and make necessary adjustments on the fly.

ROI Measurement:

For both influencers and brands, analyzing the return on investment (ROI) is crucial. By closely monitoring the performance of each campaign, it becomes possible to identify the strategies that yield the most significant impact and revenue, facilitating effective resource allocation.

Content Optimization:

Analyzing content performance aids influencers in refining their content strategy. By understanding which types of content resonate most with the audience, influencers can create more compelling and shareable content.

Fraud Detection:

Data analytics plays a crucial role in identifying fraudulent practices, such as fake followers or engagement bots. Early detection of these issues allows brands to avoid partnerships with dishonest influencers, preserving their credibility.

Predictive Analytics:

Utilizing historical data, predictive analytics forecasts future trends and behaviors. In influencer marketing, it assists influencers and brands in anticipating audience preferences, enabling them to tailor their content accordingly.

Reporting and Transparency:

Analytics provides a foundation of transparency and accountability in influencer marketing campaigns. Both influencers and brands can utilize data-driven reports to showcase the impact of their collaboration to stakeholders and partners.

7.4 Challenges and Considerations

While data analytics provides significant advantages, influencers and brands must also address data privacy and ethical considerations. The collection and utilization of data should adhere to privacy regulations and prioritize user consent.

The landscape of influencer marketing is undergoing a transformation through data analytics, evolving it into a strategic and data-driven approach. The capacity to gain insights into audience behavior, campaign performance, and content effectiveness empowers both influencers and brands to make informed decisions and optimize their strategies. As influencer marketing continues to progress, those who recognize data analytics as an integral element of their campaigns are more likely to attain their marketing objectives, forge stronger partnerships, and deliver impactful content to their target audience. In the digital era, data is not merely information; it serves as the key to success in influencer marketing [22–26].

8 Future Trends: Emerging Technologies in Influencer Marketing

8.1 Influencer Marketing Trends

Given the sustained enthusiasm for influencer marketing, it's unsurprising that a forecast for 2024 suggests marketers will invest a combined total exceeding $7.14 billion in influencers. With the rising popularity, demand, and budget allocation for influencer marketing, it becomes crucial to take note of the upcoming trends in this field for 2024 [27]. This way, one can ensure that one's attention and advertising expenditures are directed toward the most pertinent aspects [31–33].

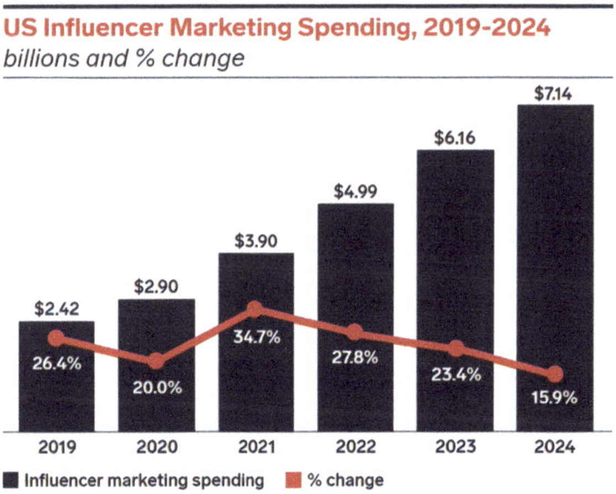

Source: insiderintelligence.com

In this year's examination of influencer marketing trends, certain enduring favorites from the previous list have been omitted. Notably, the emphasis on escalating video content and the significance of micro-influencers has been amplified. This adjustment is not due to a diminished importance in influencer marketing but stems from the recognition that these aspects are now considered standard practices. The creation and sharing of video content on social platforms have become widespread among influencers and the general public alike. After eight years of inclusion in this list, it seemed redundant to label the increased use of micro-influencers as a "trend." Instead, collaborating with micro- and nano-influencers is now regarded as the norm and an indispensable component of a well-rounded influencer marketing strategy [28–30].

1. **Rise of AI in Influencer Marketing**:

AI is progressively becoming an integral part of influencer marketing technology. An instance of this integration is evident in the influencer marketing platform Upfluence, which has incorporated ChatGPT to introduce advanced messaging features aimed at transforming influencer recruitment. Industry experts predict that the utilization of AI for selecting potential influencer partners will become increasingly prevalent in 2024 and beyond. A notable trend accompanying this is the emergence of virtual influencers—digital characters created using computer graphics software. Endowed with personalities and identities, these virtual influencers engage on social media platforms, boasting engagement rates nearly three times higher than real influencers. Major brands are already engaging with virtual influencers, and this trend is expected to gain momentum [31–34].

2. **Cross-Platform Emphasis on Preferred Creators**:

While many creators may have a preferred social network, the shift now is toward creators operating across multiple platforms. What distinguishes successful creators is the presence of "super fans" who interact with them across all platforms seamlessly. Unlike being confined to a single platform, these creators identify as specialists showcasing their work to a diverse audience across the web. They strategically choose platforms based on specific purposes, emphasizing an integrated approach rather than being exclusively associated with one social media channel.

3. **Influencers as Key Players in Affiliate Programs**:

Traditionally viewed as separate activities, influencer marketing and affiliate marketing are increasingly interconnected. Platforms like Refersion recognize this synergy, serving both as an affiliate and influencer platform. The distinction between influencers and affiliates is diminishing as influencer platforms integrate more with eCommerce, evolving beyond simple awareness campaigns to actively drive sales. Influencers are now recommending products they personally use or trust, shifting away from the traditional affiliate model. This change allows brands to benefit from influencers as genuine advocates who provide personal testimonials, demonstrations, and calls to action. Influencers guide their followers through the entire sales funnel authentically, from awareness to consideration and even purchase. Brands find this approach more cost-effective, and consumers benefit from the convenience of shopping through affiliate links shared by their favourite influencers. This paradigm shift encourages influencers to deliver results, fostering a mutually beneficial relationship between influencers and brands.

4. **Continued Significance of Live Shopping in Influencer Campaigns**:

The prominence of live shopping, a trend that has gained traction in recent years, is set to persist throughout 2024. According to Marijana Kay from Shopify, "Live commerce is here to stay." Influencers and notable thought leaders will continue to endorse products for their communities through live video and interactive content on their favorite social channels. Live shopping is becoming the preferred option for the majority of consumers, pushing platforms such as Amazon, Facebook, TikTok, and Instagram to create livestream purchasing features and make partnerships. Instagram, for example, offers Live Shopping, which allows users to buy things directly from Instagram Live sessions. The attractiveness of this format stems from the fact that it offers shoppers an immersive environment in which to find and purchase things [35, 36]. Notably, several influencers have successfully collaborated with brands on live shopping experiences, and large stores such as Walmart have teamed up with TikTok producers to offer live shopping events. Expect a boom in collaborative activities in the coming year.

5. **Heightened Focus on Diversity and Representation in Influencer Marketing**:

The cultural emphasis on diversity and representation has a huge impact on influencer marketing. The top LGBTQ + influencers on sites such as YouTube have roughly 20 million followers. Specialized influencer agencies are emerging, with a focus on expanding representation. For example, REFLECT focuses on underrepresented groups in media campaigns, working exclusively with businesses that share their commitment to broadening media and implementing diversity and inclusion into their brand strategy and marketing efforts. Other agencies, such as Shade and Black Girl Digital, specialize in representing black and brown creators. For instance, Black Girl Digital specifically focuses on empowering black females, aiming to inspire brands to establish meaningful partnerships with this group of influencers and support their pursuit of a sustainable business. In the current landscape, influencers and consumers scrutinize branded messages and campaigns to ensure inclusive representation. Looking ahead, brands will find it increasingly crucial to reshape their corporate cultures, prioritizing inclusivity and eliminating any remnants of exclusivity.

9 Conclusion

9.1 *Importance of a Data-Driven Approach in Influencer Marketing*

Social media stands as one of the world's largest repositories of data, with users generating an impressive amount of content every minute. On platforms like Facebook, 1.7 million pieces of content are posted, YouTube witnesses the upload of 500 h of video, and Instagram sees the sharing of 66,000 photos. Each interaction, be it a like, share, follow, view, or click, contributes to a real-time database of digital behavior.

Marketers, using big data analytics, have turned this massive pool of social data into a lucrative source of information. This plethora of information allows them to develop data-driven strategies and make informed judgements. How do top brands use this data to shape their influencer marketing strategy and offer exact content to the appropriate audience at the right time?

9.2 *Utilizing Data for Strategic Insights*

The volume of content and different topics covered by social media influencers make for an invaluable data source. This enables marketers to develop effective plans and gather insights from a variety of sectors. To do this, a thorough understanding of the "who," "what," "when," "where," and "why" (the five W's) of your product category, industry, and target audience is required. Leveraging data insights allows you to have a better understanding of your audience's interests, behaviours, and preferences. This, in turn, influences decisions on influencer selection, creative execution, timing, delivery, and messaging.

9.3 *Developing Influencer Strategies Based on the Five W's/ Constructing Influencer Strategies Anchored in the Five W's*

The five W's provide a straightforward framework for marketers to deconstruct complex and complicated business concerns. This framework helps them to ask essential questions, identify key difficulties, and develop successful solutions that are consistent with their marketing objectives. The application of the five W's adapts and evolves based on the marketing channel, relying on available data and actionable insights.

The Who:

Identifying the "who" entails defining your target audience, including their demographics, geolocation, interests, and affinities. Understanding the "who" is critical to ensuring that one's messaging effectively reaches the desired demographic, resulting in increased brand engagement.

The What:

Recognising the types of content that resonate most with your audience is critical to your campaign plan. Because not all viewers respond in the same way, having a clear grasp of one's target demographic is critical to your overall strategy. For example, a brief 30 s cosmetics tutorial on TikTok may generate considerable engagement from women under 25, whereas a 15 min video on YouTube may be the best option for reaching women over 30.

The When:

Given that social platforms are highly competitive digital spaces, pinpointing the optimal time to connect with the community is crucial for campaign success. Being cognizant of the "when" ensures that one's campaign capitalizes on the most relevant moments.

The Where:

The consideration of where to activate one's campaigns is as pivotal as the campaign's content. For example, if one's brand's community is more engaged on Instagram and content typically receives three times the engagement as on other platforms, this information should lead one's campaign placement.

The Why:

Finally, clarity on the purpose behind your campaign is essential. Influencer marketing strategies can support various aspects of your digital marketing strategy, and understanding why one is running a campaign is critical. Analyzing competitor mentions garnered through influencer activation can aid in determining the strategic reasons behind hiring influencers.

References

1. Zietek, N.: Influencer Marketing: the characteristics and components of fashion influencer marketing (2016)
2. Brown, D., Hayes, N.: Influencer marketing. Routledge (2008)
3. Leung, F.F., Gu, F.F., Palmatier, R.W.: Online influencer marketing. J. Acad. Market. Sci. 1–26 (2022)
4. Grafström, J., Jakobsson, L., Wiede, P.: The impact of influencer marketing on consumers' attitudes (2018)
5. Gómez, A.R.: Digital fame and fortune in the age of social media: a classification of social media influencers. aDResearch: Int. J. Commun. Res. (19), 8–29 (2019)

6. Zhou, F., Su, C., Xu, S., Lü, L.: Influence fast or later: two types of influencers in social networks. Chin. Phys. B **31**(6), 068901 (2022)
7. Ouvrein, G., Pabian, S., Giles, D., Hudders, L., De Backer, C.: The web of influencers. A marketing-audience classification of (potential) social media influencers. J. Market. Manag. **37**(13–14), 1313–1342 (2021)
8. The Metropolitan Museum of Art: Josiah Wedgwood and Sons I Queen Charlotte I British, Etruria, Staffordshire. The Metropolitan Museum of Art (n.d.). Retrieved from https://www.metmuseum.org/art/collection/search/199822
9. Standage, T.: A History of Influencers: From Shakespeare to Instagram. The Economist. Retrieved from https://www.economist.com/1843/2019/07/03/a-history-of-influencers-from-shakespeare-to-instagram (2019)
10. Stever, G. S., Lawson, K.: Twitter as a way for celebrities to communicate with fans: implications for the study of parasocial interaction. N. Am. J. Psychol. **15**(2), 339–354 (2013)
11. Glucksman, M.: The rise of social media influencer marketing on lifestyle branding: a case study of Lucie Fink. Elon J. Undergraduate Res. Commun. **8**(2), 77–87 (2017)
12. Bognar, Z.B., Puljic, N.P., Kadezabek, D.: Impact of influencer marketing on consumer behaviour. In: Economic and Social Development: Book of Proceedings, pp. 301–309 (2019)
13. Kanaveedu, A., Kalapurackal, J.J.: Influencer marketing and consumer behaviour: a systematic literature review. Vision, 09722629221114607 (20220
14. Chopra, A., Avhad, V., Jaju, A.S.: Influencer marketing: an exploratory study to identify antecedents of consumer behavior of millennial. Bus. Perspect. Res. **9**(1), 77–91 (2021)
15. Zhou, S., Blazquez, M., McCormick, H., Barnes, L.: How social media influencers' narrative strategies benefit cultivating influencer marketing: tackling issues of cultural barriers, commercialized content, and sponsorship disclosure. J. Bus. Res. **134**, 122–142 (2021)
16. Ahuja, Y., Loura, I.: An investigative study of influencer marketing: nuances, challenges and impact. J. Global Bus. Adv. **13**(5), 561–583 (2020)
17. https://www.linkedin.com/pulse/influencer-marketing-case-studies-real-world-examples-success-arbif/
18. Bansal, D., Chaudhary, M.: Blockchain: The Chain for the Changing Marketing Sphere, pp. 89–103 (2023)
19. Han, X., Wang, L., Fan, W.: Cost-effective social media influencer marketing. Informs J. Comput. **35**(1), 138–157 (2023)
20. Zak, S., Hasprova, M.: The role of influencers in the consumer decision-making process. In: SHS Web of Conferences, vol. 74, p. 03014. EDP Sciences (20200
21. Barbosa, B., Saura, J.R., Zekan, S.B., Ribeiro-Soriano, D.: Defining content marketing and its influence on online user behavior: a data-driven prescriptive analytics method. Ann. Oper. Res. 1–26 (2023)
22. Bansal, D., Bhattacharya, N., Shandilya, P.: Digital entrepreneurship and its transformative impacts on business and society. In: Entrepreneurship and Creativity in the Metaverse, pp. 251–263. IGI Global (2024)
23. Koski, A.M.: Increasing Brand Awareness Through Social Media Marketing for Software as a Service of EAS (2022)
24. Bansal, D., Kumar, A.: Big Data: A Road-Map for Successful Digital Marketing, pp. 85–97 (2023)
25. Brag, G.: Predicting the Future of Influencer Marketing: How Public Relations Experts perceive and measure the impacts of Influencer Marketing in their organizations. A Critical Study reflecting the discussions and future thinking of this particular phenomenon (2019)
26. Sharma, T., Kaushik, S., Yu, Y., Ahmed, S.I., Wang, Y.: User perceptions and experiences of targeted ads on social media platforms: learning from Bangladesh and India. In: Proceedings of the 2023 CHI Conference on Human Factors in Computing Systems, pp. 1–15 (2023)
27. Nguyen, T.M., Le, D., Quach, S., Thaichon, P., Ratten, V.: The current trends and future direction of digital and relationship marketing: a business perspective. In: Developing Digital Marketing: Relationship Perspectives, pp. 191–200. Emerald Publishing Limited (2021)

28. Ibrahim, S.S.: The future of marketing: emerging trends and technologies. EPRA Int. J. Res. Dev. (IJRD) **8**(5), 419–423 (2023)
29. Le, D., Nguyen, T.M., Quach, S., Thaichon, P., Ratten, V.: The development and current trends of digital marketing and relationship marketing research. In: Developing Digital Marketing, pp. 1–18. Emerald Publishing Limited (2021)
30. Joshi, Y., Lim, W.M., Jagani, K., Kumar, S.: Social media influencer marketing: foundations, trends, and ways forward. Electron. Commerce Res. 1–55 (2023)
31. Aw, E.C.X., Agnihotri, R.: Influencer marketing research: review and future research agenda. J. Market. Theory Practice 1–14 (2023)
32. Dimitrieska, S., Efremova, T.: The effectiveness of the influencer marketing. Econ. Manag. **18**(1), 109–118 (2021)
33. Gabhane, D., Varalaxmi, P., Rathod, U., Hamida, A.G.B., Anand, B.: Digital marketing trends: analyzing the evolution of consumer behavior in the online space. Boletin de Literatura Oral Literary J. **10**(1), 462–473 (2023)
34. Vlahov, M.M., Vlahov, A.: Travel influencers as a new strategic partners in tourism. In: Proceedings of FEB Zagreb International Odyssey Conference on Economics and Business, vol. 3, No. 1, pp. 1249–1265. University of Zagreb, Faculty of Economics and Business (2021)
35. https://www.statista.com/topics/2496/influence-marketing/#topicOverview
36. Putri, M.K., Dermawan, R.: Influence of content marketing and influencer marketing on the purchase intention of somethinc products on TikTok Gen Z users in Surabaya. Indonesian J. Bus. Anal. **3**(5), 1663–1672 (2023)

Data Analytics and AI for Influencer Marketing

Nurul Asmak Md. Lazim and Sharifah Hafizah Syed Ariffin

Abstract Before the age of modern influencer marketing, most business organizations utilized traditional marketing where the marketing was made through the advertisement that came out from pop-up ads on websites, on the TV, and also the radio. However, this form of advertisement is no longer effective due to many potential customers simply ignoring the ad and clicking out of it or most of the time choosing to skip it. This kind of phenomenon is somehow blocking the growth of the advertisement. It is reported that one out of five people worldwide will set the ad blocking from their phones and this amount increases tremendously in the past year alone. People tend to dislike this kind of ads due to many reasons such as annoying, interruption to their focus, disturbance and many more. Thus, there is the demand for the new type of marketing universally. With more than billion active users, Instagram, one of the social media platforms, is the most popular social network in the world. Many users followed the trends to fully maximize their strengths and skills in becoming the internet celebrities which in turn found their way to make profit from their fame. This group is called an influencer. Influencer Marketing Hub is known as the people with the power of affecting the buying habits of customers through social media platforms and this kind of group was defined as influencer marketing as a brand working activities either to market the products or services through online as well as improving the brand reputation for the respective products. Influencers can be anyone in the world that has more creative freedom than celebrities to attract, introduce and influence the followers who are keen to see the platform in their free time to buy or view the products. These kinds of platforms like Instagram and Tiktok have their potential as informal and transparent ways of marketing. Meanwhile, data analytics has become an exclusively important tool for decision making especially

N. A. Md. Lazim (✉) · S. H. S. Ariffin
Department of Bioprocess Engineering and Polymer, Faculty of Chemical and Energy Engineering, Johor Bahru, Johor, Malaysia
e-mail: nurulasmak89@gmail.com

Faculty of Electrical Engineering, Universiti Teknologi Malaysia, 81310 Johor Bahru, Johor, Malaysia

S. H. S. Ariffin
e-mail: shafizah@utm.my

S. Dutta et al. (eds.), *Advances in Data Analytics for Influencer Marketing: An Interdisciplinary Approach*, Information Systems Engineering and Management 9,
https://doi.org/10.1007/978-3-031-65727-6_6

in the 21th century. This includes businesses, organizations as well as individuals. With the good exploitation of advanced technology and increasing data availability has become authoritative for organizations to make data-driven decisions. The mass amounts of data generated in the modern era required versatile techniques for their analysis which where the data analytics play its role worldwide. Data analytics is the process to examine and transform raw data into meaningful and valuable information that can help with decision making. Businesses and organizations will rely on data analytics to understand their customers, competitors, and market trends and make decisions for the growth and sustainable operations. Nowadays, artificial intelligence (AI) was reported to become a part of a commercial entity for the entire world in the long term in future. The new trends in AI-driven automation can reflect the change in the AI landscape. This includes reconfigured ideas, interests and advertisements in the field of AI adoption by enterprise. This technology is very versatile as it can recognize the faces and objects which have various business applications. For security objectives, facial recognition can distinguish individuals while for the objects, it can be detected by image analysis. Thus, this kind of service can meet more customers' preferences and facial recognition will allow the business owner to diagnose their customers' moods which in turn can help with product recommendations and satisfaction. Therefore, this chapter aims to discuss in detail the actual role of data analytics and AI in marketing. The applications of each issue are also concerning. This includes ethics, privacy, security, and many more are discussed. Future recommendations for these platforms are also analyzed.

Keywords Marketing · Influencer · Data analytics · Artificial intelligence · Customer needs

1 Introduction

In a wide range of industries, organizations must contend with challenges like declining funds, growing costs, growing complexity, and expanding service requirements. To get around this, a lot of businesses use knowledge management (KM) to make better decisions. A corporation with global reach is not the same as one that sells goods all over the world; a marketer's expertise is limited to their home nation. It inspired several knowledge management practices and strategies to boost productivity, foster innovation, and accomplish corporate objectives. If businesses want to establish and maintain a competitive advantage over time, knowledge management is crucial. They have grown significantly now that consumers are willing to pay a high price for these data, which could enable them to target particular market segments [28, p. 1].

Prior to the advent of contemporary influencer marketing, many commercial enterprises employed traditional marketing strategies, which involved advertising through pop-up windows on websites, television, and radio. Unfortunately, because so many potential buyers choose to ignore the advertisement and click away, or simply ignore it altogether, it is no longer an efficient type of marketing. The expansion of the advertisement is being impeded in some way by this kind of phenomenon. According to reports, one in five individuals globally will block advertisements on their phones, and this number has significantly increased in only the last year. People often detest these kinds of advertisements for a variety of reasons, including being bothersome, disrupting, and obtrusive. Consequently, there is a need for the new kind of marketing on a global scale [18, p. 744]. One of the most important metrics for assessing the success and results of marketing initiatives is marketing performance. It helps businesses to make better decisions and keep an eye on marketing activities, both of which are crucial to a company's growth and competitiveness [15, p. 3].

The evolution of information technology over the past few years has altered the features of websites, making them more user-centric, participatory, and interactive social media. "Forms of electronic communication through which users create online communities to share information, ideas, personal messages, and other content" is the definition of social media. Applications that let users generate and share content include social networking sites, blogs, content communities, forums, bulletin boards, and other interactive apps. These are referred to as social media. Web 2.0 technologies, or the most recent iterations of social media, are projected to play a major role in marketing in the future due to their immense popularity on the Internet. The rapid expansion of social media platforms has changed how many businesses engage with their customers and how they draw in and hold on to them. According to a research, American firms were more likely to contact with consumers or prospects through social media channels than through Google. According to an annual research, marketers have been using Facebook and Twitter as their primary social media tools for the last two years. Due to the immersive character of entertainment products (like movies), which necessitates value co-creation by both the producer and the user, these kinds of promotional methods are especially crucial [13, p. 1].

However, nowadays, Instagram, a social networking platform, has over a billion active members, making it the most popular social network globally. Many users followed the trends in order to fully utilize their abilities and qualities in order to become online celebrities, who then figured out how to cash in on their notoriety. We refer to this group as influencers. Influencer marketing hubs are groups of people who have the ability to influence consumers' purchasing decisions via social media platforms. Influencer marketing is a term used to describe a brand's efforts to market products or services online while also enhancing the reputation of the associated brand. Anyone in the world with greater creative flexibility than superstars can be an influencer if they can draw in and entice followers who are eager to use the platform to browse or purchase things during their leisure time. These kinds of sites, such as Tiktok and Instagram, have the potential to be casual and transparent marketing channels [21, p. 3, 18, p. 744].

Meanwhile, in the modern networked corporate world, the availability and multiplicity of data, along with the methods and resources for gathering, processing, and evaluating it, are regarded as the main disruptive factors. It should come as no surprise that more businesses are spending money on data analytics to produce insights that will boost productivity and competitiveness. Surveys conducted by practitioners attest to the rise in these kinds of investments over the past few years among businesses. The scientific literature has examined and shown the beneficial effects of data analytics on a variety of firm level skills and performance outcomes, reflecting practitioners' efforts in the field [23, p. 1]. In the 21st century in particular, data analytics has emerged as a crucial instrument for decision making. Both persons and corporations are included in this. It is now authoritative for firms to make data-driven decisions thanks to the effective use of cutting-edge technology and growing data availability. Modern times have produced vast volumes of data that need to be analyzed using a variety of approaches, which is where data analytics comes into play globally. The act of analyzing and turning unstructured data into insightful knowledge that may support decision-making is known as data analytics. To make selections for their expansion and long-term operations, businesses and organizations will need to use data analytics to comprehend their clients, rivals, and industry trends [17, p. 1].

Artificial intelligence (AI) is being used by brands more frequently to automate service delivery, and artificial advertising is becoming more and more popular. Because of this, digital people or at the very least, AI-powered conversational interfaces may soon be seen in retail settings. Artificial intelligence (AI) is already being used to automatically write and reply to posts, which helps brands manage their social media accounts more effectively [2, p. 3]. Artificial intelligence (AI) was recently claimed to be integrated into a global commercial entity in the long run. The shift in the AI environment may be reflected in the new trends in AI-driven automation. This covers reframed concepts, pursuits, and marketing related to enterprise AI adoption. Given its ability to identify faces and objects-a feature that has numerous business applications- this technology is incredibly adaptable. While image analysis can detect objects, facial recognition can distinguish persons for security purposes. This means that more consumers' tastes can be satisfied by this type of service, and facial recognition technology will enable the company owner to identify the emotions of their clients, which will aid in product recommendations and customer contentment [8, p. 119].

Since the start of the decade, deep learning techniques which are aided by the performance of powerful GPU clusters and the vast amount of unstructured data created on the Internet have led to improvements in image processing approaches. It is now feasible to identify objects in images or videos with a fair degree of accuracy thanks to image recognition. Tens of thousands of images have been automatically tagged to estimate the average ER for common object categories (person, dog, cake, etc.) using a conventional neural network trained on the Microsoft COCO data set [14]. This neural network is based on the YOLO method for object detection [20].

Thus, the purpose of this chapter is to go into great detail on the real use of AI and data analytics in marketing. Applications are also each concerning issue's. This covers the topics of ethics, security, privacy, and a host of other things. Future suggestions for these platforms are examined as well.

2 The Role of Influencer in Marketing

The last ten years have seen a significant change in social media advertising, driven primarily by influencer marketing. The market for influencer marketing was estimated to be worth $9.7 billion in 2020 and is expected to grow to $15 billion by 2023. The market, which was worth barely $500,000 in 2015, has grown at a rate of more than 50% per year. Influencer marketing gained popularity as more people turned to internet resources for news, support, and leisure during the COVID-19 pandemic. As a result, influencer marketing is becoming a crucial component of digital marketing strategy [2, p. 1].

In today's more competitive social media ecosystem, long-term partnerships between brands and influencers are forming as a result of influencer marketing's rapid rise. Influencer marketing can be seen as a multi-layered type of relationship marketing that entails partnerships between businesses and consumers, influencers and consumers, and influencers and brands (Fig. 1). In order to enhance followers' engagement, marketers have realized the importance of consistently and progressively integrating their marketing messages into influencers' narratives [2, pp. 1–2].

More precisely, influencers promote a company by taking use of the trust and relationship they have already established with their audience. Because of this, trust is crucial in these kinds of interactions and is the basis of any fruitful leader–follower dynamic. Influencers are trusted by 92% of social media users more than traditional marketing channels, according to a report, highlighting the importance of trust in influencer marketing. In a 2019 worldwide consumer survey conducted by Ipsos, 46% of participants said they did not trust traditional media, including radio, TV, newspapers, or magazines. Additionally, 63% said they were more likely to heed the advice of social media influencers than to listen to commercials from businesses. Social media marketers must develop new marketing competencies in order to succeed in this new environment. These competencies must be focused on building and preserving trust between online opinion leaders and their virtual audiences [2, p. 2].

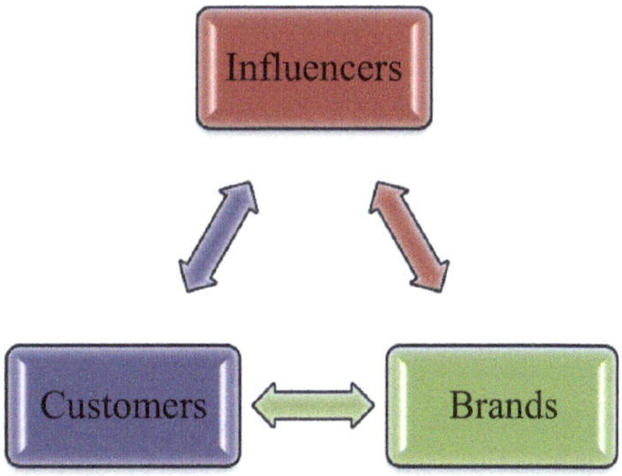

Fig. 1 Relationship between influencers, brands, customers

This is in agreement with the study by [21, p. 3]. Although influencers utilize comparable status and achievement criteria to celebrities, they are also approachable and regarded as genuine in a way similar to a typical customer. Influencers are characterized by certain qualities that set them apart from other people. They become well-known by effectively positioning themselves as authorities on social media platforms, they typically target niche audiences; they communicate with their audience in both directions; and because they share details of their personal lives, they are approachable. Crucially, research indicates that influencers and traditional endorsers have favorable brand results and boundary conditions that are comparable.

Because of these extreme levels, the average size of community on each social network is strongly biased. For marketing efforts, influencers marketing (IM) professionals typically require that influencers have communities with thousands to many more thousands of followers, which is the case for the majority of our influencers. In an effort to better categorize influencers based on the size of their communities, business-oriented categories have recently been proposed [29].

- Micro-influencer: less than 15,000 followers
- Regular-influencer: between 15,000 and 50,000 followers
- Rising-influencer: between 50,000 and 100,000 followers
- Mid-influencer: between 100,000 and 500,000 followers
- Macro-influencer: between 500,000 and 1,000,000 followers
- Mega-influencer: more than 1,000,000 followers

The needs of the business are focused on Regular and Rising Influencers. Large community influencers, also known as macro and mega influencers, are in fact well-known and have personal ties to reputable companies. Which means that too few followers will not bring much benefit, but too many followers will "overshadow" the company profile.

Influencers fill in for models, photographers, and occasionally even editors, sparing the business the expense of hiring these staff members. Since most of them operate from home, the fee they charge includes all of this, making it significantly less than what you would pay for a conventional professional photo shoot. To put this into perspective, influencer marketing costs three times as much as social media marketing that is posted by the brand. In addition, the market for influencers has grown and will continue to develop quickly, as many young individuals aspire to become well-known online. On Instagram, the count is currently 500,000 and rising [18, pp. 749–750].

It is now the responsibility of marketers to gain the trust of consumers in the digital age. Companies need to understand the interests of their customers and what motivates them to purchase a product. The brand can only start aggressively marketing itself once the buyer has established trust and is really curious. Because followers look up to influencers, an influencer whose beliefs and image fit a product will also make followers wonder if they match the product as well. Owing to their widespread online recognition, they are regarded by their followers in social networks as authoritative sources and thought leaders. Perceived competence and reliability are the foundation of influencers' credibility. The degree to which their followers are "sticky," or devoted to the influencer, determines their capacity to turn them into brand clients [18, p. 750].

The pseudo-social bonds that fans develop with these microcelebrities are what give these relationships their "stickiness." This phenomenon bears resemblance to the way a bond forms with media characters through regular exposure. There is no need for direct social contact for this process, known as the parasocial relationship (PSR), as it is developed through social appeal. By subscribing to their channels or following their accounts on social media, users establish a PSR. Because of these connections and because there is not much commenting and replying on the internet, followers see them as friends as well as conversation partners. Three primary characteristics are crucial for defining influencers: physical attractiveness, social attractiveness, and attitude homophily [18, p. 751].

3 Data Analytics in Marketing Growth

Our everyday use of social media platforms has increased dramatically, which has resulted in a rapid collection of unstructured and heterogeneous data that shares many characteristics that define big data. Researchers from a variety of fields have attempted to use a variety of methodologies for data analysis in order to obtain new benefits and opportunities. Social Media Data Analytics (SMDA) is a topical and developing field of study these days. Traditional analytical procedures have become inefficient and sometimes impossible to handle this massive volume of data, which is the fundamental issue facing analysts and practitioners. This has given rise to new, as of yet underutilized potential for the analysis of Big Data with the goal of finding and extracting knowledge that can bring value across a variety of domains. The practice of gathering and analyzing social media data in order to address certain issues and derive valuable insights is known as social media data analysis, or SMDA. The idea of SMDA is being used extensively in a variety of fields these days, most notably marketing [4, p. 96].

The correct analytical methods must be used in order to efficiently examine this massive amount of data and produce consistent results. The primary SMDA techniques are statistical analysis, social network analysis, sentiment analysis, content analysis, and trend analysis, according to [22]. These approaches encompass a multitude of strategies that can be employed singly or in combination as needed. Machine learning, modeling, social network analysis, optimization, visualization, simulation, data mining, web mining, text mining, forecasting, and statistics are the eleven categories into which [6] have mentioned the methodologies. Applying these methods to provide meaningful data has grown difficult for academics across a range of fields, particularly marketing [4, p. 98].

A significant task for researchers is to use SMDA to enhance marketing strategies. In fact, over the past ten years, a number of studies that use the SMDA to accomplish the goals of marketing strategies have been published. Kim et al. [11] used lexicon-based sentiment analysis and natural language processing to evaluate textual data from Twitter in order to uncover market knowledge and to take advantage of competitive intelligence. Pournarakis et al. [19] used sentiment analysis, latent dichlet allocation (LDA), and data visualization (Word cloud) to analyze Twitter data in order to determine inventory performance for business-to-business enterprises and to assist in enhancing logistics strategy. SMDA was utilized by [27] to determine valuable customer knowledge. Data mining was utilized by Marine-Roig et al. [16] to examine information taken from travel websites with the goal of enhancing branding and positioning tactics related to travel and destinations. Airbnb, an online marketplace, employed artificial intelligence and machine learning approaches to assist hosts in optimizing their pricing tactics related to their property [5]. Yunus et al. [26] employed Krippendorf's Content Analysis to examine two Instagram profiles of Grab, the most well-known online transportation service app in Southeast Asia, in order to assess marketing strategies through social media.

As the number of uses for the Internet grows, so does the need for precise evaluation of for example, a website's performance (website analytics). Web Metrics, a quantitative measurement of such things, can help organizations determine how a website is doing in terms of a certain attribute or activity. Measurements are used to assess and monitor the effectiveness of a website or a person's online information-seeking behaviors. On the Web, metrics like page views, page transitions, and average session durations are frequently utilized. It is hard to separate important indications because there is so much data and information available. The number of visits a website consistently receives can be used to measure the effectiveness of a company's marketing and determine how well a website is performing. The success of the marketing initiatives will depend on this, even though the healthy ranges themselves will differ. This will be an important indicator for evaluating the site's and the business's overall health. A visit is a waste of time if visitors leave without looking at the products or services the owner offers. A website's "bounce rate" is the percentage of users who browse only one page before leaving. The greater the likelihood that a visitor will remain on the site and continue exploring it, the lower the percentage of visitors that depart the site right away. In general, a lower bounce rate is preferable, particularly for business and e-commerce websites. A high bounce rate makes sense in certain situations, such when a website's main goal is to provide users with all they need on a single page. The "dropout rate" is the proportion of website visitors that depart without completing a purchase. Its principles can shed light on the appeal of a business's web material and the caliber of its image [1, p. 564].

4 Artificial Intelligence Applications in Marketing

In the marketing sector, virtual influencers also referred to as AI-generated influencers or digital avatars have become more prevalent. These digital models or computer-generated characters are made to appear and behave like actual people. They also have social media accounts where they share information produced by studios, agencies, or marketers. Virtual influencer marketing has emerged as a successful strategy for firms to interact and establish a connection with consumers on social media. These virtual influencers resemble actual people in terms of appearance and personality. An interesting trend that has emerged is the employment of virtual influencers to promote products on social media. Their perceived superior predictability and dependability over human social media. Influencers is one of the main factors contributing to their attractiveness. Notably, virtual influencers are controlled by underlying algorithms that can be adjusted to the brand's specifications, whereas real influencers may on occasion display bias and irrationality. Brands therefore have more power and influence over their sponsored content. As a result, in comparison to real social media influencers, virtual influencers are much more affordable. Nevertheless, not much is known about how virtual influencers affect customer behavior, particularly with regard to perceived trust, intention to buy, and customer preference for virtual versus human influencers [7, p. 1].

Virtual influences are artificially created representations of social media influencers that can be any size, shape, or style following that straightforward demeanor. These virtual influencers are mostly generated by artificial intelligence technology, and a significant amount of human labor goes into choosing their appearance, content, and communication style [7, p. 1].

One notable feature of virtual influencers is that they are not constrained by time or geography. They do not age like humans do, they do not get tired or need to relax, and they do not have to deal with death's inevitable conclusion. Influencers that operate virtually can follow their passions without having to pay for anything in real life, which makes them more economical over time. Virtual influencers are also less vulnerable to scandals and controversies than their real-world counterparts, which makes them even more appealing to companies looking for a dependable and low-risk influencer marketing plan. Purchase intention among consumers has increased as a result of businesses using social media influencers. The fundamental cause of this phenomena is thought to be the influencers' capacity to boost their influence on purchase intention by creating a sense of trust and similarity with the target audience. The results of [10], who also looked into the beneficial influence of social media influencers on customer purchase decisions, are in line with this.

On the other hand, 60% of Singaporean respondents to [9] study on virtual influencers said they felt that virtual influencers had no bearing on their decisions to buy. The respondents felt that real social media influencers had a greater influence on their purchasing decisions than virtual ones.

5 Ethics, Privacy, and Security Concern of Customer Data to Business

Since social media platforms are often financed by online advertising and the sale of user data, users can access them for free. Social media companies obtain a lot of data about their users through their website visits, search terms, interests, and demographics. Consumers willingly give up a lot of information when they visit websites, make online profiles, "like," and "share" content. It is possible to create a comprehensive profile of an individual by merging different data sets. Social networking sites "aggregate information from various Internet locations and comprise the virtual identity of the user in the process". However, there are signs that users are growing more aware of privacy concerns on social media [12, p. 7].

Laws like the California Consumer Privacy Act (CCPA), China Cyber Security Law, and General Data Protection Regulation (GDPR) were created in response to concerns about data privacy. These laws give people control over the collection, disclosure, erasure, and protection of their data from automated decision-making. The CCPA ensures data processing transparency with a focus on for-profit businesses operating in California. Using data from other organizations can help address class imbalance and expanding databases. However, it is challenging to exchange, for example in the case of medical records due to privacy, technological, and legal obstacles. Google uses federated learning to train machine learning algorithms on local datasets without explicitly sharing data. Local nodes trade parameters to build a single global model capable of handling diverse datasets and unreliable clients. In centralized federated learning, algorithmic stages and participating nodes are managed by a central server. On the other hand, decentralized federated learning transfers model updates to networked nodes without the aid of a central server, hence removing the potential for a single point of failure. Bottlenecks in update transmission are avoided by employing this technique. When distributed model training is combined with federated learning and AI in a hybrid method, healthcare systems for instance, are better able to manage limitations and privacy concerns [3, pp. 7–8].

6 Recommendations for Future Attempt

Further research on such topics is necessary to gain a better grasp of how data analytics and AI affects every aspect of marketing. Regarding this research, we have only touched on a few topics because it was exploratory in nature. For example, we looked into companies who prefer to outsource their marketing functions in regions where we could not find enough prior study. We suggest using a quantitative research approach to test this in a subsequent study. In terms of the research's conclusions, we found that businesses are concentrating more on developing their brand identities as a result of the growing usage of data analytics and AI assistants in the future [24].

This is true because, while utilizing AI assistants, consumers will receive a variety of recommendations depending on their preferences; yet, out of all the recommendations, they will be most likely to return to the company they trust due to its strong brand image and devoted following. Since this was one of our research's findings, we were only able to discuss a small portion of it. Still, a more thorough investigation into this may be carried out in the future. Because of this, it is anticipated that AI assistants will significantly alter the market in the near future. However, little is known about the ramifications of this development and the potential disruptions. We recommend that more research be done in the future on the several approaches that could help move the present distribution channels in favor of AI helpers. The last topic that may be studied is the set of abilities that marketing managers should possess in order to succeed and live up to future expectations. The study's conclusions indicate that in the years to come, there will be a greater need for people with marketing expertise as well as technological know-how, creative mindsets, and emotional intelligence [24].

As a part of business organizations like higher education institutions (HEIs), for example, the same methods (social media, data analytics and AI) were also used to promote the institutions in order to attract people to apply as students and as employees. Furthermore, soft skills are becoming more prevalent in social media due to influencers' delivery of relevant content and their followers' responses to their leader profiles and personas. These factors can impact an organization's overall effectiveness. Academics and administrators in higher education acknowledge that social media has an impact on their students and their institutions these days. Some regulations prohibit its use in education, but others have to allow it for recruiting and marketing objectives, as well as, to some extent, for retention and improving teaching and learning. The conflicting viewpoints on each problem have been investigated and several research have been conducted. The effectiveness of social media initiatives is still, at best, difficult to measure, and using social media frequently is thought to be either extremely difficult or impossible in higher education institutions [12, p. 5]. However, because of the advantages and services that technology provides, people are becoming more and more likely to increase the reach of their personal brands. Some contend that creating a brand identity is the first step in the process of developing a personal brand, and it considers an individual's strengths, values, personality, professional goals, and brand promise. After that, it proceeds to the stage of brand positioning before coming to a close with brand image setting. In summary, there are several ways to develop strategic management and personal branding. Higher education became more widely accessible, and higher education providers turned into "private business entities" in order to survive and make a profit. These days, the education industry is a "commodity". The liberalization of Malaysian education laws has led to the emergence of a growing number of private higher education institutions (PHEIs) that provide more opportunities and space to potential students. This has created new challenges for those involved in the industry. Too many universities are fighting for the same limited number of slots by offering the same courses. Therefore, in order to attract and retain students while retaining profitability, colleges must improve their corporate image or brand and elevate the quality of their services.

This can be done with respective social media, data analytics, and AI [3-unpublish article]. The authors are currently monitoring this subject.

7 Conclusion

Overall, the study, high engagement content is consistent with, as evidenced by specific postings that combine specific textual and visual patterns. When unstructured data is present, convolutional neural networks, high-level deep learning features like word embedding, and hashtags can be used to identify this type of content. When combined with more easily accessible influencer data (following, engagement, country, and gender), it provides a more contextualized perspective of the material that the influencer has created.

Customers' purchasing patterns are greatly influenced by these aspects since well-known accounts individuals play a key role in the decision-making process regarding purchases made by customers. Positive consumer perceptions of a certain brand or product might be influenced by well-known reports and personalities endorsing them. Influencer marketers help to reach target audiences that were previously unreachable or unconvinced by their actions. Increasing brand and product awareness, reaching a wider range of accounts' target consumers, and making sure that more people view material related to brands and products can all help with this. When you start doing influencer marketing on multiple accounts and websites, your online visibility will start to increase. Consequently, companies need to invest in artificial intelligence technologies and influencer-based marketing techniques.

Studying some of the subtle changes made by AI to conventional sales and marketing operational models has enormous potential. AI will eventually surpass human capacity for creative thought as it continues to develop to become smarter and cleverer to support human thought. It will also be ready for further humanization. As AI develops, security issues will likely grow, and the continuing privacy vs personalization argument will likely enlarge to include softer parts of marketing. Future marketing efforts will be channeled and focused more effectively thanks to AI's ongoing learning and interpretation of customer emotions and buying intents, which will result in extreme automation and customisation. This field of study is projected to evolve enormously, even if there will be ongoing, and live arguments over deep reasoning, clever AI, continuous/catastrophic learning, and many other aspects of human AI delegation. Furthermore, it is anticipated that the sustainability frontier of AI in marketing would expand and deepen, gradually producing research inputs with more robust and useful insights.

Acknowledgements The authors would like to express their gratitude to Universiti Teknologi Malaysia (UTM) for research support and we thank all our supporters.

References

1. Al Adwan, A., Kokash, H., Al Adwan, R., Khattak, A.: Data analytics in digital marketing for tracking the effectiveness of campaigns and inform strategy. Int. J. Data Netw. Sci. **7**, 563–574 (2023)
2. Alboqami, H.: Trust me, I'm an influencer!—causal recipes for customer trust in artificial intelligence influencers in the retail industry. J. Retail. Consum. Serv. **72**, 103242 (2023). https://doi.org/10.1016/j.jretconser.2022.103242
3. Arif, S.M., Lazim, N.A.M.: Chemically engineered devices for skin analysis: a comprehensive mini-review. J. Hum. Centered Technol. **3**(2), 1–9 (2024). https://doi.org/10.11113/humentech.v3n2.72
4. Benslama, T., Jallouli, R.: Social media data analytics for marketing strategies. the path from data to value. J. Telecommun. Dig. Econ. **10**(2), 96–110 (2022). https://doi.org/10.18080/jtde.v10n1.523
5. Campbell, C., Sands, S., Ferraro, C., Tsao, H.-Y., Mavrommatis, A.: From data to action: how marketers can leverage AI. Bus. Horiz. **63**(2), 227–243 (2020). https://doi.org/10.1016/j.bushor.2019.12.002
6. Galetsi, P., Katsaliaki, K., Kumar, S.: Big data analytics in health sector: theoretical framework, techniques and prospects. Int. J. Inf. Manag. **50**, 206–216 (2020). https://doi.org/10.1016/j.ijinfomgt.2019.05.003
7. Gerlich, M.: The power of virtual influencers: impact on consumer behaviour and attitudes in the age of AI. Adm. Sci. **13**, 178 (2023). https://doi.org/10.3390/admsci13080178
8. Haleem, A., Javaid, M., Qadri, M.A., Singh, R.P., Suman, R.: Artificial intelligence (AI) applications for marketing: a literature-based study. Int. J. Intell. Netw. **3**, 119–132 (2022). https://doi.org/10.1016/j.ijin.2022.08.005
9. Hirschmann, R.: Virtual Influencers' Impact on Purchasing Decisions Singapore 2020 (2021). Available online: https://www.statista.com/statistics/1103884/singapore-impact-of-virtual-influencerson-purchasing-decision/. Accessed 22 April 2024
10. Jiménez-Castillo, D., Sánchez-Fernández, R.: The role of digital influencers in brand recommendation: examining their impact on engagement, expected value and purchase intention. Int. J. Inf. Manag. **49**, 366–376 (2019)
11. Kim, Y., Dwivedi, R., Zhang, J., Jeong, S.R.: Competitive intelligence in social media Twitter: IPhone 6 versus Galaxy S5. Online Inf. Rev. **40**, 42–61 (2016). https://doi.org/10.1108/OIR-03-2015-0068
12. Lazim, N.A.M., Ariffin, S.H.S.: Influence of social media platform trends on academic management and leadership in higher education institutions: a systematic review. Malays. J. Qualitative Res. **9**(2), 172–182 (2023)
13. Liao, L., Huang, T.: The effect of different social media marketing channels and events on movie box office: an elaboration likelihood model perspective. Inf. Manag. **58**, 103481 (2021). https://doi.org/10.1016/j.im.2021.103481
14. Lin, T.Y., Maire, M., Belongie, S., Hays, J., Perona, P., Ramanan, D., Dollár, P., Zitnick, C.L.: Microsoft coco: Common objects in context. In: European Conference on Computer Vision, pp. 740–755. Springer, Berlin (2014)
15. Luo, Z., Guo, J., Benitez, J., Scaringella, L., Lin, J.: How do organizations leverage social media to enhance marketing performance? unveiling the power of social CRM capability and guanxi. Decis. Support. Syst. **178**, 114123 (2024). https://doi.org/10.1016/j.dss.2023.114123
16. Marine-Roig, E., Anton Clavé, S.: Tourism analytics with massive user-generated content: a case study of Barcelona. J. Destin. Mark. Manag. **4**(3), 162–172 (2015). https://doi.org/10.1016/j.jdmm.2015.06.004
17. Mohammadiyeh, S.A., Purhasani, H.: Data analytics in the 21st century: the importance of statistical data analysis. In: 1st International Conference on Statistical Data Analysis, pp. 1–3 (2023)
18. Peng, C.: Influencer Marketing: Statistics and Skepticism. Open J. Bus. Manag. **11**, 744–754 (2023). https://doi.org/10.4236/ojbm.2023.112040

19. Pournarakis, D.E., Sotiropoulos, D.N., Giaglis, G.M.: A computational model for mining consumer perceptions in social media. Decis. Support Syst. **93**(C), 98–110 (2017). https://doi.org/10.1016/j.dss.2016.09.018
20. Redmon, J., Divvala, S., Girshick, R., Farhadi, A.: You only look once: Unified, real-time object detection. In: Proceedings of the IEEE Conference on Computer Vision and Pattern Recognition, pp. 779–788 (2016)
21. Sands, S., Campbell, C.L., Plangger, K., Ferraro, C.: Unreal influence: leveraging AI in influencer marketing. Euro. J. Market. (2022). https://doi.org/10.1108/EJM-12-2019-0949
22. Stieglitz, S., Mirbabaie, M., Ross, B., Neuberger, C.: Social media analytics–challenges in topic discover, ty, data collection, and data preparation. Int. J. Inf. Manag. **39**, 156–168 (2018). https://doi.org/10.1016/j.ijinfomgt.2017.12.002
23. Sáenz, J., Ortiz de Guinea, A., Peñalba-Aguirrezabalaga, C.: Value creation through marketing data analytics: the distinct contribution of data analytics assets and capabilities to unit and firm performance. Inf. Manag. **59**, 103724 (2022). https://doi.org/10.1016/j.im.2022.103724
24. Tahir Iqbal Md Nazmul Khan: The Impact of Artificial Intelligence (AI) on CRM and Role of Marketing Managers. University of Gävle, Master degree (2021)
25. Yunus, E., Susilo, D., Riyadi, S., Indrasari, M., Putranto, T.D.: The effectiveness marketing strategy for ride-sharing transportation: Intersecting social media, technology, and innovation. Entrepreneurship Sustain. Issues **7**(2), 1424–1434 (2019). https://doi.org/10.9770/jesi.2019.7.2(44)
26. Yunus, E., Susilo, D., Riyadi, S., Indrasari, M., & Putranto, T. D.: The effectiveness marketing strategy for ride-sharing transportation: intersecting social media, technology, and innovation. Entrepreneurship Sustain. Issues, **7**(2), 1424–1434 (2019). https://doi.org/10.9770/jesi.2019.7.2(44)
27. Zhang, Z., He, Q., Zhu, S.: Potentials of using social media to infer the longitudinal travel behavior: a sequential model-based clustering method. Transp. Res. Part C: Emerg. Technol. **85**, 396–414 (2017). https://doi.org/10.1016/j.trc.2017.10.005
28. Zhang, H., Zang, Z., Zhu, H., Uddin, M.I., Amin, M.A.: Big data-assisted social media analytics for business model for business decision making system competitive analysis. Inf. Process. Manag. **59**, 102762 (2022). https://doi.org/10.1016/j.ipm.2021.102762
29. Zietek, N.: Influencer marketing: the characteristics and components of fashion influencer marketing (2016)

Nurul Asmak Md. Lazim is a well-known author in the field of the journal scope. She obtained her highest degree from the prestigious university and worked in the respective field for almost three years. She has published many chapters in the finest publications in various area. Her research interests include (but not limited to) Biology, Bioprocess Engineering, Drug delivery, and Science Social.

Sharifah Hafizah Syed Ariffin received Ph.D 2006 from Universiti Teknologi Malaysia, and Queen Mary, University of London, London. She is now a professor in Universiti Teknologi Malaysia. Her current research interests (not limited to) are on Internet of Things, Ubiquitous computing and smart devices, Wireless sensor networks, IPV6, Handoff Management, Network and Mobile Computing System. She had published 116 papers, 17 copyrights, 1 integrated circuit, and 1 trademark.

Interdisciplinary Data Analytics Transforming Influencer Marketing Strategies

Pawan Whig, Jhansi Bharathi Madavarapu, Nikhitha Yathiraju, and Ramya Thatikonda

Abstract This book chapter delves into the dynamic intersection of interdisciplinary data analytics and its transformative impact on influencer marketing strategies. In an era characterized by information overload and rapidly evolving digital landscapes, businesses and marketers face unprecedented challenges in identifying, engaging, and maximizing the impact of influencers. This chapter explores how the integration of diverse data analytics approaches, spanning fields such as machine learning, social network analysis, and sentiment analysis, is reshaping the landscape of influencer marketing. The chapter begins by providing an overview of the current state of influencer marketing and the inherent complexities associated with it. It then navigates through various interdisciplinary data analytics methodologies, emphasizing their applicability in extracting valuable insights from vast and heterogeneous datasets. By harnessing the power of advanced analytics, businesses can gain a deeper understanding of audience behavior, preferences, and engagement patterns, facilitating the identification of authentic influencers who resonate with target demographics. Furthermore, the chapter highlights case studies and practical implementations of interdisciplinary data analytics in influencer marketing campaigns. It discusses the challenges and opportunities presented by these innovative approaches, emphasizing the need for a strategic blend of quantitative analysis and qualitative understanding. The synergy between data-driven decision-making and the creative nuances of influencer collaborations is explored, showcasing how organizations can optimize their marketing efforts..

Keywords Influencer marketing · Data analytics · Interdisciplinary approach · Campaign measurement · Collaboration efficacy · Digital landscape

P. Whig (✉) · N. Yathiraju
Vivekananda Institute of Professional Studies-TC, New Delhi, India
e-mail: pawanwhig@gmail.com

J. B. Madavarapu · R. Thatikonda
University of The Cumberlands, Williamsburg, KY, USA

S. Dutta et al. (eds.), *Advances in Data Analytics for Influencer Marketing: An Interdisciplinary Approach*, Information Systems Engineering and Management 9,
https://doi.org/10.1007/978-3-031-65727-6_7

1 Introduction

In the rapidly evolving landscape of digital marketing, influencer marketing has emerged as a pivotal strategy for brands seeking to connect with their target audiences in authentic and engaging ways [1]. As businesses navigate the complexities of this dynamic field, the integration of interdisciplinary data analytics has become a game-changer, reshaping the strategies employed in identifying, collaborating with, and maximizing the impact of influencers [2]. This introduction explores the symbiotic relationship between interdisciplinary data analytics and influencer marketing, providing insights into how this convergence is revolutionizing the industry[3].

To comprehend the significance of interdisciplinary data analytics in influencer marketing, it's essential to reflect on the evolution of this marketing approach [4–6]. Influencer marketing has transitioned from a niche strategy to a mainstream marketing channel, fueled by the rise of social media platforms. Today, influencers wield considerable influence over their followers, and brands recognize the potential of leveraging these authentic connections [7–9]. However, as the influencer marketing landscape expands, so do the challenges associated with identifying the right influencers, measuring the impact of campaigns, and ensuring authenticity in brand partnerships. One of the fundamental challenges in influencer marketing lies in the identification of influencers who align with a brand's values and resonate with the target demographic. Traditional methods of manual selection prove increasingly insufficient as the pool of influencers diversifies [10]. This is where interdisciplinary data analytics steps in, offering sophisticated tools and algorithms that analyze vast datasets to identify influencers whose content, engagement patterns, and audience demographics align strategically with a brand's objectives. By combining machine learning algorithms and social network analysis, businesses can gain a nuanced understanding of the influencer landscape, facilitating more informed and targeted decision-making [11–14].

The marriage of data analytics and influencer marketing extends beyond identification to inform strategic collaborations. Brands are now equipped with actionable insights derived from data analytics, providing a comprehensive understanding of audience behavior, preferences, and trends [15]. This wealth of information enables brands to tailor their influencer partnerships to maximize impact. Through sentiment analysis, marketers can gauge the authenticity of an influencer's engagement with their audience, ensuring a more genuine and effective collaboration that resonates with consumers [16–18].

Measuring the success of influencer marketing campaigns has historically been a challenge. However, with the integration of interdisciplinary data analytics, businesses can now quantify the impact of their influencer collaborations with greater precision. Advanced analytics tools allow for the tracking of key performance indicators (KPIs) such as reach, engagement, and conversion rates. This data-driven approach not only facilitates real-time campaign optimization but also enables marketers to demonstrate the return on investment (ROI) of influencer marketing initiatives to stakeholders. While interdisciplinary data analytics presents immense opportunities for enhancing influencer marketing strategies, it is not without its challenges. The ethical considerations of data usage, privacy concerns, and the need for transparent collaborations are critical factors that demand attention. Striking the right balance between data-driven decision-making and the creative nuances of influencer marketing remains an ongoing challenge. However, it is precisely in navigating these challenges that the potential for innovation and improvement lies.

This chapter serves as a comprehensive exploration of the transformative impact of interdisciplinary data analytics on influencer marketing. It will delve into case studies and practical implementations, showcasing how businesses across industries have successfully harnessed the power of data to drive meaningful influencer collaborations. By examining the synergy between quantitative analysis and qualitative understanding, this chapter aims to provide a roadmap for industry professionals, researchers, and students navigating the evolving landscape of influencer marketing [19]. The convergence of interdisciplinary data analytics and influencer marketing represents a paradigm shift in how brands approach audience engagement. This introduction sets the stage for an in-depth exploration of the strategies, applications, and implications of this transformative partnership, highlighting the potential for innovation, growth, and authenticity in influencer marketing campaigns. Literature review with research gap is shown in Table 1.

2 The Emergence of Influencer Marketing

In the ever-evolving landscape of digital marketing, the emergence of influencer marketing has marked a significant shift in how brands connect with their target audiences. Influencers, individuals who have developed a substantial and dedicated following on various online platforms, have become powerful voices capable of shaping consumer opinions and behaviors. This section explores the origins and growth of influencer marketing, tracing its roots to the rise of social media and the transformative impact it has had on traditional advertising models. In the early days of social media, individuals with a passion or expertise in specific niches began attracting followers drawn to their authentic and relatable content. These early influencers paved the way for a new form of marketing, where brands could leverage the trust and influence these individuals had built with their audience. Platforms like YouTube, Instagram, and later TikTok provided the perfect stage for influencers to share their experiences, recommendations, and opinions, creating a direct line of communication with consumers.

The emergence of influencer marketing is closely tied to a shift in consumer behavior. Traditional advertising channels faced a decline in effectiveness as consumers became adept at tuning out or actively avoiding ads. Influencers, on the other hand, offered a more personalized and engaging way for brands to connect with audiences. Consumers began to view influencers as authentic and trustworthy sources, valuing their recommendations and insights over traditional advertising messages. As the popularity of influencer marketing grew, so did the diversity

Table 1 Literature review with research gap

References	Title	Key findings	Research gap
[1]	Interdisciplinary research in services marketing	Explores interdisciplinary approaches in services marketing	Limited exploration of specific interdisciplinary methodologies employed
[2]	The value of influencer marketing for business	Provides a bibliometric analysis of influencer marketing	Lack of focus on the practical implications for businesses in the analysis
[3]	Digital Marketing and Analytics Education	Systematic review of digital marketing and analytics education	Insufficient attention to emerging trends and technologies in digital education
[4]	Digital transformation: A multidisciplinary reflection	Reflects on digital transformation with a multidisciplinary perspective	Need for more concrete recommendations for businesses undergoing digital transformation
[5]	Artificial Intelligence (AI): Multidisciplinary perspectives	Explores various perspectives on AI from a multidisciplinary angle	Limited discussion on the ethical implications of AI
[6]	AI in Sustainable Fashion Marketing	Examines AI's role in sustainable fashion marketing	Lack of analysis on the scalability and adoption challenges in the fashion industry
[11]	Cancel Culture in Influencer Marketing	Investigates strategies and crisis management in influencer marketing	Limited exploration of preventive measures and influencers' perspectives
[13]	Digital transformation in tourism sector	Analyzes digital transformation trends in the tourism sector	Inadequate coverage of the potential socio-cultural impacts of digital transformation in tourism
[14]	AI for Social Good	Explores social transformations for health equity beyond AI applications	Need for more concrete examples and case studies of successful implementations
[18]	Industry 4.0 and marketing 4.0	Discusses the impact of Industry 4.0 on marketing	Lack of in-depth analysis on the challenges and opportunities for businesses in adapting to Industry 4.0

within the influencer ecosystem. Micro-influencers, nano-influencers, and macro-influencers each carved out their niche, catering to different audience sizes and demographics. This section explores the various types of influencers and how brands strategically choose influencers based on their target market and campaign objectives.

2.1 Evolution of Data Analytics in Marketing

Simultaneously, the evolution of data analytics in marketing has played a pivotal role in shaping and refining influencer marketing strategies as shown in Fig. 1. This section traces the evolution of data analytics, from its early applications to the sophisticated and interdisciplinary approaches employed today. The integration of data analytics has not only enhanced the efficiency of marketing campaigns but has also provided a data-driven foundation for decision-making in the complex landscape of influencer collaborations.

The evolution of data analytics in marketing began with basic metrics such as impressions, click-through rates, and conversion tracking. Marketers used these insights to measure the success of their campaigns and make data-informed adjustments. The advent of digital marketing channels facilitated the collection of valuable data, laying the groundwork for more sophisticated analytics methodologies. With the proliferation of digital touchpoints and the increasing volume of data generated, marketers turned to advanced analytics and machine learning to extract meaningful insights. This section explores how machine learning algorithms analyze large datasets to identify patterns, predict consumer behavior, and optimize marketing strategies. The marriage of data analytics and marketing has empowered businesses to make strategic decisions based on predictive models and real-time analysis.

The evolution of data analytics in marketing has seen the integration of interdisciplinary approaches, combining insights from fields such as psychology, sociology,

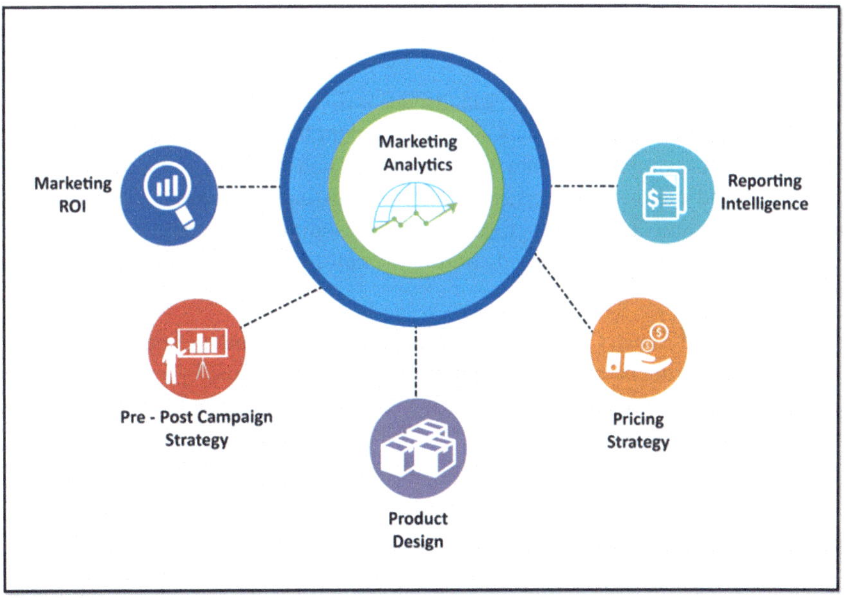

Fig. 1 Marketing analytics

and economics. This holistic perspective allows marketers to not only understand consumer behavior but also to delve into the underlying motivations and emotions that drive purchasing decisions. The section highlights the benefits of interdisciplinary approaches in creating more nuanced and effective marketing strategies. The emergence of influencer marketing and the evolution of data analytics in marketing are intricately linked, representing a transformative force in the way brands engage with consumers. As we delve deeper into these topics, it becomes evident that the synergy between influencers and data-driven insights is reshaping the marketing landscape, offering new possibilities and challenges for businesses seeking to connect meaningfully with their audiences.

3 Understanding Influencer Marketing

In the contemporary realm of digital marketing, influencer marketing has risen to prominence as a dynamic and impactful strategy. This section delves into the core concepts and dynamics of influencer marketing, exploring the fundamental elements that define this approach and examining its role in the broader marketing landscape.

At its essence, influencer marketing involves collaborating with individuals who possess a significant and dedicated following on various online platforms. These individuals, known as influencers, leverage their authenticity, expertise, or relatability to engage with their audience. Brands strategically partner with influencers to promote their products or services, harnessing the influencers' sway over their followers to amplify their messaging. This section provides a comprehensive definition of influencer marketing and outlines its key components.

Types of Influencers: The influencer landscape is diverse, encompassing individuals with varying audience sizes, niches, and engagement levels as shown in Fig. 2. This section categorizes influencers into different types, such as macro-influencers, micro-influencers, and nano-influencers, each offering distinct advantages depending on campaign goals and target demographics. Understanding the characteristics of each influencer type is crucial for brands aiming to tailor their strategies for maximum impact.

Authenticity and Trust: Central to the success of influencer marketing is the concept of authenticity. Influencers build a rapport with their audience by sharing genuine experiences, recommendations, and insights. This section explores how authenticity is a cornerstone of influencer marketing, contributing to the trust that followers place in influencers. Brands seeking successful collaborations must prioritize authenticity to ensure the resonance of their message with the influencer's audience.

The Impact of Social Media Platforms: Influencer marketing has become closely intertwined with the rise of social media platforms. This section examines how platforms like Instagram, YouTube, TikTok, and others have become influential stages

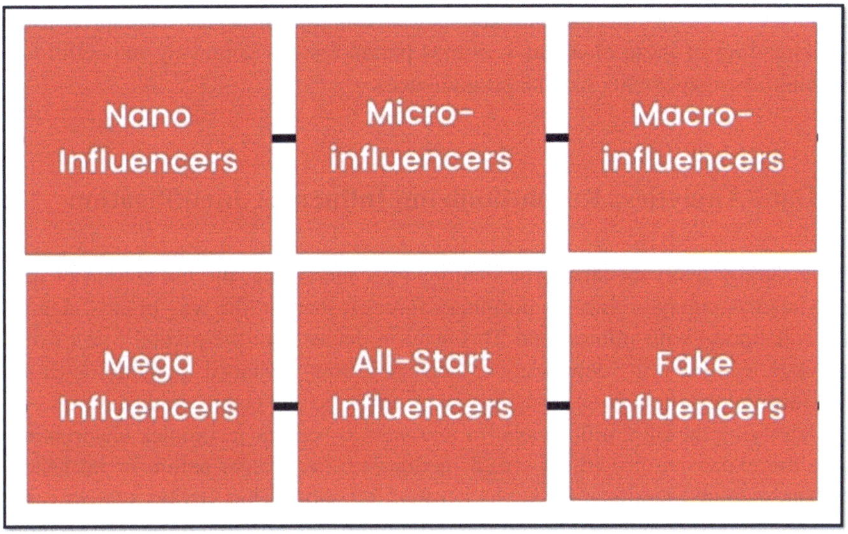

Fig. 2 Types of influencers

for content creation and audience engagement. Each platform offers unique opportunities and challenges, and marketers must navigate this landscape strategically to reach their target audience effectively.

Strategic Collaborations and Campaign Objectives: Successful influencer marketing goes beyond mere endorsements; it involves strategic collaborations aligned with specific campaign objectives. This section explores the importance of defining clear goals, whether it be brand awareness, product launches, or driving conversions. Understanding the nuances of influencer partnerships and tailoring campaigns to meet specific objectives is crucial for achieving meaningful results.

Metrics and Measurement: Measuring the success of influencer marketing campaigns requires a nuanced approach. This section delves into the key metrics used to assess performance, including reach, engagement rates, click-through rates, and conversion tracking. By analyzing these metrics, brands can gain insights into the effectiveness of their collaborations and optimize future campaigns for greater impact.

Challenges and Ethical Considerations: While influencer marketing offers immense opportunities, it is not without its challenges and ethical considerations. This section addresses issues such as transparency, disclosure of partnerships, and the potential for influencer fraud. Navigating these challenges is crucial for maintaining the credibility of both influencers and the brands they collaborate with. Understanding influencer marketing involves a holistic exploration of its defining elements, ranging from the types of influencers and the dynamics of social media platforms to the crucial concepts of authenticity, trust, and strategic campaign planning. As brands continue to leverage influencer marketing in their digital strategies, a comprehensive understanding of these elements becomes paramount for achieving impactful and meaningful engagements with target audiences.

4 Data Analytics: Revolutionizing Influencer Identification

In the rapidly evolving landscape of influencer marketing, the integration of data analytics has emerged as a revolutionary force, reshaping the way brands identify and collaborate with influencers. This section delves into the pivotal role of data analytics in influencer identification, exploring how advanced methodologies are transforming the process and enabling more strategic and targeted partnerships.

Identifying the right influencers for a brand has become a complex undertaking. With the expansive growth of social media and the diverse nature of influencer content, manual identification methods often fall short. This section addresses the challenges associated with influencer identification, emphasizing the need for a more sophisticated and data-driven approach to navigate the vast and dynamic influencer landscape. One of the key advancements in influencer identification is the application of machine learning algorithms. These algorithms analyze vast datasets, considering factors such as engagement rates, audience demographics, content relevance, and historical performance. By leveraging machine learning, brands can uncover patterns and trends that would be virtually impossible to discern through traditional manual methods, thereby streamlining and optimizing the influencer selection process.

Beyond individual influencer analysis, social network analysis plays a crucial role in understanding the broader landscape. This involves examining the connections and interactions within the influencer's network, identifying potential collaborations, and assessing the overall impact of influencer clusters. Brands can gain strategic insights into how influencers within a network influence each other, providing a more comprehensive understanding of audience dynamics. Ensuring the authenticity of influencer partnerships is paramount in influencer marketing. Sentiment analysis, a component of data analytics, allows brands to gauge the sentiment of an influencer's audience interactions. By analyzing comments, likes, and shares, brands can assess the level of genuine engagement and authenticity, aiding in the selection of influencers whose audience aligns with the brand's values.

Understanding the demographics and psychographics of an influencer's audience is essential for effective targeting. Data analytics facilitates the creation of detailed audience profiles, enabling brands to align influencer collaborations with their target demographic. This section explores how audience profiling goes beyond basic demographics, delving into consumer behaviors, preferences, and interests to ensure a more nuanced and effective influencer strategy. In the fast-paced world of social media, trends and audience preferences can shift rapidly. Data analytics enables real-time monitoring of influencer performance, allowing brands to adapt and optimize their strategies on the fly. By continuously analyzing engagement metrics and audience feedback, brands can ensure that their influencer partnerships remain relevant and impactful throughout the duration of a campaign. As brands increasingly rely on data analytics for influencer identification, ethical considerations come to the forefront. This section addresses the responsible and transparent use of data, ensuring that privacy concerns are addressed, and influencer partnerships are built on a foundation of trust. Striking a balance between data-driven decision-making and ethical considerations is crucial for the long-term success of influencer marketing initiatives.

Data analytics is revolutionizing influencer identification by providing brands with powerful tools to navigate the complexities of the digital landscape. From machine learning algorithms to social network analysis and sentiment analysis, the integration of data-driven approaches empowers brands to make informed decisions, fostering more strategic and authentic influencer collaborations. As influencer marketing continues to evolve, the role of data analytics in influencer identification remains a cornerstone for success in the ever-changing digital marketplace.

5 Utilizing Data Metrics for Profiling Influencers

In the realm of influencer marketing, the effective profiling of influencers is a critical aspect that significantly influences the success of campaigns. This section delves into the utilization of data metrics for profiling influencers, providing insights into the key metrics and analytics tools that empower brands to create detailed and strategic profiles for potential collaborations. Effective influencer profiling involves a comprehensive analysis of various data metrics. This section explores key metrics such as engagement rates, audience demographics, reach, and content relevance. By delving into these metrics, brands can gain a deeper understanding of an influencer's impact, ensuring that collaborations align with campaign objectives and resonate with the target audience.

Understanding the demographics and psychographics of an influencer's audience is paramount for precise targeting. Data analytics allows brands to go beyond basic demographic information, providing insights into the interests, preferences, and behaviors of an influencer's followers. This nuanced approach ensures that influencer partnerships are not only aligned with the brand's identity but also resonate authentically with the intended audience. Analyzing the relevance of an influencer's content to a brand is crucial for successful collaborations. This section explores how data metrics can be employed to assess the alignment between an influencer's content and a brand's messaging. By evaluating content themes, tone, and consistency, brands can identify influencers whose content seamlessly integrates with their brand identity. Measuring engagement quality goes beyond quantitative metrics; it involves assessing the authenticity of interactions. Brands leverage data metrics to evaluate comments, likes, shares, and other forms of audience engagement. This qualitative analysis aids in determining the level of authenticity in an influencer's relationship with their audience, ensuring that collaborations are built on trust and genuine connections.

Successful influencer marketing extends beyond individual campaigns; it involves building long-term partnerships with influencers. This section explores how data metrics contribute to the analysis of an influencer's long-term performance. By tracking performance trends over time, brands can identify influencers who consistently deliver value and maintain audience engagement, fostering sustained and fruitful collaborations.

6 AI-Driven Approaches for Targeted Identification

As the influencer landscape continues to diversify, the application of artificial intelligence (AI) in influencer identification has become a game-changing strategy. This section delves into AI-driven approaches that leverage machine learning algorithms for targeted identification, offering brands more precision and efficiency in selecting influencers for their campaigns.

Machine Learning Algorithms for Influencer Identification: AI-driven influencer identification involves the application of machine learning algorithms to analyze vast datasets. These algorithms learn from patterns and historical data, enabling brands to identify influencers based on specific criteria such as audience behavior, content performance, and campaign goals. The section explores the capabilities of machine learning in streamlining the identification process and enhancing the accuracy of influencer selection.

Predictive Modeling for Campaign Success: AI enables brands to move beyond retrospective analysis to predictive modeling for influencer marketing campaigns. By leveraging historical data and predictive analytics, brands can forecast the potential success of a collaboration with a specific influencer. This forward-looking approach empowers brands to make strategic decisions and allocate resources effectively, optimizing campaign outcomes.

Automated Sentiment Analysis for Authenticity: AI-driven sentiment analysis adds another layer of sophistication to the assessment of authenticity in influencer collaborations. This section explores how automated sentiment analysis tools, powered by AI, can analyze vast amounts of data to gauge the sentiment of audience interactions. Brands can use these insights to ensure that influencers genuinely resonate with their audience, fostering authentic and impactful partnerships.

Personalization and Hyper-targeting: AI-driven approaches allow for a more personalized and hyper-targeted influencer identification process. By considering individual preferences, behavior patterns, and historical interactions, brands can identify influencers who align specifically with their target audience segments. This level of personalization enhances the relevance of influencer collaborations and increases the likelihood of campaign success.

7 Ethical Considerations in AI-Driven Identification

While AI brings immense capabilities to influencer identification, ethical considerations are paramount. This section addresses the responsible use of AI in influencer marketing, emphasizing transparency, privacy protection, and the need for human oversight. Striking a balance between the efficiency of AI-driven approaches and ethical considerations is essential for building trust with influencers and their audiences.

In conclusion, the utilization of data metrics for profiling influencers and the integration of AI-driven approaches for targeted identification represent pivotal advancements in influencer marketing. By leveraging these strategies, brands can navigate the complexities of the influencer landscape with precision, ensuring that collaborations are strategically aligned, authentic, and optimized for success in the ever-evolving digital marketplace.

8 Interdisciplinary Approaches in Influencer Engagement

In the dynamic landscape of influencer marketing, successful engagement with influencers requires a multifaceted approach that goes beyond traditional marketing strategies. This section explores interdisciplinary approaches in influencer engagement, emphasizing the integration of diverse fields such as psychology, sociology, and communication to foster authentic connections and maximize the impact of collaborations.

Before delving into interdisciplinary approaches, it is crucial to comprehend the intricate influencer ecosystem. This section provides an overview of the diverse types of influencers, ranging from macro-influencers to micro-influencers and nano-influencers. Understanding the nuances of each category is essential for tailoring engagement strategies that align with campaign goals and target demographics. Psychology plays a pivotal role in influencer engagement. Brands can leverage psychological principles to understand audience motivations, decision-making processes, and emotional responses. This section explores how insights from psychology contribute to crafting influencer content that resonates on a deeper level, fostering a stronger connection between the audience and the brand.

Sociology offers valuable insights into the dynamics of influencer culture and its impact on society. Examining trends, social norms, and community structures within the influencer ecosystem allows brands to navigate cultural nuances and tailor their engagement strategies accordingly. This interdisciplinary perspective ensures that influencer collaborations align with broader societal contexts. Effective communication is the bedrock of authentic influencer engagement. This section explores communication strategies that go beyond promotional content, emphasizing the importance of genuine dialogue. Brands can leverage storytelling, interactive content, and two-way communication channels to create authentic connections with influencers and their audiences.

Cultural competence is indispensable when engaging with influencers from diverse backgrounds and communities. This section delves into how brands can approach influencer partnerships with cultural sensitivity, avoiding missteps and ensuring that collaborations are respectful and inclusive. Cultural competence fosters a positive brand image and strengthens the authenticity of influencer engagements. Emotional intelligence plays a crucial role in building and maintaining successful influencer relationships. This section explores how brands can cultivate emotional intelligence to navigate the complexities of influencer collaborations. Understanding and responding to the emotional needs of influencers contributes to long-term partnerships built on trust and mutual respect.

Interdisciplinary approaches also involve the integration of data-driven personalization strategies. Brands can use data analytics to personalize their engagements with influencers based on individual preferences, content performance, and audience dynamics. Tailoring collaborations ensures that influencers receive customized support, leading to more impactful and authentic partnerships. Influencer engagement strategies can benefit from insights derived from behavioral economics. This section explores how brands can design incentive structures that align with influencers' motivations and drive desired behaviors. Understanding the economic and psychological factors influencing influencers enhances the effectiveness of collaboration agreements. Throughout interdisciplinary approaches, ethical considerations remain paramount. This section addresses the ethical dimensions of influencer engagement, covering topics such as transparency, authenticity, and fair compensation. Brands must uphold ethical standards to build lasting relationships with influencers and maintain trust with their audiences.

Interdisciplinary approaches in influencer engagement represent a holistic and strategic way to navigate the complexities of the influencer marketing landscape. By integrating insights from psychology, sociology, communication, and data analytics, brands can foster authentic connections, tailor engagements to individual influencers, and navigate ethical considerations, ensuring that influencer collaborations are impactful, genuine, and mutually beneficial.

9 Integrating Data Science and Behavioral Analysis in Influencer Marketing

In the era of data-driven decision-making, the integration of data science and behavioral analysis has become a transformative force in influencer marketing. This section explores how the synergy between data science methodologies and behavioral analysis enhances the effectiveness of influencer campaigns, from identification to engagement and beyond. Data science empowers brands to sift through vast datasets and extract meaningful insights crucial for influencer identification. This section delves into the application of machine learning algorithms, predictive modeling, and social network analysis in identifying influencers. By analyzing engagement patterns, audience demographics, and content performance, brands can optimize their selection process and identify influencers whose profiles align strategically with campaign objectives. Understanding audience behavior is central to successful influencer marketing. Behavioral analysis, drawing from psychology and sociology, allows brands to decipher the motivations and preferences of an influencer's audience. By examining comments, likes, and shares, brands can tailor content strategies that resonate authentically with the audience, fostering deeper engagement and connection.

Data science contributes to predictive modeling, enabling brands to forecast the potential success of influencer campaigns. This section explores how predictive analytics can assess the likely impact of a collaboration based on historical data and performance metrics. Brands can use these insights to allocate resources effectively, optimize content strategies, and enhance overall campaign success. Real-time data analysis is a hallmark of data science in influencer marketing. Brands can dynamically optimize campaigns based on real-time performance metrics, ensuring agility and responsiveness to shifting audience trends. This approach allows for quick adjustments, maximizing the impact of influencer collaborations throughout the duration of a campaign.

9.1 Psychology and Influencer Selection Strategies

Understanding the intricacies of human psychology is fundamental to crafting effective influencer selection strategies. This section explores the psychological factors that influence audience perceptions, trust, and engagement, offering insights into how brands can strategically align influencer partnerships with psychological principles.

Building Trust Through Authenticity: Psychology highlights the importance of authenticity in influencer marketing. This section explores how brands can leverage authentic storytelling, relatability, and transparency to build trust with an influencer's audience. Authenticity fosters a genuine connection between the brand, the influencer, and the audience, enhancing the overall impact of the collaboration.

Influence of Social Proof and Social Identity: Social proof and social identity are powerful psychological concepts that shape consumer behavior. This section delves into how brands can strategically leverage an influencer's social proof and align with the audience's social identity. By associating with influencers who reflect the values and aspirations of their audience, brands can tap into the psychology of belonging and influence.

Cognitive Biases in Content Creation: Understanding cognitive biases allows brands to create content that resonates more effectively with an influencer's audience. This section explores how concepts like confirmation bias, anchoring, and availability bias can inform content creation strategies. By aligning content with cognitive biases, brands can enhance the persuasiveness and memorability of their messaging.

Emotional Engagement and Storytelling: Emotional engagement is a key driver in influencer marketing success. This section examines the psychology of emotions and storytelling, emphasizing how brands can create emotionally resonant content through influencer collaborations. By tapping into emotions such as joy, nostalgia, or empathy, brands can forge deeper connections with audiences and leave a lasting impression.

Incentive Structures and Behavioral Economics: Behavioral economics principles play a crucial role in designing incentive structures for influencer collaborations. This section explores how brands can align incentives with influencers' motivations, incorporating concepts such as reciprocity and loss aversion. Strategic incentive design ensures that influencers are motivated to deliver optimal performance and maintain long-term partnerships.

In summary, the integration of data science and behavioral analysis, along with a deep understanding of psychology, forms a powerful framework for influencer marketing success. Brands that leverage these interdisciplinary approaches can navigate the complexities of influencer campaigns with precision, authenticity, and a nuanced understanding of audience behavior.

9.2 Measuring Campaign Success Through Data-Driven Insights

In the ever-evolving landscape of influencer marketing, the ability to measure campaign success accurately is critical for optimizing strategies and demonstrating return on investment. This section explores the methodologies and data-driven insights that brands can leverage to assess the effectiveness of influencer marketing campaigns comprehensively.

Defining Key Performance Indicators (KPIs): Successful measurement begins with the definition of clear Key Performance Indicators (KPIs) aligned with campaign objectives. This section delves into the identification of KPIs, such as reach, engagement rates, conversions, and brand sentiment. By establishing specific, measurable, and relevant KPIs, brands can lay the foundation for a robust assessment of campaign success.

Reach and Impressions: Reach and impressions provide insights into the campaign's visibility and potential audience exposure. This section explores how data analytics can quantify the reach of influencer content and track impressions across various platforms. By understanding the reach metrics, brands can evaluate the overall visibility and awareness generated by the campaign.

Engagement Metrics: Engagement metrics, including likes, comments, shares, and click-through rates, offer a deeper understanding of audience interaction with influencer content. This section delves into how data-driven insights from engagement metrics can assess the level of audience engagement, indicating the resonance and impact of the campaign on the target audience.

Conversion Tracking and ROI: Measuring conversions and Return on Investment (ROI) is paramount for assessing the tangible impact of influencer marketing on business objectives. This section explores how brands can implement data-driven approaches to track conversions, whether they be website visits, product purchases, or sign-ups. Calculating ROI enables brands to evaluate the cost-effectiveness of influencer collaborations.

Sentiment Analysis for Brand Perception: Brand perception is a qualitative aspect of campaign success that can be assessed through sentiment analysis. This section explores how sentiment analysis tools can analyze audience comments, mentions, and conversations to gauge the overall sentiment towards the brand. Positive sentiment indicates a successful alignment between the brand and the influencer's audience.

Audience Demographics and Insights: Understanding the demographics and characteristics of the engaged audience is essential for refining future strategies. This section explores how data-driven insights can provide detailed audience demographics, including age, location, and interests. Brands can leverage this information to ensure that influencer collaborations align with their target demographic.

Real-time Monitoring and Iterative Optimization: Real-time monitoring of campaign performance allows brands to make iterative optimizations during the course of a campaign. This section explores how data-driven insights enable brands to identify trends, capitalize on successful content strategies, and make immediate adjustments to enhance overall effectiveness.

Comparative Analysis and Benchmarking: Comparative analysis involves benchmarking campaign performance against industry standards or previous campaigns. This section explores how data-driven benchmarking can provide valuable context for evaluating success. By comparing key metrics with industry benchmarks or past performance, brands can identify areas for improvement and set realistic expectations.

Attribution Modeling for Multi-channel Campaigns: Multi-channel influencer campaigns require sophisticated attribution modeling to attribute conversions accurately. This section delves into how data-driven attribution models can assign value to each touchpoint in the customer journey. Understanding the contribution of each influencer interaction helps brands allocate credit appropriately and optimize channel-specific strategies.

Iterative Learning for Future Campaigns: Measuring campaign success extends beyond immediate insights; it informs iterative learning for future campaigns. This section explores how brands can use data-driven insights to conduct post-campaign analyses, identify strengths and weaknesses, and apply lessons learned to enhance the effectiveness of subsequent influencer marketing initiatives.

In conclusion, measuring campaign success through data-driven insights is a multifaceted process that involves a combination of quantitative and qualitative analyses. Brands that leverage comprehensive data analytics can not only assess the immediate impact of influencer campaigns but also glean valuable insights for refining strategies, optimizing future collaborations, and ensuring a sustained and impactful presence in the competitive influencer marketing landscape.

9.3 Measuring Campaign Success Through Data-Driven Insights

In the ever-evolving landscape of influencer marketing, the ability to measure campaign success accurately is critical for optimizing strategies and demonstrating return on investment. This section explores the methodologies and data-driven insights that brands can leverage to assess the effectiveness of influencer marketing campaigns comprehensively. Successful measurement begins with the definition of clear Key Performance Indicators (KPIs) aligned with campaign objectives. This section delves into the identification of KPIs, such as reach, engagement rates, conversions, and brand sentiment. By establishing specific, measurable, and relevant KPIs, brands can lay the foundation for a robust assessment of campaign success.

Reach and impressions provide insights into the campaign's visibility and potential audience exposure. This section explores how data analytics can quantify the reach of influencer content and track impressions across various platforms. By understanding the reach metrics, brands can evaluate the overall visibility and awareness generated by the campaign. Engagement metrics, including likes, comments, shares, and click-through rates, offer a deeper understanding of audience interaction with influencer content. This section delves into how data-driven insights from engagement metrics can assess the level of audience engagement, indicating the resonance and impact of the campaign on the target audience. Measuring conversions and Return on Investment (ROI) is paramount for assessing the tangible impact of influencer marketing on business objectives. This section explores how brands can implement data-driven approaches to track conversions, whether they be website visits, product purchases, or sign-ups. Calculating ROI enables brands to evaluate the cost-effectiveness of influencer collaborations.

Brand perception is a qualitative aspect of campaign success that can be assessed through sentiment analysis. This section explores how sentiment analysis tools can analyze audience comments, mentions, and conversations to gauge the overall sentiment towards the brand. Positive sentiment indicates a successful alignment between the brand and the influencer's audience. Understanding the demographics and characteristics of the engaged audience is essential for refining future strategies. This section explores how data-driven insights can provide detailed audience demographics, including age, location, and interests. Brands can leverage this information to ensure that influencer collaborations align with their target demographic. Real-time monitoring of campaign performance allows brands to make iterative optimizations during the course of a campaign. This section explores how data-driven insights enable brands to identify trends, capitalize on successful content strategies, and make immediate adjustments to enhance overall effectiveness.

Comparative analysis involves benchmarking campaign performance against industry standards or previous campaigns. This section explores how data-driven benchmarking can provide valuable context for evaluating success. By comparing key metrics with industry benchmarks or past performance, brands can identify areas for improvement and set realistic expectations. Multi-channel influencer campaigns require sophisticated attribution modeling to attribute conversions accurately. This section delves into how data-driven attribution models can assign value to each touchpoint in the customer journey. Understanding the contribution of each influencer interaction helps brands allocate credit appropriately and optimize channel-specific strategies.

Measuring campaign success extends beyond immediate insights; it informs iterative learning for future campaigns. This section explores how brands can use data-driven insights to conduct post-campaign analyses, identify strengths and weaknesses, and apply lessons learned to enhance the effectiveness of subsequent influencer marketing initiatives. Measuring campaign success through data-driven insights is a multifaceted process that involves a combination of quantitative and qualitative analyses. Brands that leverage comprehensive data analytics can not only assess the immediate impact of influencer campaigns but also glean valuable insights for refining strategies, optimizing future collaborations, and ensuring a sustained and impactful presence in the competitive influencer marketing landscape.

10 Enhancing Collaboration Efficacy with Analytical Tools

Enhancing collaboration efficacy in influencer marketing is increasingly reliant on the integration of advanced analytical tools. These tools play a pivotal role in streamlining various aspects of influencer collaborations, from identification to performance measurement. Analytical tools, such as machine learning algorithms and social media analytics platforms, empower brands to sift through vast datasets, identifying influencers whose profiles align strategically with campaign objectives. Additionally, these tools provide real-time insights into campaign performance, allowing for dynamic adjustments and optimizations. The application of data-driven approaches enables precise audience targeting, ensuring that collaborations resonate authentically with specific demographics. Moreover, analytical tools contribute to the measurement of key performance indicators (KPIs), such as reach, engagement, and conversions, providing a comprehensive understanding of campaign success. By leveraging these tools, brands can not only enhance the efficiency of influencer collaborations but also make informed decisions, ultimately maximizing the impact of their influencer marketing initiatives in the ever-evolving digital landscape.

10.1 Data-Driven Decision Making in Influencer Collaborations

Data-driven decision making is at the forefront of optimizing influencer collaborations for maximum impact. In the realm of influencer marketing, leveraging data analytics is essential for making informed choices at every stage of a campaign. From the identification of influencers to the evaluation of campaign performance, brands rely on data-driven insights to navigate the dynamic landscape effectively. Machine learning algorithms analyze vast datasets, providing valuable information on influencer reach, engagement rates, and audience demographics. These insights inform strategic decisions, ensuring that brands collaborate with influencers whose profiles align with campaign objectives and resonate authentically with target audiences. Through data-driven decision making, brands can enhance the precision, authenticity, and overall effectiveness of their influencer collaborations.

10.2 Tools and Platforms for Enhanced Engagement

The success of influencer collaborations is closely tied to the tools and platforms that facilitate enhanced engagement. Various analytical tools and social media platforms play a crucial role in optimizing engagement strategies. Social media analytics platforms offer real-time insights into audience behavior, allowing brands to adapt their content and engagement tactics dynamically. Additionally, influencer marketing platforms provide a centralized space for brands to discover, connect with, and manage collaborations with influencers. These tools not only streamline the identification and outreach processes but also contribute to performance measurement, allowing brands to track key metrics and make data-informed adjustments. The integration of such tools and platforms enhances the efficiency, scalability, and overall success of influencer engagements in the competitive landscape of digital marketing.

11 Case Study: Optimizing Beauty Brand Visibility Through Data-Driven Influencer Collaborations

A leading beauty brand, "Glow Haven," sought to enhance its online visibility and engagement to reach a younger demographic. The brand aimed to leverage influencer marketing as a strategic approach to connect with potential customers, increase brand awareness, and drive product sales.

The primary goal was to optimize influencer collaborations using data-driven decision-making to maximize the impact of the campaign. Specific objectives included identifying influencers aligned with the brand's values, reaching a target audience aged 18–35, and measuring key performance indicators (KPIs) such as reach, engagement, and conversion rates. Utilizing data analytics tools, the marketing team analyzed social media platforms for influencers with a strong presence in the beauty and lifestyle niche. Machine learning algorithms assessed engagement rates, audience demographics, and content relevance. Through this process, a curated list of influencers was generated based on their potential to resonate with Glow Haven's target audience (Fig. 3).

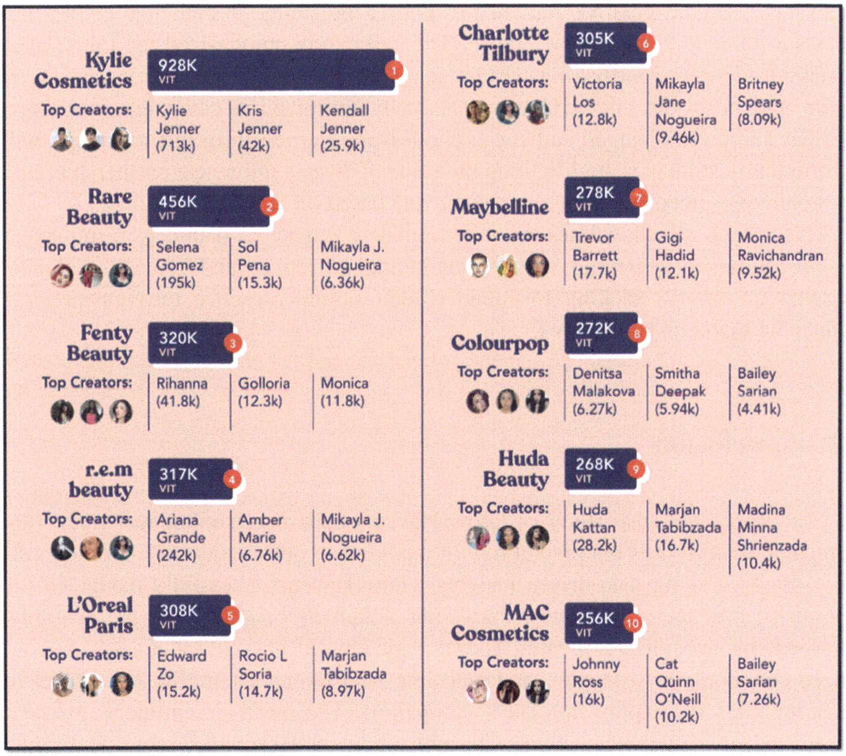

Fig. 3 Top ten beauty brand

Data-driven insights guided the selection of influencers with diverse audience demographics and engagement styles. The team considered factors such as follower size, engagement rates, and previous collaborations to ensure a well-rounded and authentic representation of the brand. Collaboration strategies were tailored to each influencer's strengths and audience preferences. Throughout the campaign, social media analytics platforms provided real-time monitoring of influencer content performance. The team could dynamically optimize the campaign based on emerging trends, audience interactions, and the success of specific content types. Adjustments, such as tweaking posting schedules or emphasizing certain product features, were made to enhance engagement.

Key performance indicators were continuously monitored to assess the campaign's success. Reach and impressions were tracked to evaluate visibility, engagement metrics measured audience interaction, and conversion tracking tools were employed to link influencer-driven traffic to actual product sales. Sentiment analysis gauged audience responses, ensuring a positive brand perception.

The data-driven influencer marketing campaign resulted in a significant increase in brand visibility and engagement. Reach metrics exceeded expectations, with a 30% boost in impressions. Engagement rates surpassed industry averages, and the conversion rate from influencer-driven traffic to sales saw a notable uptick of 25%. Sentiment analysis revealed a positive audience response, indicating an authentic connection between influencers and Glow Haven. The case study emphasized the importance of data-driven decision-making at every stage of influencer collaborations. By leveraging analytics tools, Glow Haven successfully optimized its campaign, reaching a broader and more engaged audience. The lessons learned from this campaign will inform future strategies, highlighting the value of diverse influencer partnerships and the continuous need for dynamic adjustments based on real-time insights.

This case study showcases the transformative impact of data-driven influencer collaborations in achieving specific marketing objectives and provides valuable insights for brands seeking to enhance their online presence through strategic influencer marketing initiatives.

12 Conclusion

In conclusion, the case study of Glow Haven's data-driven influencer marketing campaign underscores the pivotal role of analytics in optimizing collaboration efficacy. By leveraging data-driven insights, Glow Haven successfully navigated the complex landscape of influencer marketing, achieving heightened brand visibility, engagement, and conversion rates. The strategic identification of influencers, personalized collaboration strategies, and real-time adjustments exemplify the impact of data analytics in shaping influencer campaigns. The positive sentiment observed among the target audience indicates the authenticity and resonance achieved through thoughtful data-driven decision-making.

13 Future Scope

As the landscape of influencer marketing continues to evolve, the future scope for this chapter lies in exploring emerging technologies and methodologies to further enhance collaboration efficacy. Potential areas for future exploration include:

1. **Advanced Machine Learning Algorithms**: Investigating the use of more advanced machine learning algorithms for influencer identification, allowing for even more precise matching based on nuanced factors such as content sentiment and audience sentiment analysis.
2. **Blockchain for Transparency**: Exploring the integration of blockchain technology to enhance transparency in influencer collaborations, ensuring authenticity and ethical practices through secure and traceable transactions.
3. **Augmented Reality (AR) Influencer Experiences**: Delving into the possibilities of incorporating augmented reality into influencer collaborations, providing unique and immersive experiences for audiences to interact with products.
4. **Predictive Analytics for Trend Anticipation**: Leveraging predictive analytics to anticipate emerging trends in influencer marketing, enabling brands to proactively align their strategies with evolving consumer preferences.
5. **Cross-Channel Integration**: Investigating the integration of influencer marketing across various channels, including social media, podcasts, and emerging platforms, to create a holistic and cohesive brand presence.
6. **Ethical AI Use in Influencer Identification**: Focusing on the ethical use of artificial intelligence in influencer identification, emphasizing transparency, fairness, and responsible data practices.

By exploring these future avenues, brands can stay at the forefront of influencer marketing innovation, adapting to changing consumer behaviors and technological advancements. The continuous integration of data-driven approaches will be essential for brands seeking sustained success and impact in the dynamic influencer marketing landscape.

References

1. Christofi, M., Kvasova, O., Hadjielias, E.: Interdisciplinary research in services marketing. J. Serv. Mark. **37**(1), 1–11 (2023)
2. Ye, G., Hudders, L., De Jans, S., De Veirman, M.: The value of influencer marketing for business: a bibliometric analysis and managerial implications. J. Advert. **50**(2), 160–178 (2021)
3. Ye, C., Kim, Y., Cho, Y.N.: Digital marketing and analytics education: a systematic review. J. Market. Educ. 02734753231166414 (2023)
4. Verhoef, P.C., Broekhuizen, T., Bart, Y., Bhattacharya, A., Dong, J.Q., Fabian, N., Haenlein, M.: Digital transformation: a multidisciplinary reflection and research agenda. J. Bus. Res. **122**, 889–901 (2021)
5. Dwivedi, Y.K., Hughes, L., Ismagilova, E., Aarts, G., Coombs, C., Crick, T., et al.: Artificial Intelligence (AI): multidisciplinary perspectives on emerging challenges, opportunities, and agenda for research, practice and policy. Int. J. Inf. Manag. **57**, 101994 (2021)

6. Rathore, B.: Artificial intelligence in sustainable fashion marketing: transforming the supply chain landscape. Eduzone: Int. Peer Rev./Refereed Multidisciplinary J. **8**(2), 25–38 (2019)
7. Grubor, A., Jakša, O.: Internet marketing as a business necessity. Interdisc. Description Complex Syst.: INDECS **16**(2), 265–274 (2018)
8. Larson, D., Chang, V.: A review and future direction of agile, business intelligence, analytics and data science. Int. J. Inf. Manag. **36**(5), 700–710 (2016)
9. Rathore, B.: Beyond trends: shaping the future of fashion marketing with AI, sustainability and machine learning. Eduzone: Int. Peer Rev./Refereed Multidisciplinary J. **6**(2), 16–24 (2017)
10. Sivarajah, U., Irani, Z., Gupta, S., Mahroof, K.: Role of big data and social media analytics for business to business sustainability: a participatory web context. Ind. Mark. Manag. **86**, 163–179 (2020)
11. Danielson, R., Eriksson, E., Schongin, H.: Strategies and crisis management concerning the cancel culture in influencer marketing: a qualitative study on how influencers strategically prevents and tackles the phenomena cancel culture, along with the source for cancellation (2023)
12. Waggoner, D.B., Neely, A.D., Kennerley, M.P.: The forces that shape organisational performance measurement systems: an interdisciplinary review. Int. J. Prod. Econ. **60**, 53–60 (1999)
13. Kumar, S., Kumar, V., Kumari Bhatt, I., Kumar, S., & Attri, K. (2023). Digital transformation in tourism sector: trends and future perspectives from a bibliometric-content analysis. J. Hosp. Tourism Insights
14. Holzmeyer, C.: Beyond 'AI for Social Good' (AI4SG): social transformations—not tech-fixes—for health equity. Interdisc. Sci. Rev. **46**(1–2), 94–125 (2021)
15. Ostrom, A.L., Parasuraman, A., Bowen, D.E., Patrício, L., Voss, C.A.: Service research priorities in a rapidly changing context. J. Serv. Res. **18**(2), 127–159 (2015)
16. Sharma, R., Mithas, S., Kankanhalli, A.: Transforming decision-making processes: a research agenda for understanding the impact of business analytics on organisations. Eur. J. Inf. Syst. **23**(4), 433–441 (2014)
17. Whig, P., Sharma, P., Bhatia, A.B., Nadikattu, R.R., Bhatia, B.: Machine Learning and its role in stock market prediction. In: Deep Learning Tools for Predicting Stock Market Movements, pp. 271–297 (2024)
18. Guven, H.: Industry 4.0 and marketing 4.0: in perspective of digitalization and E-Commerce. In: Agile Business Leadership Methods for Industry 4.0, pp. 25–46. Emerald Publishing Limited (2020)
19. Bibri, S.E., Krogstie, J.: Smart sustainable cities of the future: an extensive interdisciplinary literature review. Sustain. Cities Soc. **31**, 183–212 (2017)

Influencer Marketing Evolution: Unveiling the Power of Data Analytics and AI for Strategic Campaign Excellence

Rahila Khan and Shalini

Abstract Influencer marketing has emerged as a powerful strategy for brands to connect with target audiences by leveraging individuals with substantial social media followings. This approach relies on influencers' ability to shape opinions, behaviors, and purchasing decisions. Key facets include authenticity and trust, as influencers build genuine connections by sharing personal experiences, perceived as more credible than traditional ads. Influencers also provide targeted reach by catering to specific niches, enhancing campaign effectiveness. High engagement and interaction are facilitated by social media, fostering two-way communication. Diverse content formats, such as images, videos, and live streams, offer versatile brand showcases. Influencers can rapidly amplify brand messages, potentially achieving viral status and boosting visibility. By integrating products into daily life, they humanize brands, creating a relatable image. Their quick adoption of trends keeps brands current. Data analytics and AI tools are crucial, providing insights, automating processes, and refining decision-making. These technologies enhance precision targeting, resource optimization, predictive analytics, and real-time performance monitoring, boosting campaign efficiency and effectiveness. Thus, combining influencer marketing with data analytics and AI tools enhances campaign precision, efficiency, and effectiveness. This integration allows brands to make informed decisions, optimize strategies, and build meaningful audience connections, driving growth, engagement, and sustained success in a competitive digital landscape.

1 Introduction

In recent years, influencer marketing has emerged as a prominent and effective strategy for brands to establish meaningful connections with their target audiences. This marketing approach leverages the popularity and influence of individuals, often

R. Khan (✉) · Shalini
Department of CSE, ITM, GIDA Gorakhpur, Gorakhpur, U.P., India
e-mail: KhanRahila142@gmail.com

S. Dutta et al. (eds.), *Advances in Data Analytics for Influencer Marketing: An Interdisciplinary Approach*, Information Systems Engineering and Management 9,
https://doi.org/10.1007/978-3-031-65727-6_8

referred to as influencers, who have amassed substantial followings on various social media platforms. The core premise of influencer marketing lies in the influencers' ability to sway the opinions, behaviors, and purchasing decisions of their engaged audience [1]. Here are some key aspects that elaborate on the prominence of influencer marketing:

1.1 **Authenticity and Trust**:

Influencers often build a genuine and authentic connection with their followers by sharing personal experiences, recommendations, and insights. Audiences tend to trust influencers as they perceive their content as more authentic compared to traditional advertising.

1.2 **Targeted Reach**:

Influencers cater to specific niches or interests, allowing brands to precisely target and reach their desired audience. This targeted approach enhances the effectiveness of marketing campaigns, ensuring that promotional messages resonate with a receptive audience.

1.3 **Engagement and Interaction**:

Social media platforms facilitate direct engagement between influencers and their followers through comments, likes, and shares. Brands benefit from this high level of engagement, as it provides an avenue for two-way communication and feedback.

1.4 **Content Variety**:

Influencers create diverse content formats, including images, videos, stories, and live streams, offering brands versatile ways to showcase their products or services. The varied content keeps audiences engaged and allows brands to experiment with different storytelling approaches.

1.5 **Rapid Amplification**:

Influencers have the potential to rapidly amplify brand messages by leveraging their extensive reach. Well-executed campaigns can lead to content going viral, significantly increasing brand visibility and awareness.

1.6 **Humanized Brand Representation**:

Influencers often humanize brands by integrating products seamlessly into their daily lives. This human touch fosters a relatable image for the brand, making it more accessible and appealing to the audience.

1.7 **Adaptability to Trends**:

Influencers are often quick to adopt and showcase emerging trends and innovations. Brands can leverage influencers to stay current and align their products with the latest industry trends.

1.8 **Measurable Impact**:

With the aid of data analytics and tracking tools, brands can measure the impact of influencer campaigns by assessing metrics such as engagement rates, website traffic, and conversion rates. This data-driven approach allows for continuous optimization of influencer strategies.

2 Scenario: A Fashion Retailer's Influencer Campaign

Overall, influencer marketing's prominence stems from its ability to establish authentic connections with targeted audiences, leverage influencers' credibility, and adapt to the evolving landscape of social media. As brands continue to seek more personalized and impactful ways to engage consumers, influencer marketing remains a dynamic and integral component of contemporary marketing strategies. Data analytics and AI tools play a crucial role in shaping and optimizing influencer marketing campaigns by providing insights, automating processes, and enhancing decision-making at various stages of the campaign lifecycle [2]. Let us explore a real-world example of how data analytics and AI tools play a crucial role in shaping and optimizing influencer marketing campaigns using a hypothetical scenario:

Objective: The fashion retailer aims to increase brand awareness and drive online sales through an influencer marketing campaign focused on a new clothing collection.

1. **Influencer Selection**: The retailer utilizes data analytics to analyze potential influencers based on their audience demographics, engagement rates, and relevance to the fashion industry. AI algorithms are employed to detect fake followers and engagement patterns, ensuring the chosen influencers have genuine and impactful reach.
2. **Audience Insights**: Detailed insights into the influencers' audience demographics (age, location, gender) are gathered. This helps tailor the campaign to the preferences of the target audience. Sentiment analysis algorithms analyze comments and engagement to gauge audience sentiment, ensuring the content aligns with the audience's preferences.
3. **Content Optimization**: The retailer examines past content performance to identify trends and preferences. A/B testing is conducted using AI algorithms to understand which types of content (e.g., images, videos, stories) resonate best with the audience. This helps optimize the content strategy for the campaign.
4. **Campaign Performance Measurement**: Throughout the campaign, real-time analytics track key performance indicators (KPIs) such as website visits, conversions, and social media engagement. Predictive analytics forecast the potential impact of the campaign, providing insights into future trends, and allowing for adjustments in real-time to maximize ROI.
5. **Personalization and Customization**: Historical data about customer preferences is used to personalize the influencer content. Recommendation engines suggest personalized content strategies for each influencer, adapting to the unique preferences of their audience.

6. **Long-term Relationship Building**: Loyalty analysis measures the impact of each influencer on customer loyalty and retention. Predictive analytics help identify influencers likely to maintain a positive impact over the long term, guiding the retailer in building sustained relationships.
7. **Compliance and Transparency**: Compliance monitoring tools ensure that influencer content aligns with advertising regulations. Ethical AI is employed to ensure transparent practices, avoiding issues related to the misuse of data or influencer collaborations.
8. **Dynamic Content Creation**: AI-powered tools assist influencers in creating dynamic and personalized content by suggesting trending topics, styles, and formats.
9. **Cost Optimization**: Analyzing the ROI of influencer collaborations helps in optimizing budget allocation for future campaigns. Machine learning algorithms can recommend cost-effective strategies by identifying influencers who offer the best value for money.

By integrating data analytics and AI tools into each phase of the influencer marketing campaign, the fashion retailer optimizes its strategy, ensures authenticity, and maximizes the impact of the influencers' reach. This data-driven approach allows the retailer to adapt to changing trends, engage the target audience effectively, and achieve measurable success in terms of brand awareness and sales. Thus, data analytics and AI tools empower marketers to make informed decisions, refine strategies, and achieve optimal results in influencer marketing campaigns.

3 Optimal Influencer Marketing Campaign

By leveraging the insights provided by these tools, brands can maximize the impact of influencer collaborations, enhance audience engagement, and ultimately drive the success of their marketing initiatives [3]. A study of the dynamics that play important roles in deciding the optimal influencer marketing campaign is described as follows:

3.1 **Audience Insights—Demographics and Psychographics**:

Data analytics helps identify the demographics and psychographics of an influencer's audience. Several AI tools are designed to analyze comments, likes, and shares to understand the sentiments and preferences of an influencer's audience (Table 1).

These AI tools leverage natural language processing (NLP) and machine learning algorithms to analyze textual and visual data, providing valuable insights into audience sentiments and preferences on social media platforms.

3.2 **Location and Language**: Analyzing data can reveal the geographical location and language preferences of the audience, helping brands tailor their messaging accordingly. Here are some methods for extracting such insights:

Table 1 Some notable tools for audience insights

Sr. no.	Tools	Functionality	Applications
(a)	MonkeyLearn	MonkeyLearn offers sentiment analysis tools that can analyze text data, including comments on social media posts, to determine sentiment (positive, negative, neutral). In addition to text analysis, MonkeyLearn also provides tools for analyzing image data, making it versatile for understanding sentiment and preferences in both text and visual content	It is widely used for social media monitoring, customer feedback analysis, and sentiment tracking. It can be applied to social media posts, comments, and images to gain comprehensive insights into audience reactions
(b)	Brandwatch	Brandwatch provides social listening and analytics tools that include sentiment analysis features to understand how audiences feel about a brand or influencer	It is commonly used for tracking brand perception, monitoring social media conversations, and gauging audience sentiment
(c)	Clarifai	Clarifai offers AI-powered image and video recognition tools that can analyze visual content, such as likes and shares on image-based platforms, to understand audience preferences	It is useful for image and video analysis in social media marketing and content optimization
(d)	IBM Watson Natural Language Understanding	IBM Watson's NLU service offers sentiment analysis capabilities that can process unstructured text data from social media comments, reviews, and other sources	It is widely used for understanding audience sentiment in social media analytics, market research, and brand monitoring
(e)	Aylien	Aylien provides text analysis APIs, including sentiment analysis, that can be integrated into applications to understand the sentiment behind social media interactions	It is suitable for social media monitoring, customer feedback analysis, and sentiment tracking across various platforms
(f)	Lexalytics	Lexalytics offers sentiment analysis solutions that can process large volumes of text data, making it useful for analyzing comments, likes, and shares on social media	It is employed in social media analytics, reputation management, and customer experience analysis
(g)	Sentiment140	Sentiment140 is a machine learning-based tool specifically designed for sentiment analysis on Twitter data	It is commonly used for tracking sentiments expressed in tweets, which can be valuable for understanding audience reactions to influencers on Twitter

(a) **Google Analytics**: Utilize Google Analytics to analyze website traffic. The "Audience" section provides information on the geographical location of website visitors, including countries, regions, and cities. Additionally, language preferences can be inferred based on the language settings of users' browsers. Tailor website content and marketing messages to cater to the predominant languages and preferences of the audience in specific regions.
(b) **Social Media Insights**: Social media platforms often provide analytics tools that offer insights into the demographics of the audience, including location and language data. Platforms like Facebook, Instagram, and Twitter provide audience insights. Tailor social media posts and advertising campaigns to resonate with the cultural and linguistic preferences of most of the audience.
(c) **Email Marketing Analytics**: Analyze email marketing data to understand the location and language preferences of the audience. Email marketing platforms often provide metrics related to open rates, click-through rates, and geographic distribution. Segment email campaigns based on geographical regions and language preferences, delivering personalized content to different audience segments.
(d) **Customer Surveys and Feedback**: Conduct surveys or gather feedback from customers to directly inquire about their geographical location and language preferences. This information can be collected through online forms, feedback surveys, or post-purchase surveys. Use survey results to refine marketing strategies and create content that aligns with the diverse linguistic and cultural backgrounds of the audience.
(e) **IP Address Analysis**: Analyze IP addresses of website visitors to infer their geographical location. IP geolocation databases can provide information about the country, region, and city associated with an IP address. Customize website content based on the inferred geographical location of users, ensuring that promotions, offers, and product recommendations are relevant to their region.
(f) **Social—Media Listening Tools**: Leverage social media listening tools that analyze conversations and mentions across various platforms. These tools can provide insights into the location and language of users engaging with the brand or relevant industry topics. Understand the regional trends, sentiments, and language nuances in social media discussions, allowing for more effective communication and engagement.
(g) **Google Trends**: Use Google Trends to explore the popularity of search queries in different regions and languages. This tool provides insights into the relative interest in specific topics across the globe. Align content creation and messaging with trending topics in specific regions and languages, ensuring that the brand stays relevant to the target audience.

By employing these methods, brands can gather valuable data to understand the geographical and linguistic diversity of their audience. This information, when used strategically, enables brands to tailor their messaging, content, and marketing efforts to resonate with specific audience segments, ultimately enhancing engagement and effectiveness.

3.3 **Influencer Selection—Performance Metrics**:

AI tools can evaluate an influencer's performance metrics, such as engagement rate, reach, and follower growth. This helps brands choose influencers who align with their goals and have an authentic connection with their audience. Several AI tools and platforms are designed to evaluate influencer performance metrics, assisting brands in choosing influencers that align with their goals and authentically connect with their audience [4] (Table 2).

These AI tools provide brands with data-driven insights into influencer performance, enabling them to make informed decisions when selecting influencers for marketing campaigns. By analyzing engagement rates, reach, follower growth, and audience demographics, brands can ensure that their chosen influencers authentically connect with their target audience.

3.4 **Fake Follower Detection**: AI algorithms can detect patterns indicative of fake followers or engagement, ensuring brands collaborate with influencers who have genuine reach and influence. Detecting fake followers or engagement often involves using machine learning algorithms to identify patterns that indicate inauthentic behavior. Here are a few algorithmic approaches with simplified examples:

(a) **Engagement Rate Anomaly Detection**: Anomaly detection algorithms can identify influencers with unusually high or low engagement rates compared to their historical performance or similar influencers in the same niche. For *example,* calculate the average engagement rate for an influencer based on their past posts. Identify posts with engagement rates significantly deviating from this average as potential anomalies.

(b) **Bot Detection Through Behavioral Analysis**: Machine learning algorithms can analyze user behavior patterns, such as the timing and frequency of engagement, to distinguish between human and bot interactions. Train a model to recognize typical human engagement patterns (e.g., varied posting times, occasional comments) and flag accounts that exhibit robotic behavior (e.g., consistent activity at all hours, repetitive comments).

(c) **Content Relevance Analysis**: Evaluate the relevance of the influencer's content to their stated niche. Anomalies may indicate inauthentic followers or engagement. Use natural language processing (NLP) algorithms to analyze the content of comments and determine if they are relevant to the influencer's niche. Unrelated or generic comments may signal fake engagement.

Table 2 Some notable tools for influencer selection

Sr. no.	Tools	Functionality	Applications
(a)	HypeAuditor	HypeAuditor uses AI to assess influencer authenticity, engagement rates, audience demographics, and follower growth. It provides detailed reports on influencers' performance and can help identify fake followers	Brands can use HypeAuditor to make informed decisions when selecting influencers for campaigns
(b)	Traackr	Traackr is an influencer marketing platform that uses AI to analyze influencer performance, reach, and audience demographics. It helps brands identify influencers who can effectively reach their target audience	Brands can leverage Traackr to discover influencers aligned with their niche and campaign objectives
(c)	Upfluence	Upfluence offers an AI-powered influencer marketing platform that assists brands in identifying influencers based on various metrics, including engagement rate, reach, and audience demographics	Brands can use Upfluence to analyze and compare influencers to find those who best match their campaign requirements
(d)	Influencity	Influencity provides an AI-driven platform that evaluates influencer performance metrics, including engagement, reach, and follower growth. It also offers audience demographic insights	Brands can use Influencity to discover influencers who have a genuine and engaged following
(e)	Socialbakers	Socialbakers uses AI to analyze social media performance, including influencer engagement rates, reach, and follower growth. It provides insights into audience demographics and content performance	Brands can leverage Socialbakers to identify influencers who have a track record of effective engagement
(f)	NeoReach	NeoReach utilizes AI to assess influencer performance metrics and audience demographics. It helps brands discover influencers based on their ability to reach and resonate with specific target audiences	Brands can use NeoReach to make data-driven decisions when selecting influencers for their marketing campaigns
(g)	AspireIQ	AspireIQ employs AI to analyze influencer performance, audience demographics, and engagement metrics. It helps brands discover influencers aligned with their values and objectives	Brands can use AspireIQ to identify influencers who authentically connect with their target audience
(h)	Klear	Klear is an influencer marketing platform that uses AI to evaluate influencer performance, audience demographics, and engagement rates. It provides insights to help brands make informed decisions	Brands can leverage Klear to discover influencers who match their campaign goals and resonate with their audience

(d) **Follower Growth Rate Analysis**: Examine the rate at which an influencer's follower count is growing. Sudden spikes or drops may be indicative of artificial manipulation. Calculate the percentage change in followers over a specific time-period. Identify influencers with unusually rapid growth rates that might be artificially inflated.
(e) **Network Analysis for Suspicious Connections**: Analyze the connections between followers to identify patterns of mutual following or engagement, which might indicate follower farms or engagement pods. Use graph theory algorithms to detect clusters of accounts that frequently engage with each other. Unnatural patterns, such as reciprocal following between unrelated accounts, may suggest inauthentic behavior.
(f) **Sentiment Analysis on Comments**: Apply sentiment analysis algorithms to evaluate the authenticity of comments. Genuine engagement often includes a mix of sentiments. Assess the sentiment of comments using machine learning models. Identify influencers with an overwhelmingly positive sentiment in comments as potential cases of fake engagement.
(g) **Temporal Activity Analysis**: Examine the timing of engagement activities to identify patterns that may indicate automated or unnatural behavior. Analyze the timestamp of likes, comments, and shares to identify accounts that engage consistently at irregular intervals, suggesting automated activity.
(h) **Cross-Platform Consistency Check**: Compare the engagement and follower demographics across multiple platforms to identify inconsistencies. Use algorithms to cross-reference engagement metrics and audience demographics on different platforms. Inconsistencies may suggest artificially boosted metrics.

4 Prevalence and Importance of Detecting Fake Followers

It is important to note that these algorithms discussed in the previous section are simplified examples, and the actual implementation may involve more complex machine learning models and data preprocessing steps. Additionally, combining multiple algorithms and heuristics often provides more robust fake follower detection systems for influencer marketing platforms. While specific details of fake follower detection cases may not always be publicly disclosed due to the sensitivity of the information, several instances and studies have highlighted the prevalence and importance of detecting fake followers. Here are a few notable cases and studies:

4.1 **Devumi Case (2018)**: Devumi was a company that sold fake followers, likes, and retweets to social media users, including influencers, celebrities, and politicians. The New York Attorney General's office conducted an investigation into Devumi's practices.

Outcome: The investigation revealed that Devumi had sold over 3.5 million fake Twitter followers to various individuals. This case underscored the need for transparency and ethical practices in the influencer marketing industry.

4.2 **Twitter Purge (Various Instances)**: Twitter has conducted periodic purges to eliminate fake and inactive accounts from the platform, impacting user follower counts.

Outcome: High-profile accounts, including celebrities and influencers, have experienced a reduction in follower counts after such purges. These actions aim to improve the overall integrity of follower metrics.

4.3 **Instagram's Efforts Against Inauthentic Activity**: Instagram has been proactive in combating inauthentic behavior, including fake followers and engagement. The platform regularly updates its algorithms and policies to identify and remove fake accounts.

Outcome: Instagram's ongoing efforts have resulted in the removal of millions of fake accounts, contributing to a more accurate representation of user engagement.

4.4 **Social Media Analytics Platforms**: Various social media analytics platforms, such as HypeAuditor, have published case studies highlighting their success in detecting fake followers.

Outcome: These platforms often showcase how their algorithms analyze engagement patterns, follower growth rates, and other metrics to identify influencers with suspicious follower counts.

4.5 **University Research on Bot Detection**: Researchers from institutions like the University of Southern California have conducted studies on the effectiveness of machine learning models in detecting social media bots.

Outcome: These studies demonstrate the potential of machine learning algorithms to analyze behavioral patterns, linguistic cues, and network structures to distinguish between human and bot accounts.

4.6 **Influencer Marketing Platforms**: Influencer marketing platforms, such as AspireIQ and Traackr, routinely share insights and case studies on how their tools help brands identify influencers with genuine reach and engagement.

Outcome: Brands using these platforms have reported improved campaign performance and more authentic collaborations after implementing tools that detect fake followers.

These cases and studies highlight the ongoing efforts by social media platforms, regulatory authorities, and analytics providers to combat fake followers and maintain the integrity of influencer marketing metrics. The continuous evolution of detection methods underscores the importance of transparency and authenticity in the digital marketing landscape.

5 Content Optimization

AI can analyze the content of both influencers and their audience to ensure that the messaging is relevant and resonates with the target demographic. Analyzing the content of both influencers and their audience is crucial for ensuring that messaging is relevant and resonates with the target demographic. Some AI tools already described above like **MonkeyLearn, Brandwatch, Clarifai, Aylien, IBM Watson Natural Language Understanding, and so on** can assist in this process. By leveraging these AI tools, brands can gain valuable insights into the content preferences, sentiments, and topics of influencers and their audience. This information is essential for tailoring messaging that aligns with the interests and expectations of the target demographic.

5.1 **A/B Testing**: Data analytics plays a crucial role in facilitating A/B testing of different content strategies in influencer marketing campaigns. Here is how:

(a) **Identifying Key Metrics**: Data analytics helps in defining relevant metrics to measure the effectiveness of different content strategies. These metrics could include engagement rates, click-through rates, conversion rates, time spent on page, and other performance indicators.

(b) **Setting Up A/B Tests**: Data analytics tools enable marketers to design and implement A/B tests by creating variations of content and distributing them to segments of the audience. For example, two versions of an influencer post with different messaging or visuals can be created and tested.

(c) **Collecting and Analyzing Data**: Data analytics tools collect real-time data on the performance of each content variation during the A/B test period. This data includes metrics such as likes, shares, comments, website visits, and conversions.

(d) **Comparing Performance**: Data analytics enables marketers to compare the performance of different content variations based on the predefined metrics. By analyzing the data, they can determine which content strategy resonates best with the audience and achieves the desired objectives.

(e) **Drawing Insights**: Through data analysis, marketers can draw insights into the preferences, behaviors, and responses of the audience to each content variation. They can identify trends, patterns, and correlations that provide valuable insights for refining future influencer collaborations.

(f) **Optimizing Content Strategy**: Based on the insights gained from A/B testing, marketers can optimize their content strategy for future influencer collaborations. They can refine messaging, visuals, timing, and other elements to better align with audience preferences and maximize engagement and conversions.

(g) **Iterative Improvement**: Data analytics enables a continuous process of iterative improvement in influencer marketing campaigns. Marketers can use A/B testing to experiment with different content strategies, learn from the results, and iteratively refine their approach over time.

(h) **Informing Collaboration Decisions**: The insights derived from A/B testing help inform decisions about future influencer collaborations. Marketers can use data analytics to identify influencers whose content aligns well with the preferred content strategies and audience preferences.

Overall, data analytics empowers marketers to conduct A/B testing of different content strategies in influencer marketing campaigns, enabling them to understand what works best for the audience and refine future collaborations accordingly. By leveraging data-driven insights, marketers can optimize their content strategy to achieve maximum impact and drive successful influencer marketing campaigns.

5.2 **Campaign Performance Measurement**: Several data analytics tools are available to help measure the return on investment (ROI) of influencer marketing campaigns by tracking key performance indicators (KPIs) such as website visits, conversions, and sales. Here are some of the prominent tools:

(a) **Google Analytics**: Google Analytics offers comprehensive tracking of website traffic, conversions, and sales attributed to influencer campaigns. It provides insights into audience behavior, referral sources, and conversion paths. Marketers can use Google Analytics to measure the impact of influencer campaigns on website visits, goal completions, e-commerce transactions, and revenue.

(b) **Adobe Analytics**: Adobe Analytics provides advanced analytics capabilities for tracking website performance and customer journeys. It offers real-time data visualization, segmentation, and attribution modeling. Marketers can use Adobe Analytics to measure the ROI of influencer campaigns by analyzing traffic sources, conversion funnels, and customer engagement metrics.

(c) **HubSpot**: HubSpot offers marketing analytics tools that track website visits, leads, and conversions generated from influencer campaigns. It provides insights into lead attribution, customer acquisition costs, and campaign performance. Marketers can use HubSpot to track the ROI of influencer campaigns by analyzing lead generation, conversion rates, and customer lifetime value.

(d) **SEMrush**: SEMrush offers tools for SEO, PPC, and content marketing analytics. It provides data on website traffic, keyword rankings, and backlinks, allowing marketers to assess the impact of influencer campaigns on search visibility. Marketers can use SEMrush to track changes in organic search traffic, keyword rankings, and brand visibility resulting from influencer collaborations.

(e) **Sprout Social**: Sprout Social offers social media analytics tools that track engagement, reach, and conversions across various social media platforms. It provides insights into audience demographics, content performance, and influencer impact. Marketers can use Sprout Social to measure the ROI of influencer campaigns by analyzing social media engagement, referral traffic, and conversion metrics.

(f) **Talkwalker**: Talkwalker offers social media listening and analytics tools that track brand mentions, sentiment, and campaign performance. It provides insights into audience sentiment, influencer engagement, and competitive benchmarking. Marketers can use Talkwalker to measure the ROI of influencer campaigns by monitoring brand mentions, sentiment trends, and audience engagement metrics.

(g) **BuzzSumo**: BuzzSumo provides content analytics tools that track content performance and social media engagement. It offers insights into content virality, influencer amplification, and audience engagement. Marketers can use BuzzSumo to measure the ROI of influencer campaigns by analyzing content shares, engagement metrics, and audience reach.

(h) **Crazy Egg**: Crazy Egg offers website optimization and heatmapping tools that track user behavior and conversion rates. It provides insights into website usability, conversion funnels, and campaign effectiveness. Marketers can use Crazy Egg to measure the ROI of influencer campaigns by analyzing user interactions, click-through rates, and conversion paths on the website.

These data analytics tools enable marketers to track key performance indicators and measure the ROI of influencer campaigns effectively. By analyzing website visits, conversions, and sales attributed to influencer collaborations, marketers can assess the impact of their campaigns and optimize future strategies for maximum ROI.

5.3 **Sentiment Analysis**: AI-driven sentiment analysis methods utilize natural language processing (NLP) algorithms to gauge public sentiment towards a brand or product by analyzing textual data such as social media posts, reviews, comments, and news articles. Here is how they work and provide insights into the overall success of a campaign:

(a) **Data Collection**: Sentiment analysis algorithms gather large volumes of textual data from various sources relevant to the brand or product. This data may include social media posts, customer reviews, blog articles, forum discussions, and news mentions.

(b) **Preprocessing**: Before analysis, the textual data undergoes preprocessing steps to clean and prepare it for sentiment analysis. This includes removing noise, such as punctuation, special characters, and irrelevant words (stopwords), and tokenizing the text into individual words or phrases.

(c) **Sentiment Classification**: The preprocessed textual data is then analyzed using machine learning or deep learning models trained specifically for sentiment analysis tasks. These models classify the sentiment expressed in each piece of text as positive, negative, or neutral.

(d) **Feature Extraction**: Sentiment analysis algorithms extract features from the textual data to identify sentiment-bearing elements such as words, phrases, emotions, and context. These features help the algorithms understand the nuanced expressions of sentiment within the text.

(e) **Sentiment Scoring**: After feature extraction, sentiment analysis algorithms assign sentiment scores to each piece of text based on the overall sentiment expressed. The scores may range from highly positive to highly negative, with neutral sentiment falling in between.
(f) **Aggregate Analysis**: Sentiment analysis methods aggregate the sentiment scores from individual pieces of text to provide an overall assessment of public sentiment towards the brand or product. This can be done by calculating average sentiment scores, sentiment distributions, or sentiment trends over time.
(g) **Insight Generation**: Based on the aggregated sentiment analysis results, insights are generated to assess the overall success of the campaign. Positive sentiment indicates that the campaign has resonated well with the target audience, generated positive buzz, and enhanced brand perception. Conversely, negative sentiment may suggest areas of improvement or issues that need to be addressed.
(h) **Actionable Recommendations**: Sentiment analysis insights inform strategic decision-making by providing actionable recommendations for optimizing future campaigns. Marketers can identify successful campaign elements, understand audience preferences, address negative feedback, and refine messaging strategies accordingly.
(i) **Monitoring and Iteration**: Sentiment analysis is an ongoing process that involves continuous monitoring of public sentiment towards the brand or product. Marketers use sentiment analysis tools to track sentiment trends over time, monitor campaign performance, and iterate on strategies to maintain positive sentiment and drive success.

Overall, AI-driven sentiment analysis methods provide valuable insights into public sentiment towards a brand or product, helping marketers evaluate the overall success of a campaign, identify areas for improvement, and make data-driven decisions to optimize future marketing efforts.

5.4 **Personalization and Customization**:

Several AI-powered recommendation engines specialize in suggesting influencers based on various criteria such as the brand's target audience, industry, and campaign objectives (Table 3).

These AI-powered recommendation engines enable brands to discover influencers that align with their target audience, industry niche, and campaign objectives, facilitating personalized and effective influencer marketing strategies.

Table 3 Some notable tools for personalization and customization

Sr. no.	Tools	Functionality	Applications
1	Upfluence	Upfluence offers an AI-driven influencer marketing platform that uses machine learning algorithms to match brands with relevant influencers based on audience demographics, industry relevance, and campaign goals	Brands can use Upfluence to discover influencers tailored to their specific requirements and personalize influencer marketing strategies accordingly
2	Traackr	Traackr provides an influencer relationship management platform powered by AI algorithms. It analyzes influencer data, audience demographics, and content relevance to recommend influencers aligned with brand objectives	Marketers can leverage Traackr to identify influencers who resonate with their target audience and industry niche, enabling personalized influencer marketing strategies
3	Influencity	Influencity offers an AI-powered influencer marketing platform that matches brands with influencers based on audience demographics, engagement metrics, and campaign objectives. It provides personalized recommendations for influencer collaborations	Brands can use Influencity to discover influencers that fit their target audience profile and campaign goals, facilitating personalized influencer marketing strategies
4	AspireIQ	AspireIQ utilizes AI algorithms to analyze influencer data, audience insights, and content relevance. It recommends influencers based on brand preferences, target audience characteristics, and campaign objectives	Marketers can use AspireIQ to discover influencers who align with their brand values and objectives, enabling personalized influencer marketing strategies
5	HypeAuditor	HypeAuditor offers an AI-powered influencer discovery platform that assesses influencers' authenticity, audience demographics, and content relevance. It recommends influencers based on brand requirements and campaign goals	Brands can leverage HypeAuditor to identify authentic influencers with a genuine reach and resonance with their target audience, facilitating personalized influencer marketing strategies
6	NeoReach	NeoReach provides an AI-driven influencer marketing platform that analyzes influencer data, audience behavior, and content performance. It recommends influencers based on brand affinity, industry relevance, and campaign objectives	Marketers can use NeoReach to discover influencers who have the potential to drive meaningful engagement and conversions among their target audience, enabling personalized influencer marketing strategies
7	CreatorIQ	CreatorIQ offers an AI-powered influencer marketing platform that evaluates influencers' audience demographics, engagement metrics, and content authenticity. It recommends influencers based on brand fit and campaign objectives	Brands can utilize CreatorIQ to identify influencers with a strong affinity for their brand and a genuine connection with their target audience, facilitating personalized influencer marketing strategies

6 Case Study: AI-Powered Content Personalization for Influencers

AI tools can assist influencers in creating dynamic and personalized content, adapting to the preferences and behaviors of the target audience. Here is a case study illustrating how AI tools can assist influencers in creating dynamic and personalized content, adapting to the preferences and behaviors of the target audience:

Background: Imagine a scenario where a fashion influencer, let us call her Emily, wants to create engaging content for her audience of young, trendy individuals interested in sustainable fashion. Emily aims to increase audience engagement and brand affinity by delivering personalized content tailored to her followers' preferences and behaviors.

Challenge: Emily faces the challenge of understanding her audience's diverse preferences and delivering relevant content that resonates with them. She needs a solution to analyze her audience's behaviors, preferences, and engagement patterns to create dynamic and personalized content.

Solution: Emily decides to leverage AI-powered content personalization tools to analyze her audience data and tailor her content accordingly. She partners with an AI platform that offers advanced analytics and content optimization features.

Implementation:

(a) **Audience Analysis**: The AI platform collects and analyzes data from Emily's social media accounts, including demographics, engagement metrics, content interactions, and user preferences.

(b) **Behavioral Insights**: Using machine learning algorithms, the platform identifies patterns in audience behavior, such as preferred content types, topics of interest, peak engagement times, and purchasing behaviors.

(c) **Content Recommendations**: Based on the audience analysis, the AI platform generates personalized content recommendations for Emily. These recommendations include content topics, formats (e.g., videos, images, stories), caption styles, and posting schedules tailored to her audience's preferences.

(d) **Dynamic Content Creation**: Emily uses the AI platform's content creation tools to develop dynamic and personalized content aligned with the recommendations. She incorporates insights from the platform to craft engaging posts, stories, and videos that resonate with her audience.

(e) **Performance Monitoring**: After publishing the content, Emily monitors its performance using the AI platform's analytics dashboard. She tracks metrics such as engagement rates, click-through rates, conversion rates, and audience sentiment to measure the effectiveness of her personalized content.

Results: By leveraging AI-powered content personalization tools, Emily achieves the following outcomes:

(a) **Increased Engagement**: Emily's personalized content resonates with her audience, resulting in higher engagement rates and interaction levels.
(b) **Improved Relevance**: The tailored content recommendations ensure that Emily delivers relevant and timely content that aligns with her audience's interests and preferences.
(c) **Enhanced Brand Affinity**: By delivering dynamic and personalized content, Emily strengthens her connection with her audience, leading to increased brand affinity and loyalty.
(d) **Optimized Performance**: Emily continuously optimizes her content strategy based on real-time performance data, driving ongoing improvements in audience engagement and campaign effectiveness.

Conclusion: In this case study, AI-powered content personalization tools enable influencers like Emily to create dynamic and personalized content that adapts to the preferences and behaviors of their target audience. By leveraging audience insights and recommendations, influencers can enhance engagement, relevance, and brand affinity, ultimately driving success in their influencer marketing campaigns.

7 Long-Term Relationship Building

Assessing the long-term impact of influencer collaborations on audience loyalty and retention requires sophisticated data analytics methods that track and analyze audience behavior over time. Here are some data analytics methods that can help in this regard:

7.1 **Customer Lifetime Value (CLV) Analysis**: CLV analysis calculates the net present value of a customer's future purchases, providing insights into the long-term revenue generated by customers acquired through influencer marketing. By segmenting customers based on their acquisition source (e.g., influencer campaigns), marketers can assess the CLV of customers acquired through influencer collaborations and compare it with other acquisition channels.

7.2 **Cohort Analysis**: Cohort analysis tracks the behavior of specific groups (cohorts) of customers over time, allowing marketers to measure retention, repeat purchase rates, and engagement levels. Marketers can create cohorts of customers acquired through influencer campaigns and analyze their behavior compared to cohorts from other acquisition channels. This helps in understanding the long-term impact of influencer collaborations on audience loyalty and retention.

7.3 **Churn Analysis**: Churn analysis identifies customers who have stopped engaging with a brand or making purchases over time. It helps in understanding retention rates and predicting future churn. Marketers can analyze churn rates among customers acquired through influencer marketing to assess their long-term loyalty. Lower churn rates indicate higher audience retention and loyalty stemming from influencer collaborations.

7.4 **Repeat Purchase Analysis**: Repeat purchase analysis tracks the percentage of customers who make multiple purchases over time, indicating their loyalty and engagement with the brand. By analyzing the repeat purchase rates of customers acquired through influencer campaigns, marketers can gauge the effectiveness of influencer collaborations in driving long-term audience loyalty.

7.5 **Customer Segmentation and RFM Analysis**: Customer segmentation divides the audience into groups based on demographic, behavioral, or transactional attributes. RFM (Recency, Frequency, Monetary) analysis categorizes customers based on their transactional behavior. Marketers can segment customers acquired through influencer marketing and conduct RFM analysis to identify high-value segments with strong loyalty and retention. This helps in tailoring marketing strategies to nurture and retain these valuable customers.

7.6 **Surveys and Feedback Analysis**: Surveys and feedback collection methods gather qualitative data on customer satisfaction, brand perception, and loyalty. Marketers can use surveys and feedback analysis to gather insights directly from customers acquired through influencer collaborations. Understanding their satisfaction levels and brand affinity provides valuable insights into long-term audience loyalty.

7.7 **Social-Media Listening and Sentiment Analysis**: Social-media listening tools track brand mentions, sentiment, and conversations across social media platforms. Marketers can use social media listening and sentiment analysis to monitor audience sentiment towards the brand over time. Positive sentiment and ongoing engagement indicate sustained audience loyalty resulting from influencer collaborations.

By leveraging these data analytics methods, marketers can assess the long-term impact of influencer collaborations by measuring audience loyalty and retention. Understanding how influencer marketing contributes to customer lifetime value, repeat purchases, churn rates, and audience sentiment provides valuable insights for optimizing future influencer strategies and maximizing long-term brand success.

8 Predictive Analytics

AI can play a significant role in predicting which influencers are likely to maintain a positive impact on the brand over the long term, thereby aiding in building sustained relationships. Here's how AI can accomplish this:

8.1 **Influencer Performance Analysis**: AI algorithms can analyze historical data on influencer performance, including engagement metrics, audience demographics, content relevance, and campaign outcomes. By identifying patterns and trends in past performance, AI can predict which influencers are likely to consistently deliver positive results for the brand.

8.2 **Audience Engagement Prediction**: AI can analyze audience engagement patterns and sentiment towards influencers to predict which influencers are likely to maintain strong audience connections over time. By examining factors such as audience growth rates, interaction levels, and sentiment trends, AI can forecast which influencers will continue to resonate with their audience in the long term.
8.3 **Content Relevance Assessment**: AI-powered content analysis tools can evaluate the relevance and authenticity of influencer content to the brand and its target audience. By assessing content quality, alignment with brand values, and audience receptiveness, AI can predict which influencers are capable of consistently producing content that positively impacts the brand.
8.4 **Social Listening and Sentiment Analysis**: AI-driven social listening tools can monitor conversations, mentions, and sentiment surrounding influencers and the brand across social media platforms. By analyzing sentiment trends, audience feedback, and brand mentions, AI can identify influencers who are perceived positively by their audience and are likely to maintain a favorable impact on the brand over time.
8.5 **Predictive Modeling and Machine Learning**: AI algorithms can utilize predictive modeling and machine learning techniques to forecast the future performance of influencers based on historical data and current trends. By training models on a dataset of influencer characteristics and campaign outcomes, AI can predict which influencers are most likely to drive sustained positive impact for the brand.
8.6 **Relationship Mapping and Network Analysis**: AI-powered network analysis tools can map relationships between influencers, brands, and their respective audiences. By analyzing network dynamics, influencer collaborations, and audience overlap, AI can identify influencers who are embedded within relevant networks and are likely to maintain long-term partnerships with the brand.
8.7 **Real-time Monitoring and Adaptive Learning**: AI can continuously monitor influencer performance, audience engagement, and brand sentiment in real-time. By adapting to changing trends and dynamics in the influencer landscape, AI algorithms can refine their predictions and recommendations over time, ensuring that the brand builds sustained relationships with the most impactful influencers.

By leveraging AI-driven predictive analytics and machine learning capabilities, brands can identify influencers who are likely to maintain a positive impact over the long term, enabling them to build sustained relationships that drive continuous value and success for the brand.

9 Compliance and Transparency

Several AI tools can monitor influencer content for compliance with advertising regulations, ensuring transparency, and preventing potential legal issues (Table 4).

By utilizing these AI tools, brands can effectively monitor influencer content for compliance with advertising regulations, ensure transparency in sponsored content, and mitigate legal risks associated with non-compliant posts, thereby safeguarding their brand reputation and integrity in influencer marketing campaigns.

10 Ethical AI

Ensuring ethical AI practices in influencer marketing is crucial for maintaining trust with the audience and preventing the misuse of data. Here are some methods to promote ethical AI practices in influencer marketing:

11.1 **Transparency and Disclosure**: Ensure transparency in the use of AI algorithms and data analytics tools in influencer marketing campaigns. Disclose to the audience how AI is utilized to analyze data and make recommendations for influencer collaborations.

11.2 **Informed Consent**: Obtain informed consent from influencers and their audience regarding the collection, processing, and utilization of data for AI-driven analysis and personalized content recommendations. Clearly communicate the purpose and scope of data usage to all parties involved.

11.3 **Data Privacy and Security**: Implement robust data privacy and security measures to protect sensitive information collected through AI-powered analytics tools. Adhere to relevant data protection regulations, such as GDPR, CCPA, and other regional privacy laws, to safeguard user data.

11.4 **Fairness and Bias Mitigation**: Ensure fairness and mitigate bias in AI algorithms used for influencer marketing. Regularly audit and assess AI models to identify and address biases in data collection, feature selection, and decision-making processes.

11.5 **Accountability and Oversight**: Establish accountability mechanisms and oversight processes to monitor the ethical use of AI in influencer marketing. Designate responsible individuals or teams to oversee AI practices, compliance with ethical guidelines, and adherence to industry standards.

11.6 **User Empowerment and Control**: Empower users, including influencers and their audience, with control over their data and AI-driven experiences. Provide options for users to opt-in or opt-out of data collection, personalization features, and targeted advertising based on their preferences.

11.7 **Education and Awareness**: Educate stakeholders, including marketers, influencers, and consumers, about the ethical implications of AI in influencer marketing. Raise awareness about data privacy, consent, transparency, and the responsible use of AI technologies.

Table 4 Some notable tools for compliance and transparency

Sr. no.	Tools	Features	Application
(a)	BrandBastion	BrandBastion offers AI-powered content moderation and compliance monitoring for social media platforms. It detects sponsored content, disclosures, and compliance violations, helping brands ensure transparency in influencer marketing campaigns	Brands can use BrandBastion to monitor influencer content for adherence to advertising regulations, identify non-compliant posts, and take corrective actions to mitigate legal risks
(b)	CompliSpace	CompliSpace provides AI-driven compliance monitoring solutions for advertising and marketing content. It analyzes influencer posts for compliance with regulatory requirements, including disclosure of sponsored content and endorsements	Brands can leverage CompliSpace to automate the monitoring of influencer content, detect potential legal issues, and ensure adherence to advertising regulations across various platforms
(c)	Adverity	Adverity offers AI-powered marketing analytics and compliance monitoring tools. It analyzes influencer content for transparency, disclosure of paid partnerships, and compliance with advertising standards	Brands can use Adverity to monitor influencer campaigns, track compliance with advertising regulations, and generate reports to demonstrate transparency and regulatory compliance
(d)	AdGuard	AdGuard provides AI-driven ad monitoring and compliance tools for influencer marketing campaigns. It identifies sponsored content, affiliate links, and non-compliant disclosures to ensure transparency and compliance with advertising regulations	Brands can deploy AdGuard to monitor influencer content across platforms, detect potential legal issues, and enforce compliance with advertising guidelines and disclosure requirements
(e)	Influencity Compliance	Influencity Compliance is an AI-powered platform specifically designed for monitoring influencer content for compliance with advertising regulations. It analyzes influencer posts for proper disclosures, sponsored content, and compliance with FTC guidelines	Brands can utilize Influencity Compliance to automate the monitoring of influencer campaigns, ensure transparency in sponsored content, and mitigate legal risks associated with non-compliant posts

(continued)

Table 4 (continued)

Sr. no.	Tools	Features	Application
(f)	Open Influence	Open Influence offers AI-driven content moderation and compliance tools for influencer marketing campaigns. It monitors influencer content for adherence to advertising regulations, FTC guidelines, and platform policies	Brands can partner with Open Influence to ensure transparency and compliance in influencer marketing campaigns, detect potential legal issues, and take proactive measures to address compliance violations
(g)	Zefr	Zefr provides AI-powered brand safety and compliance solutions for influencer marketing. It analyzes influencer content for brand alignment, compliance with advertising regulations, and protection against harmful content	Brands can leverage Zefr to monitor influencer campaigns, ensure transparency and compliance with advertising standards, and safeguard against potential legal issues arising from non-compliant content

11.8 **Continuous Improvement and Feedback**: Foster a culture of continuous improvement by soliciting feedback from stakeholders and incorporating ethical considerations into AI development and deployment processes. Regularly review and update ethical guidelines and best practices based on evolving industry standards and regulatory requirements.

11.9 **Ethics Committees and Advisory Boards**: Establish ethics committees or advisory boards comprised of diverse stakeholders, including experts in AI, ethics, privacy, and influencer marketing. These committees can provide guidance, review ethical issues, and offer recommendations for ethical AI practices.

11.10 **Public Accountability and Reporting**: Maintain transparency and accountability by publicly disclosing AI practices, data usage policies, and ethical guidelines for influencer marketing. Publish regular reports or statements detailing compliance efforts, ethical considerations, and measures taken to address any ethical concerns or incidents.

By adopting these methods, brands, influencers, and AI practitioners can promote ethical AI practices in influencer marketing, uphold trust with the audience, and mitigate the risks of data misuse or ethical lapses.

11 Conclusion

In conclusion, the marriage of influencer marketing with data analytics and AI tools enhances the precision, efficiency, and effectiveness of campaigns enabling Personalized Content Creation, Real-time Performance Monitoring, Efficient Resource Allocation, Predictive Analytics for Future Success, with enhanced Audience Insights, building Trust and Authenticity. By leveraging these technologies, brands can make informed decisions, optimize their strategies, build meaningful connections with their target audience through the insights gained from these technologies and unlock new opportunities for growth, drive meaningful engagement, thereby achieving sustainable success in an increasingly competitive digital landscape.

References

1. Tafesse, W., Wood, B.P.: Followers' engagement with instagram influencers: the role of influencers' content and engagement strategy. J. Retail. Consum. Serv. **58**, 102303 (2021)
2. Agustian, K., et al.: The influence of influencer marketing in increasing brand awareness and sales for SMEs. Technol. Soc. Perspect. (TACIT) **1**(2), 68–78 (2023)
3. Leung, F.F., et al.: Influencer marketing effectiveness. J. Market. **86**(6), 93–115 (2022)
4. How Artificial Intelligence (AI) Is Changing Influencer Marketing. https://influencermarketinghub.com/ai-influencer-marketing/. Last accessed 21 May 2024

Sustainability and Influencer Marketing: Measuring the Gap Between Consumer Attitudes and Purchase Behavior

M. L. Vasita and Vineet Chouhan

Abstract Sustainability movement has started in many parts of the world in 1987. Consumer habits seems to be a major cause of sustainable aim and environmental issues. Research on sustainable consumption indicated that consumers' attitudes and behaviour must change to attain sustainable lives. Therefore, this work addresses two main research problem of examining the elements that influence urban consumers' sustainable Fast moving consumer goods purchases and to quickly describe the situational circumstances that duplicate attitudinal Sustainable values in real purchase behaviour. For the study the data of 4735 FMCG customers residing at 5 selected districts (includes Jodhpur, Udaipur, Jaipur, Bikaner and Kota) of Rajasthan state were gathered. The results with Friedman Test results and Paired Samples Test revealed that the better purchase behaviour and factors explaining Sustainable consumption have high probability to predict actual purchasing behaviour for FMCG products.

Keywords Customer's priority · Customer's attitude · Actual purchase behaviour · Sustainable FMCG product · Rajasthan

1 Introduction

Many environmental threats have arisen as a consequence of industrialisation in the previous two decades, such as greenhouse gas emissions, iceberg melting, and sea level rise. Globally, and more significantly, in terms of climate change. Sustainable activities are those that can be done to enhance the earth's living conditions

M. L. Vasita
Department of Business Administration, University of Rajasthan, Jaipur, Rajasthan, India

V. Chouhan (✉)
Institute of Law, NIRMA University, Ahmedabad, Gujarat, India
e-mail: vineet.chouhan@nirmauni.acin; vcpc2008@gmail.com

S. Dutta et al. (eds.), *Advances in Data Analytics for Influencer Marketing: An Interdisciplinary Approach*, Information Systems Engineering and Management 9,
https://doi.org/10.1007/978-3-031-65727-6_9

by decreasing the negative effects of industrialization. For minimizing carbon footprints and environmental dangers, sustainable methods are becoming more important. However, governments in various countries have established committees and reports [such as the Bretton wood Conference, and the Kyoto Protocol (1944)] to improve the climate of the planet in which we live and hold corporations accountable for their environmental pollution. There have been several research projects in this field in order to build frameworks for sustainable growth, but no framework has been developed and accepted without question by businesses throughout the world.

In 2015, the FMCG in India, was the 4 biggest in the economy, having a total market size of USD 43.08 billion. Over the previous decade, India's FMCG business has risen at an annual rate of roughly 11%. Consumer expenditure in India is predicted to reach US$ 3.6 trillion by 2020, thanks to the country's strong economic development and growing family incomes. i.e., it is predicted to increase by above 100% (ASSOCHAM 2018). Furthermore, as the urban population grows, so does disposable income, making this group important. In India, urban customers choose things with high surface values, which violate the rule of sustainable consumption. Consumers in cities spend a large portion of their money on FMCG items. As a result of its size, the FMCG industry should pay attention to environmental issues. Furthermore, manufacturers should look at creating sustainable FMCG items that employ natural ingredients rather than synthetic chemicals throughout the production, packaging, and distribution process.

In today's world, corporate stakeholders are interested in learning about the company's profit, as well as how the profit is produced. They also want to know about the company's sustainable practices, which will help to improve the environment or prevent environmental damages.

So far, sustainable reporting practices have been recorded in Canada, Australia, Europe, and the United States. However, no work is investigated in emerging countries such as India. As a consequence, a sustainable reporting framework for the Indian cement industry is required to establish a relevant and credible reporting system that fulfils the needs of stakeholders and culminates in a framework to transform the company's day-to-day operations into sustainable operations.

Consumers are becoming more aware of the causes and impacts of pollution as a result of global developments and increased usage of information technology. Consumers are now fully aware of the extent to which polyurethane usage is wreaking havoc on the environment, including damage to the soil and drainage system. As a result, the voices for sustainable consumption are getting potential environmental commitment. The reality that each purchase has an unspoken impact on resources, waste, community, and overall environmental sustainability supports this viewpoint. When consumers have an option, they choose to acquire environmentally friendly items, especially when such products are utilised in their everyday lives or for the health of their families. Furthermore, customers have a positive attitude toward sustainable items (also known as eco-friendly products), but this does not always transfer into actual purchases. Despite significant changes in understanding and overall attitudes toward environmental and social concerns, there is little reflecting similarity in consumer purchasing behaviour. Furthermore, the rapid rate of urbanisation has transformed society into a "throw-away" civilization, resulting in environmental, social, and economic decay.

In light of this, the suggested research seems to be more relevant in the present circumstances, since it aims to analyse the important elements impacting Indian urban consumers' sustainable consumption. It also makes an attempt to establish a framework for Sustainable consumption behaviour by evaluating the components involved in determining the gap between consumers' Sustainable values and their actual consumption behaviour.

As a Sustainable goal, a host of improvement and adaption strategies have been promoted and suitably executed, but the role of consumers has been largely neglected. However, consumer lifestyles are apparently responsible for the existing environment problems to a great extent. Reviews of the literature have shown that only a slight attention is given to the issue of product attributes and Sustainable purchasing behaviour. Hence this study makes an attempt to address two broad research problems: First, to delve out the factors which explain the Sustainable purchasing behaviour of urban consumers for non-durable goods; second, to explain briefly the situational factors that have dominance to repeat attitudinal Sustainable values to the actual purchase behavior.

The aims of the research paper includes.

To analyse the conflict between consumers' attitude and actual purchase behaviour for Sustainable FMCG goods.

To analyse the conflict between consumers' attitude and actual purchase behaviour for FMCG products.

2 Review of Literature

Shrivastava [1] revealed that the first tourist strategy of the Indian government was unveiled in 1982. The primary objective was to promote sustainable tourism as a means of achieving social and economic inclusivity, while also enhancing India's image as a country with a rich history, a flourishing present, and a potential future. The policies created would concentrate on six primary categories: "swagat (welcome)", "suchana (information)", "suvidha (facilitation)", "suraksha (safety)", "sahyog (cooperation)", and "Samrachana (infrastructure development)".

Saini and Jain [2] explained that Consumers are now exhibiting a heightened level of awareness about the ecological consequences of their decisions. Environmental consciousness entails recognizing the importance of one's natural environment. Therefore, it exhibits awareness and resolve to tackle environmental issues. Consumers' perspectives on sustainable living have changed. Humanity is actively striving to mitigate its detrimental impact on the world. However, this is uncommon and still in the developmental stage. In recent times, almost all countries and communities have developed a heightened awareness of the environmental issues that need attention. Compelling evidence indicates that environmentally aware consumer behavior arises from a sense of empathy towards the natural world. Nevertheless, there has been little advancement in enhancing customers' comprehension of the advantages associated with using environmentally-friendly products. Marketers have been promoting their environmentally conscious attributes to differentiate their goods and enhance customer loyalty. Conducting research on consumers' comprehension and attitude towards embracing environmentally friendly activities is essential for achieving a substantial impact with the green idea. The present research is to examine the level of environmental awareness among urban inhabitants in the Indian state of Rajasthan, specifically focusing on their attitudes towards eco-friendly products and behaviors.

Saini and Jain [3] examines the potential of integrating more environmentally friendly consumer habits into contemporary lifestyles, taking into account the increasing availability of green products in the present context. Given the need to adapt to changing demands, the concept of marketing mix must be revised, and thus, the thinking of marketing managers must also be transformed. The environmentally-friendly product, made from non-polluting materials, plays a crucial role in green marketing. The products are accompanied with branding that promotes environmental sustainability and packaging that is designed to be eco-friendly. Promoting these things via advertising may effectively raise awareness and, by emphasizing their advantages, motivate prospective consumers to conduct transactions. Green products often have a higher price due to the additional labor and resources invested in their production. However, these costs must fall within the limits of what customers are prepared to spend. This may be achieved by reducing expenditures, saving energy, optimizing the use of scraps and leftovers, and similar measures. Furthermore, it is essential to include "green principles" into the design of distribution, namely in terms of guaranteeing enough transportation and using ecologically sustainable fuels. The author obtained important secondary data by extensively researching the literature and the internet. Subsequently, a comprehensive questionnaire was devised to gather the required primary information. The study used a sample of 200 inhabitants from a prominent city in Rajasthan as the population for sample selection. To mitigate the perceived performance risk in environmental evaluation, marketers could implement standardized environmental labeling activities.

Gupta [4] explained that as globalization continues to shape the world, it has brought forth a range of challenges, with environmental issues being among the most significant. These environmental problems have gained increasing attention in recent years, prompting discussions about their negative impact on all living beings. Consequently, consumers have developed concerns about the future of their planet and are increasingly inclined to choose environmentally friendly products.

Sheoran and Kumar [5] attempted to Examine the theoretical framework and structural aspects of sustainable consumer behavior in order to create a scale for sustainable consumer behavior specifically for sustainable electrical items. This research has aimed to explain sustainable consumer behavior by examining the whole consumption cycle, including the acquisition, use, and disposal of sustainable electronic items. The theory of planned behaviour (TPB) has been used to comprehend the many aspects of sustainable consumer behavior using qualitative and quantitative methodologies. A sustainable consumer behavior scale for sustainable electronic items was evaluated and validated via pilot research and a main study to assess its factor structure, reliability, validity, and model fit.

Baldi et al. [6] Several research studies have attempted to elucidate the factors influencing choices connected to eating. This article investigates how an individual's attitude towards the environment might influence their green behavior. This study introduces a novel methodological framework in the field of food consumption, drawing on the Campbell Paradigm and developments in attitude measurements from environmental psychology literature. They considered consumer attitudes towards environmental problems. The investigation is carried out in Italy and the UK, two nations with contrasting views on tomatoes as a consumer product and water-related concerns. The study shows that the two aspects of consumer mentality have varying impacts on the willingness to spend on sustainable tomatoes, and these variations are distinct to each nation. The research indicates that policies encouraging sustainable food items should focus on consumers who are more receptive to environmental sustainability, such as naturalists or environmentalists.

ul Haque et al. [7] shown that there is a lack of understanding regarding the extent to which consumers are willing to limit their food choices in favor of sustainable options (WTRFCIFOSA). Additionally, there is scrutiny on the government's willingness to lead by ensuring that only organic food is available for purchase. This research introduces a novel concept by proposing the idea of "willingness to limit food options in favor of sustainable alternatives." The notion originated from an evaluation of consumer behavior, ethical characteristics, health characteristics, and the responsibilities of government and food manufacturers regarding naturalness. Governments can consider implementing obligatory consumption of sustainable alternatives by limiting the availability of unnatural and less ethical food options for consumers, since naturalness plays a crucial role in this context.

Thøgersen and Alfinito [8] conducted In a study on the influence of goal activation on sustainable consumer choices, 1,031 participants from Denmark and 1,028 from Brazil. The significance of different product qualities for customer preferences might be associated with country-specific factors. Hence, it was anticipated that the significance of qualities varies between Denmark and Brazil. Confirmation was provided for the characteristic of look, but not for the qualities of organic and pricing. The goal-activation adjustment as expected affected how two out of three product features impacted selections. Activating a normative objective increased the likelihood of participants in both nations choosing organic tomatoes and reduced their emphasis on the look of the tomatoes. The goal-activation adjustment did not affect the significance of the price. The results indicate that activating a normative objective in a specific scenario might effectively encourage sustainable consumer choices in various national settings.

Kumagai [9] analyzes consumer behavior in relation to product sustainability and brand luxury. The aim is to provide strategic recommendations to garment firms looking to engage in sustainable development of plastic clothing, together with insights for marketing and sustainability research.

Kumagai and Nagasawa [10] revealed that fashion firms are working to create sustainable clothing from recycled plastic bottles due to the significant societal problem of plastic waste. This societal problem is pertinent to any organization, irrespective of the brand's degree of luxury. Previous studies have not investigated how this sustainable product type influences customer behavior, but extensive sustainable development is often mentioned in luxury branding research. This research examines the impact of introducing sustainable plastic clothing on customer brand perception and likelihood to buy. Analyzed data from 440 online survey responses in Japan shows that the introduction of sustainable plastic clothing has a good impact on brand opinion, leading to purchase intention both indirectly and directly. Alternatively, their high-end brand might become a liability as plastic recycling is becoming more of a societal expectation.

Cuzdriorean et al. [11] evaluated aspects such as environment, education, social pressure, and government measures in relation to promoting sustainable behavior among students. A poll was done at a major institution in Romania. Their findings showed a substantial influence solely on characteristics such as education and environment. Their research attempts to provide a standard for judging future sustainability efforts undertaken by institutions. The current research addresses a seldom explored area in academic literature by investigating students' perspectives on sustainable development. The study's significance is mostly based on the assumptions of the Theory of Planned conduct, which connects people' attitudes to their conduct. The implications of their findings are also analyzed.

Byrd and Su [12] performed a research with the goal of this study is to evaluate the views of customers and the behavior of consumers with regard to garment labels and environmentally sustainable and socially conscious apparel.

Salmivaara and Lankoski [13] revealed that Businesses must supplement their production-side sustainability efforts with measures that promote sustainable consumption, they measured consumer choice using a field experiment involving 1,289 participants in 19 office eateries. The study aimed to determine whether the inclusion or exclusion of two injunctive norm messages on Baltic Sea protection and local food influenced the selection of a new sustainable meal, roach fish patties. The nested analysis of variance and qualitative comparative analysis indicate that although activating injunctive norms did not influence consumer choice in the general population, it might have affected specific subgroups defined by sociodemographic traits and past consumption behaviors. Moreover, there was evidence of a mutually reinforcing interaction effect when both injunctive norms were triggered together..

Quoquab et al. [14] Created a dependable and valid scale with favorable psychometric characteristics and enough levels of reliability and validity to assess sustainable consumption from the consumer's point of view.

do Paco et al. [15] have taken several green consumer behavior models to measure the challenges have arisen in creating an ideal or universally accepted model for predicting green consumer behavior. This research aims to analyze the connections between several constructs to develop a model for green consumer behavior. The model is based on multiple factors influencing purchasing behavior, including a prosocial attitude, the importance of environmental values, and green communication. The connection between these principles is crucial for determining the most effective ways to create green marketing campaigns and communication methods that may promote green behaviors.

Zalega [16] studied the sustainable consumption habits of individuals aged 65 and above in Poland. The essay aims to provide insight into sustainable consumption in the decision-making process of elderly consumers in Poland. The results are based on actual research in the form of a survey questionnaire done on a sample of 2,537 individuals aged 65 and older in 2014–2015 in 10 Polish cities of different demographics and sizes.

3 Research Design

The FMCG products, which attempt to capture the sustainability in purchase dimension. The current study included both exploratory and descriptive research methods. Exploratory research involves reviewing existing material to create hypotheses. Descriptive research technique is used to test hypotheses and draw conclusions from data analysis. This research employs a quantitative approach to problem-solving. This study will include both exploratory and conclusive research design and the details are as under.

The research framework selected at the preliminary level will be exploratory in nature in order to identify the factors influencing Sustainable purchase behaviour for FMCG products. The major activities include pilot survey through an open-ended questionnaire; survey of experts using Delphi Technique and; qualitative analysis of secondary data.

Further the Descriptive Research Design incorporates set of procedures/ framework for empirical testing of hypotheses. The targeted population for study included urban consumers of major cities (includes Jodhpur, Udaipur, Jaipur, Bikaner and Kota) of Rajasthan state in India. The sampling techniques used is called Stratified Random Sampling. Under this technique a strata of the 5 districts of Rajasthan state is developed and the data is gathered accordingly. The sample size was initially 1000 consumers from each district that was planned to be 5000 in total but the real data gathered in all the complete way stand to be 4735 which was sufficient for the analysis purpose.

A comprehensive questionnaire was used to gather all the essential information for testing the hypotheses. The questionnaire was verified for validity using a pilot survey. The questionnaire will use a Likert rating scale, as well as nominal and rank-order scales, to collect data. The scale's reliability examination is often conducted using the Co-efficient alpha (Cronbach alpha). So, Cronbach alpha test were applied to test the reliability of the scale and the reliability 0.843 was accepted as it is way above than the desired level of 0.70. Research method and mode of administration was survey method was being used and the proposed mode of administration were personal interview of consumers residing in Rajasthan state. The questionnaire was used as a major survey instrument and responses will be arrested through personal interview. For the study the data of 4735 FMCG customers Residing at 5 selected districts (includes Jodhpur, Udaipur, Jaipur, Bikaner and Kota) of Rajasthan state were gathered which was sufficient for the analysis purpose. For the current research work Friedman Test results and multiple regression analysis was used with SPSS for analysis.

4 Data Analysis

The data from the selected respondents were gathered for inclusion in the study. The demographics are presented in Table 1 as under.

The above distribution revealed that the respondents selected were more in terms of 25–34 year of age (37.4%), Post Graduate (49.8%), Upper Middle-income earner (50.2%), Unemployed/ Homemaker (52.9%) and Financial Status wise Dependent (64.1%).

Table 1 Demographical profile

	Frequency	Percent
Age wise distribution		
18–25	1309	27.6
25–34	1769	37.4
34–44	1103	23.3
Above 44	554	11.7
Qualification wise distribution		
Higher Secondary	505	10.7
Graduate	1073	22.7
Post Graduate	2360	49.8
Professional	797	16.8
Income wise distribution		
Lower Middle	589	12.4
Upper Middle	2378	50.2
High	1768	37.3
Occupation wise distribution		
Employed	1466	31.0
Unemployed/Homemaker	2507	52.9
Retired	762	16.1
Financial Status wise distribution		
Dependent	3034	64.1
Independent	1701	35.9

Further as per the first objective to measure the conflict between consumers' attitude and actual purchase behaviour for Sustainable FMCG goods the Friedman Test is conducted to measure priorities towards sustainable product to find out that which category of the FMCG product is at most of the customer's priority for the sustainable product and whether there is any difference in the priorities. To check the same the data collected were analysed with the SPSS software and the results are presented in Table 2.

The result of the Friedman test with the chi-square value of 4158.249 is found significantly different with $p < 0.05$. Further it can be reported that with the highest mean value the Packaged Food products are the first category of FMCG product in which the sustainable product is checked by the consumers in the Rajasthan state. Further the Personal care products is the second category in which the sustainability is seen by the consumers.

Table 2 Friedman test results

Descriptive statistics

	N	Mean	SD	Min	Max
EcoFrend_Purch_resp_1	4735	4.02	1.351	1.00	5.00
EcoFrend_Purch_resp_2	4735	2.5301	1.09126	1.00	5.00
EcoFrend_Purch_resp_3	4735	3.0593	1.07437	1.00	5.00
EcoFrend_Purch_resp_4	4735	3.4714	1.25783	1.00	5.00
EcoFrend_Purch_resp_5	4735	2.2152	1.48629	1.00	5.00

Ranks

Item	SPSS code	Mean rank
Packaged food products	EcoFrend_Purch_resp_1	3.97
Home care products	EcoFrend_Purch_resp_2	2.47
Personal care products	EcoFrend_Purch_resp_3	2.99
Health care products	EcoFrend_Purch_resp_4	3.42
Any other, Please specify	EcoFrend_Purch_resp_5	2.15

Test statistics[a]

N	4735
Chi-Square	4158.249
df	4
Sig	0.000

[a]Friedman Test

5 Measuring Difference Between Customer's Attitude and Actual Purchase Behaviour

As per the above test result positively explained that the customers attitude and their intention was to buy the FMCG product that is sustainable, the objective of analysing the conflict between consumers' attitude and actual purchase behaviour for FMCG products or not is now measured with Paired sample 't' test (also called dependent 't' test). Paired sample t-test is a statistical method used to compare two population means when dealing with two related samples.

Mathematically,

H0: $\mu B = \mu A$

H1: $\mu B \neq \mu A$ Where, μB and μA are the hypothesized mean for customers attitude and actual purchase behaviour of the customers of Rajasthan in respect of FMCG product's Sustainability.

Possible conditions:

If, $\mu B - \mu A = 0$, There were no variations in abdominal expansion.

$\mu B - \mu A > 0$, Increase in the abdominal expansion.

$\mu B - \mu A < 0$, Decrease in the abdominal expansion.

In terms of our study this step is developed with the following hypothesis as under:

H_2: Factors explaining Sustainable consumption have equal probability to predict actual purchasing behaviour for FMCG products.

To identify difference between customer's attitude and actual purchase behavior, Paired Samples Test has been used with SPSS-19 software presented in Table 3 as under.

The test mainly assesses the difference between respondents' attitudes and buying behavior towards sustainable FMCG items. The results generated are shown in the table above, illustrating different stages. The first section of the table presents descriptive data for factors concerning consumer attitudes and purchasing behavior. The Paired Samples Test results of the dependent t-test for both customer's attitude and purchase behaviour for sustainable FMCG products are presented in later part of table.

The table present a positive gap difference between the customer's attitude and purchase behaviour of the respondents. H1 is accepted and we can say that the customer's attitude and purchase behaviour is significantly different ($p < 0.05$) as $\mu B > \mu A$ for first 4 pairs while $\mu A > \mu B$ for the remaining pairs of the customer's attitude and purchase behaviour. Customers demonstrate a preference for environmentally safe products by reading labels, showing a willingness to pay extra for sustainable items, making an effort to purchase biodegradable products, opting for products in refillable containers, although their attitude may not always align with their actual behavior.

Table 3 Paired sample t test for difference between customer's attitude and actual purchase behaviour

Paired samples statistics					
		Mean	N	SD	SE
Pair 1	Cust_Att_1	2.2646	4735	1.02456	0.01489
	Act_Pur_beh_1	3.0906	4735	0.95639	0.01390
Pair 2	Cust_Att_2	2.1252	4735	1.32417	0.01924
	Act_Pur_beh_2	2.3521	4735	1.12844	0.01640
Pair 3	Cust_Att_3	2.1200	4735	1.22551	0.01781
	Act_Pur_beh_3	2.5054	4735	1.11464	0.01620
Pair 4	Cust_Att_4	2.0044	4731	0.93090	0.01353
	Act_Pur_beh_4	2.15	4731	1.035	0.015
Pair 5	Cust_Att_5	3.4762	4735	1.03458	0.01504
	Act_Pur_beh_5	2.1183	4735	1.15651	0.01681
Pair 6	Cust_Att_6	2.4133	4735	1.02649	0.01492
	Act_Pur_beh_6	2.2729	4735	1.14325	0.01661
Pair 7	Cust_Att_7	2.2640	4735	1.01322	0.01472
	Act_Pur_beh_7	2.1901	4735	1.33309	0.01937
Pair 8	Cust_Att_8	2.7635	4735	1.07759	0.01566
	Act_Pur_beh_8	2.3252	4735	1.06279	0.01544
Pair 9	Cust_Att_9	2.3005	4735	1.03706	0.01507
	Act_Pur_beh_9	2.2574	4735	1.10451	0.01605
Pair 10	Cust_Att_10	3.3451	4735	1.16582	0.01694
	Act_Pur_beh_10	2.7455	4735	1.17809	0.01712
Pair 11	Cust_Att_11	2.7113	4735	1.16656	0.01695

(continued)

Table 3 (continued)

Paired samples statistics					
		Mean	N	SD	SE
	Act_Pur_beh_11	2.3014	4735	1.14726	0.01667
Pair 12	Cust_Att_12	3.4218	4735	0.91284	0.01327
	Act_Pur_beh_12	2.2667	4735	1.10172	0.01601
Pair 13	Cust_Att_13	3.4699	4735	0.91385	0.01328
	Act_Pur_beh_13	2.4492	4735	1.23321	0.01792

Correlations				
Variable SPSS code		N	Correlation	Sig
While purchasing I read the labels to see if the products are environmentally safe	Cust_Att_1 & Act_Pur_beh_1	4735	0.126	0.000
I am willing to pay more for sustainable products	Cust_Att_2 & Act_Pur_beh_2	4735	0.007	0.618
I make a special effort to buy biodegradable products	Cust_Att_3 & Act_Pur_beh_3	4735	0.079	0.000
I buy products in refillable containers	Cust_Att_4 & Act_Pur_beh_4	4731	0.285	0.000
I use environmentally safe soaps and detergents	Cust_Att_5 & Act_Pur_beh_5	4735	−0.032	0.029
I buy products having biodegradable packaging	Cust_Att_6 & Act_Pur_beh_6	4735	0.142	0.000
I am interested in reading reports/articles on environmentally-friendly products	Cust_Att_7 & Act_Pur_beh_7	4735	0.076	0.000
I prefer to buy sustainable products that care about value for money	Cust_Att_8 & Act_Pur_beh_8	4735	0.197	0.000
I prefer to buy sustainable products that have leading brand names	Cust_Att_9 & Act_Pur_beh_9	4735	0.022	0.139

(continued)

Table 3 (continued)

Paired samples statistics

	Mean	N	SD	SE
I buy sustainable products if their quality is comparable to leading brands	Cust_Att_10 & Act_Pur_beh_10	4735	0.029	0.047
When I have opportunity to choose I always prefer the products that least harm to the environment	Cust_Att_11 & Act_Pur_beh_11	4735	0.059	0.000
I have switched brands for the ecological reasons only	Cust_Att_12 & Act_Pur_beh_12	4735	0.006	0.672
I usually notice advertisements about environment friendly products	Cust_Att_13 & Act_Pur_beh_13	4735	0.086	0.000

Test result

		Paired differences					t	df	Sig. (2-tailed)
		Mean	SD	SE mean	95% confidence interval for difference				
					Low	Up			
Pair 1	Cust_Att_1−Act_Pur_beh_1	−0.82598	1.31037	0.01904	−0.86331	−0.78864	−43.37	4734	0.000
Pair 2	Cust_Att_2−Act_Pur_beh_2	−0.22682	1.73354	0.02519	−0.27621	−0.17743	−9.003	4734	0.000
Pair 3	Cust_Att_3−Act_Pur_beh_3	−0.38543	1.58984	0.02310	−0.43072	−0.34013	−16.68	4734	0.000
Pair 4	Cust_Att_4−Act_Pur_beh_4	−0.14817	1.17866	0.01714	−0.18177	−0.11458	−8.647	4730	0.000
Pair 5	Cust_Att_5−Act_Pur_beh_5	1.3579	1.57600	0.02290	1.31307	1.40287	59.29	4734	0.000
Pair 6	Cust_Att_6−Act_Pur_beh_6	0.14044	1.42386	0.02069	0.09988	0.18101	6.787	4734	0.000
Pair 7	Cust_Att_7−Act_Pur_beh_7	0.07392	1.61204	0.02343	0.02799	0.11985	3.155	4734	0.002
Pair 8	Cust_Att_8−Act_Pur_beh_8	0.43823	1.35593	0.01971	0.39959	0.47686	22.24	4734	0.000

(continued)

Table 3 (continued)

Test result

		Paired differences					t	df	Sig. (2-tailed)
		Mean	SD	SE mean	95% confidence interval for difference				
					Low	Up			
Pair 9	Cust_Att_9−Act_Pur_beh_9	0.04308	1.49871	0.02178	0.00038	0.08578	1.978	4734	0.048
Pair 10	Cust_Att_10−Act_Pur_beh_10	0.59958	1.63330	0.02374	0.55304	0.64611	25.26	4734	0.000
Pair 11	Cust_Att_11−Act_Pur_beh_11	0.40993	1.58703	0.02306	0.36471	0.45514	17.77	4734	0.000
Pair 12	Cust_Att_12−Act_Pur_beh_12	1.1550	1.42643	0.02073	1.11438	1.19566	55.72	4734	0.000
Pair 13	Cust_Att_13−Act_Pur_beh_13	1.0207	1.47016	0.02137	0.97881	1.06258	47.77	4734	0.000

6 Conclusion

The result of the study revealed that the customer are using ecologically safe soaps and detergents and prefers items with biodegradable packaging, they are interested in reading studies and articles about environmentally-friendly products, prefer to buy sustainable products that care about value for money, prefer to buy sustainable products that have leading brand names, buy sustainable products if their quality is comparable to leading brands, have opportunity to choose they always prefer the products that least harm to the environment, have switched brands for the ecological reasons only and usually notice advertisements about environment friendly products the purchase behaviour is better than the customer's attitude and towards sustainable FMCG products.

Thus, the final outcome is in favour of the better purchase behaviour and Factors explaining Sustainable consumption have high probability to predict actual purchasing behaviour for FMCG products, thus the implication of the same is that the companies should made the FMCG product with considering the sustainability goal otherwise their product will become outdated and no customer will be ready to buy. Further, to make our future sustainable the Government should also work in the area of the same and also suggest the new direction for the tax where the non-sustainable product should be charged extra tax to reduce their consumption.

Disclosure statement The paper is covered under the research project financially supported by the RUSA 2.0 Programme (RUSA) on "Measuring sustainable consumer behaviour of urban consumers towards FMCG products in Rajasthan".

Data Availability Statement The data used in the paper is not freely available, it has been submitted to RUSA 2.0.

Using Third-Party Material No third part material is used in this study.

References

1. Shrivastava, P.: A Study on Consumer Awareness and Challenges associated with Dynamic Pricing in Tourism Industry with Special Reference to Rajasthan (2023)
2. Saini, M., Jain, P.: A study of the effects of environmental awareness on purchase intention of green products in urban Rajasthan. Phoenix: Int. Multidisciplinary Res. J. **1**(2), 1–5 (2023)
3. Saini, M., Jain, P.: A study of the impact of green marketing on consumer purchasing patterns and decision making in urban Rajasthan. Int. J. Res. Manag. Stud. **7**(4), 1–12 (20230
4. Gupta, A.K.: A study on consumer behavior and green marketing with reference to consumer durable products. AIJRA **II**(III), 89.1–89.7(2023)
5. Sheoran, M., Kumar, D.: Conceptualisation of sustainable consumer behaviour: converging the theory of planned behaviour and consumption cycle. Qual. Res. Organ. Manag. Int. J. **17**(1), 103–135 (2022)
6. Baldi, L., et al.: Attitude toward environmental protection and toward nature: How do they shape consumer behaviour for a sustainable tomato? Food Q. Preference **90**, 104175 (2021)
7. ul Haque, A., Yamoah, F.A., Sroka, W.: Willingness to reduce food choice in favour of sustainable alternatives: the role of government and consumer behaviour. İn: Perspectives on Consumer

Behaviour: Theoretical Aspects and Practical Applications, pp. 31–51. Springer International Publishing, Cham (2020)

8. Thøgersen, J., Alfinito, S.: Goal activation for sustainable consumer choices: a comparative study of Denmark and Brazil. J. Consum. Behav. **19**(6), 556–569 (2020)
9. Kumagai, K.: Sustainable plastic clothing and brand luxury: a discussion of contradictory consumer behaviour. Asia Pac. J. Mark. Logist. **33**(4), 994–1013 (2021)
10. Kumagai, K., Nagasawa, S.: Launch of sustainable plastic apparel: effects of brand luxury and experience on consumer behaviour. Sustainability 12(18), 7662 (2020)
11. Cuzdriorean, D.D., Fekete, S., Vladu, A.B.: Identifying the promoters of students sustainable behaviour: an empirical study. Amfiteatru Econ. 22(54), 432–446 (2020)
12. Byrd, K., Su, J.: Investigating consumer behaviour for environmental, sustainable and social apparel. Int. J. Clothing Sci. Technol. **33**(3), 336–352 (2021)
13. Salmivaara, L., Lankoski, L.: Promoting sustainable consumer behaviour through the activation of injunctive social norms: a field experiment in 19 workplace restaurants. Organ. Environ. **34**(3), 361–386 (2021)
14. Quoquab, F., Mohammad, J., Sukari, N.N.: A multiple-item scale for measuring "sustainable consumption behaviour" construct: Development and psychometric evaluation. Asia Pac. J. Market. Logistics **31**(4), 791–816 (2019)
15. Do Paco, A., Shiel, C., Alves, H.: A new model for testing green consumer behaviour. J. Clean. Prod. **207**, 998–1006 (2019)
16. Zalega, T.: Sustainable consumption in consumer behaviour of Polish seniors (report from own research). Acta Scientiarum Polonorum. Oeconomia **17**(1), 131–139 (2018)

An Audience Threshold in the Millions: Exploring Influencer Marketing Activations Online

Brandon Chicotsky and Ken Corbit

Abstract Influencer marketing campaigns often struggle to predict audience activations and their prospective financial returns. This is further hindered by a lack of statistically relevant approaches to model expected outcomes. The current analysis offers a guided approach and recommendations for estimating and predicting social media engagements. By leveraging graph theory, regression analysis, and projection methods, the study gains insights from social media data to set feasible expectations for influencer marketing campaigns. The study demonstrates that network analysis, particularly degree-centrality, works well for choosing an initial set, provided analysts iterate over the degree-centrality range choices via subsequent regression analyses. X platform data (herein referred to by its former name, Twitter) from the Higgs Boson Graph Dataset is utilized to operationalize the investigation's approach.

Keywords Influencer marketing · Social media · Digital marketing · Marketing campaigns

1 Introduction

Companies running influencer marketing campaigns experience difficulty identifying the best influencers, objectively measuring the direct impact on the audience, and projecting expected outcomes [2]. Influencer marketing has created a unique landscape where business, marketing, and data science converge with a yearn to better understand and capitalize on this dynamic marketing tool. With an ever- expanding audience and customer pool via the internet, data science allows insight into the best means to execute marketing goals and objectives [11]. The current study aims to define a statistical algorithm that identifies the most relevant influencers, the average level of engagement a marketing campaign should expect, and the projected campaign reach based on a given set of influencers. X platform data (herein referred to by its

B. Chicotsky (✉) · K. Corbit
Neeley School of Business, Texas Christian University, Fort Worth, TX 76129, USA
e-mail: b.chicotsky@tcu.edu

S. Dutta et al. (eds.), *Advances in Data Analytics for Influencer Marketing: An Interdisciplinary Approach*, Information Systems Engineering and Management 9,
https://doi.org/10.1007/978-3-031-65727-6_10

former name, Twitter), specifically the Higgs Boson Graph Dataset, is used to test and support the proposed algorithm.

Currently, a framework for influencer marketing analytics would help overcome three main hurdles across marketing, business, and data science stakeholders. The literature shows three main gaps [4, 5]: first, applied maximization and non-linear methods on influencers' data are scarce. Second, applied regression methods to estimate and generalize influencer impact could be misleading in most cases, while domain-centric and content-driven analysis (Natural Language Processing) does not allow for comparison or comments. And third, although still very useful, graph theory provides a limited descriptive contribution if not further from regression insights. This paper aims to integrate a framework for decision-making by logically filling those three identified gaps.

1.1 Background

The way in which the global human network operates has experienced a massive shift due to the developments in online life via the world wide web (web3) [13]. Web3 has transformed how consumers engage in purchasing goods, the way friends communicate, and how people spend their leisurely time [10]. The consequences of this shift are two-fold: (1) consumers have changed the way they interact with content and go about the decision-making process, and (2) brands, companies, and organizations are changing the way they communicate with potential consumers [6].

According to DataReportal, Americans screen time is an average of 7 h and 4 min per day [7]. Popular social media platforms have catapulted the change in human interaction and the e-commerce domain [19]. The commercial marketing landscape has been tasked with keeping pace with these changes [3]. As consumer behavior changes, so does the behavior of brands, marketers, and the methods used to effect desired outcomes (e.g., higher sales volume and more social media followers) [9].

With the advent of advanced targeting tools on social media, commercial actors are able to mass advertise on a wider geographical scale, as well as leverage a network of influencers with a high social media follower count. An influencer is, "... a marketing 'rising star' whose popularity has been fueled by social media. The term influencer generally refers to an individual or third party who has accumulated a certain audience size through social media platforms and who has a direct impact on customer purchasing decisions" [14]. Effective word-of-mouth marketing and reviews of products and services via web3 is a phenomenon referred to as influencer marketing [16, 17]. Influencer marketing is more important than ever when it comes to creating a brand with far reach coupled with a high activation rate [3]. Successfully executing a proper influencer marketing campaign is difficult because of a three-fold problem facing market actors:

1. Reaching influencers who are able to establish a genuine connection with the brand and have the follower base to promote this connection in an authentic and effective manner is tedious and difficult [4],
2. Current influencer marketing agencies employ tactics that are: (a) inflexible and unfriendly to high-potential yet early-stage companies, (b) indiscriminate when it comes to the quality and longevity of the businesses they are ultimately facilitating promotion of, and (c) not utilizing analytical tools at their disposal to their full potential; and
3. The current market environment is rife with competition and is ever changing [5].

A step-by-step algorithm for estimating and predicting social media engagement is proposed. Graph theory, regression analysis, and projection methods to gain insights from social media data and to set expectations for eventual influencer marketing campaigns. Finally, the current study presents a couple of scenarios and shares the conclusions that an iterative process over the steps outlined in the proposed algorithm may offer replicable statistically sound results.

2 Materials and Methods

2.1 Theoretical Framework

Graph theory is a branch of mathematics that deals with the study of graphs, which are mathematical structures used to model pairwise relationships between objects. Graphs are used to model a wide range of systems, including social networks, transportation systems, and biological systems. The application of graph theory to the Higgs Boson Graph Dataset involved analyzing the properties of graphs to identify patterns and characteristics that may indicate the presence of the Higgs boson particle. This involved the use of various graph-based algorithms and techniques, such as community detection, centrality analysis, and clustering. Graph theory provides a powerful framework for analyzing complex systems and identifying patterns and structures that may not be immediately apparent. Its application to the Higgs Boson Graph Dataset is just one example of how this approach can be used to gain new insights into fundamental questions in physics and other fields. In this case, Twitter data is the application of interest and the model specifications are provided in the sections that follow.

2.2 Operational Framework

A step-by-step algorithm that identifies minimum thresholds to achieve the desired number of [either] activations or click-throughs via influencer marketing is defined in this study. An adapted approach to modeling social media interactions is proposed by leveraging the following:

1. Graph theory for selecting the initial set of influencers,
2. Linear regression techniques for parameter estimation,
3. Non-linear parametric modeling for maximum reach estimation.

The model assumes homogeneity of individuals, which can be interpreted as similar levels of engagement as well as similar levels of followers. Furthermore, such homogeneity also stems from domain-specific networks. The algorithm assumes that there is only one metric for tracking progress on the objective outcome of interest. In particular, the model assumes no differences across the types of engagement (e.g., activations, click-through, likes, etc.). The Twitter Ads Glossary provides common terms and concepts on this social media platform [18]

The maximum reach estimation model assumes that the model discriminant must be less than zero and the linear parameter to be non-negative real numbers.

The steps of the algorithm detailed in the following sections are as follows:

1. Gather data on the social network within the specified domain.
2. Graph the edges and nodes resulting from the network.
3. Identify the major components and nodes by degree centrality.
4. Rank nodes by degree centrality.
5. Iterate over different thresholds for cut offs.
6. Regress via OLS or Maximum Likelihood engagement over activity plus activity squared.
7. Verify maximum parameters for reach.
8. Iterate by choosing different thresholds or combination of cut offs.
9. Choose a set of model parameters and the initial set of influencers.
10. Predict and project engagements.

2.3 Model Specification

Choosing the initial set of influencers follows traditional graph theory methods via degree centrality measures [1, 8, 12]. To simplify the procedure, the current analysis proposes forming a graph from where edges are sourced based on the maximum degree centrality. The number of followers drives the ranking along with the number of interactions. The chosen graph requires an arbitrary threshold for testing. For instance, the graph provided in Fig. 1 demonstrates a desired level of between fifty and sixty engagements per node, while the centrality measure ranks the nodes by the number of edges in the graph, which in turn equals the number of re-tweets. Therefore,

$$Centrality_{!} = \overset{''}{\underset{\substack{\#\$\% \\ \&'()}}{+}} Retweets_{!}$$

Social network data often plot complex, dense graphs. Figure 1 is purposely simplified by eliminating loops and direction for easier interpretation.

After selecting the initial set, the next step is the algorithm which estimates the average engagement per influencer. The heterogeneity of individuals is somewhat assuaged by the condition run on the initial set; therefore, an Ordinary Least Square regression model should demonstrate constant variance in the error term. Social

Fig. 1 Initial set graph on Twitter data. Nodes are users, edges are retweets

media interactions from Twitter data are modeled, where aggregated counts of user retweets drive the measure of total engagement per influencer. Further, Twitter activity is aggregated at the influencer level in the model.

That is,

$$Centrality_! = \underset{()}{\overset{\overset{//}{+}}{*\&+,\ \&\&+'}}\ Retweets_!$$

While,

$$Activity_! = \underset{()}{\overset{\overset{//}{+}}{\underset{,\ \&\&+'}{-}}}\ Tweets_!$$

The two aggregations at the influencer level reveal a high-level of correlation between the two variables. The Pearson correlation coefficient between activity and engagement data is 0.36. With a logarithm of base 2 to engagement data, the Pearson correlation coefficient is ~0.49.

To visualize, both variable plots are provided in Fig. 2.

This correlation is modeled in a way that can describe multiple scenarios that are conditional to the selected parameters. The stochastic model estimates the conditional probability of engagements based on the activity level. Based on the initial set,

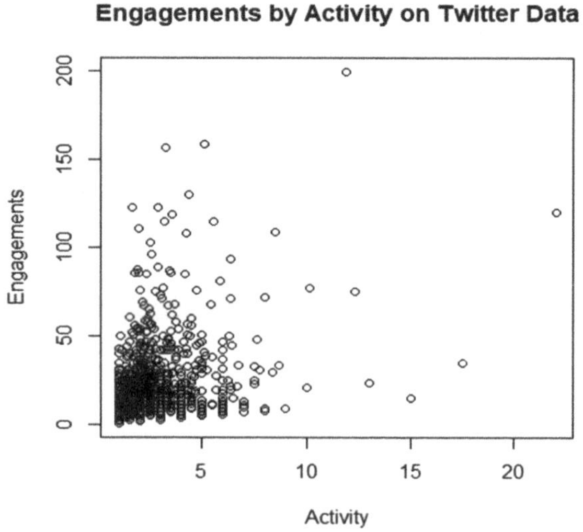

Fig. 2 Engagements by activity on Twitter data

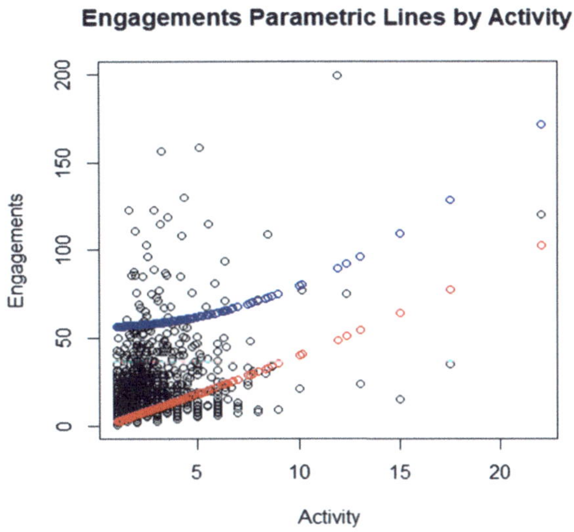

Fig. 3 Engagements parametric lines by activity

the parametric line may take multiple slopes, such as the depicted dotted-blue and dotted-red lined in Fig. 3.

2.4 *Model Algorithm*

The provided algorithm proposes that engagements are a linear function of engagements, activity, and followers as in the following equation where *Error*! is independent and normally distributed:

$$f(Engagement!) = \alpha + \beta) * Engagement! + \beta. * Activity! \\ + \beta/ *Followers! + Error!$$

To remove bias from the estimation parameter, the initial set of influencers must be restricted to a homogeneous set by its degree centrality. That means the followers variable in the model becomes a constant, which enables the study to simplify by removing it altogether. Likewise, as specified above, the model suggests a dynamic structure that often leads to serially correlated error.

The Pearson correlation between engagement and activity is ~0.49 when engagements are in the logarithm base 2 form. To avoid modeling dynamic stochastic processes and to specify the most parsimonious model, the current study leverages the empirical knowledge from the data by assuming that activity and engagements are highly collinear, which forces the choice of one of them. Thus, if Log 2 of engagement yields a 0.49 Pearson correlation coefficient, it is feasible to consider that activity squared should roughly equal engagement. Therefore, the model specification reduces to:

$$f(Engagement!) = \alpha + \beta) * Activity. + \beta. * Activity! + Error!$$

Therefore, modeling social media engagements becomes a quadratic function of activity where the dynamic term is replaced by its proxy activity squared.

Furthermore, modeling engagements in a quadratic function requires assuming that $\beta)$ is greater than zero to hold the expectation of growth over the output of the function,

$$\beta) > 0$$

and, the discriminant of the characteristic equation to be less than zero,

$$\frac{01_! \pm 3\overline{1_!{}^{!}041''5}}{0.1}$$

where,

$$\beta : -4\beta)\alpha < 0$$

As such, the model states that the expected value of engagement given different levels of activity is a stochastic function of the mean value of engagements among the initial set, plus a non-negative coefficient of the squared activity, plus activity, plus a white noise error, where the discriminant of the characteristic equation must be less than zero. The algebraic expression of the model is as follows:

$$\mathrm{E}(Engagement_! | Activity_!) = \alpha + \beta) * Activity + \beta_{.!} * Activity_! + Error_!$$

where,

$$Error_! \approx N(\mu, \sigma);$$
$$\beta) > 0;$$
$$\beta : -4\beta)\alpha < 0$$

3 Results

3.1 Initial Set

The following section provides the model outputs by adjusting the parameters for three differing influencer marketing campaign scenarios. Scenarios are defined as the model parametric line given the model fit, which implies the topic remains constant for all users/influencers and at all times. The current analysis assumes that social media users share and align their interests for a given period. The current analysis leverages Twitter data from the Higgs Boson Graph Dataset [15] with observations less than timestamp "1,341,385,159," the first peak shown in Twitter activity. Meaning, this is when the theme goes viral. Influencers are sorted in descending order by the number of engagements, activity, and number of followers. To make the graph data plottable, a subset of the highest first half of the data is used. Figure 4 demonstrates the results of 49,698 generated interactions from 7,919 Twitter users. The total network reach is estimated to be 1,161,663 Twitter users based on the sum of follower counts for all of the re-tweets produced within the network. In the figure, the initial set graph reveals a focused set of engagement driving the results, which is assumed to be influencers.

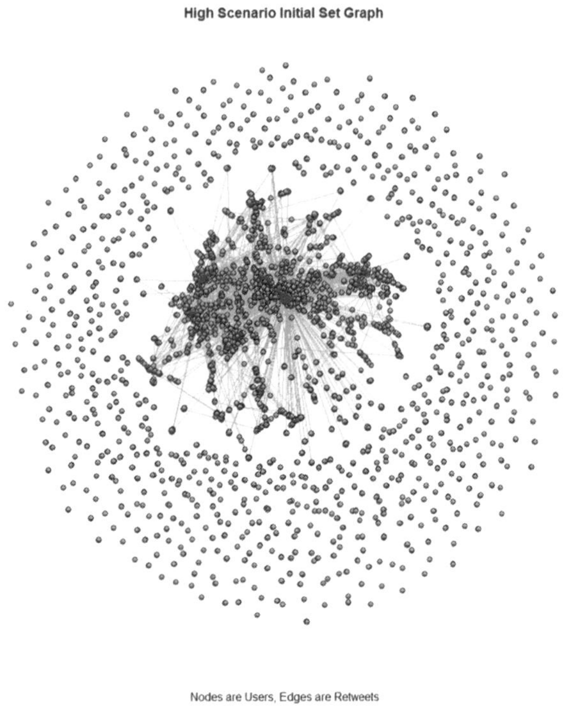

Fig. 4 High scenario initial set graph. Nodes are users, edges are retweets

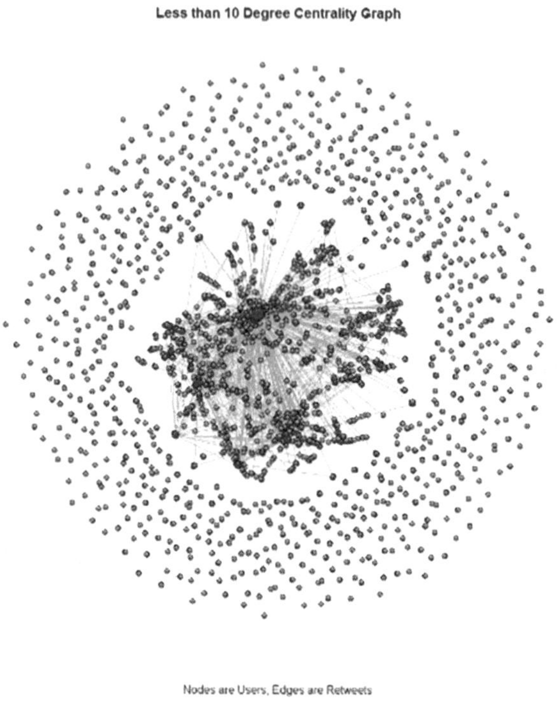

Fig. 5 Less than 10° centrality graph. Nodes are users, edges are retweets

To further explore the impact of influencers on engagement results, nodes are produced to measure the degree of centrality. The following figures show three different graphs based on three ranges of degree centrality as follows: Fig. 5 (less-than-10), Fig. 6 (between 10-and-20), and Fig. 7 (greater-than-20).

The first insight is that number of followers is not necessarily the best proxy to understand engagements. The set of Twitter users in Fig. 6 exceeds the number of engagements despite the average number of followers. On the other hand, activity seems to go accordingly with increases in engagements figure by figure. Table 1 provides the average of the activity, followers, and engagement from Twitter Data across the three figures.

The data provide insight on how to decide which set of users to choose and other parameters to consider, such as the number of followers. However, there is still a need to understand how users' activity increases engagements and projections for maximum reach Therefore, the next step in the model identifies the average effect of activity over engagements by graphs' members.

Fig. 6 Between 10 and 20° centrality graph. Nodes are users, edges are retweets

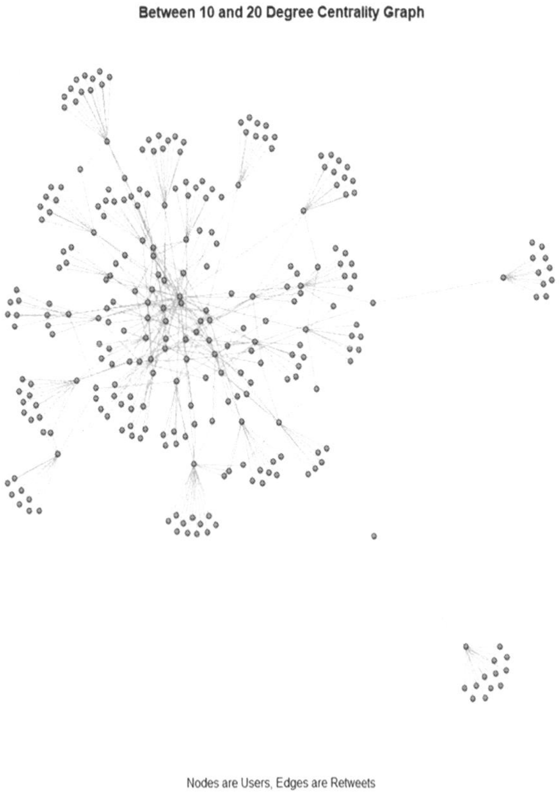

3.2 Parameter Estimation Set

In order to better understand the effects of influencer marketing, parameter adjustments led to a more granular level result. A regression model was conducted to observe engagement on both activity and the dynamic term of engagements by modeling the assumption stated in the model specification section above. Choosing parameters from the regression model illustrates the possible reach outcomes and enables inferences about differing scenarios. More precisely, one can infer the parametric model line within a degree of certainty.

Table 2 provides the regression model results. Only the graph data with less-than-10-degree centrality among the three models yield significant estimates. Neither between-10-and-20 nor greater-than-20-degree centrality nodes model yield parameters that are statistically different from zero.

The three models yield a set of different parameters, which implies different trajectories of the parametric lines. Business-wise, these models explain the expected number of interactions given a certain level of activity for a given set of Twitter users. Because an influencer can be defined differently, the model allows for one to

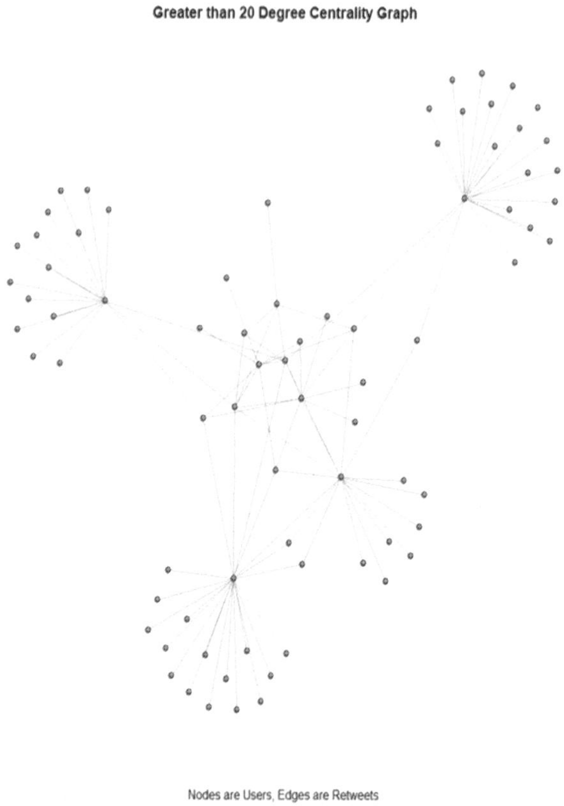

Fig. 7 Greater than 20° centrality graph. Nodes are users, edges are retweets

Table 1 Average of activity, followers and engagements from Twitter data

	Graph 3	Graph 4	Graph 5
Activity	1.854	1.938	1.146
Followers	159.8	208.4	248.8
Engagements	6.308	8.368	4.125

potentially project impact of a marketing campaign based on that definition. Figure 8 reveals the three expected trajectories of the three distinct model parameters. The less-than-10 figure model parametric line is on the blue dots, between-10-and-20 on the red, while greater-than-20 is on the green dots.

Table 2 Regression model results

	Dependent variable		
	Interactions		
	Less than 10	Between 10 and 20	Greater than 20
	(1)	(2)	(3)
Activity	1.354*** (0.622, 2.086)	19.179 (–9.567, 47.926)	–9.833 (–410.673, 391.006)
I(Activity2)	0.397*** (0.291, 0.502)	–3.492 (–9.792, 2.808)	5.000 (–148.462, 158.462)
Constant	2.083*** (1.078, 3.089)	–13.116 (–42.254, 16.021)	8.500 (–239.521, 256.521)
Observations	3,739	38	8
R^2	0.148	0.068	0.039
Adjusted R^2	0.148	0.14	−0.345
Residual Std. Error	9.047 (df = 3736)	16.316 (df = 35)	5.203 (df = 5)
F Statistic	325.410*** (df = 2; 3736)	1.271 (df = 2; 35)	0.102 (df = 2; 5)

Note *p < 0.1; **p < 0.05; ***p < 0.01

Fig. 8 Engagements parametric lines by activity

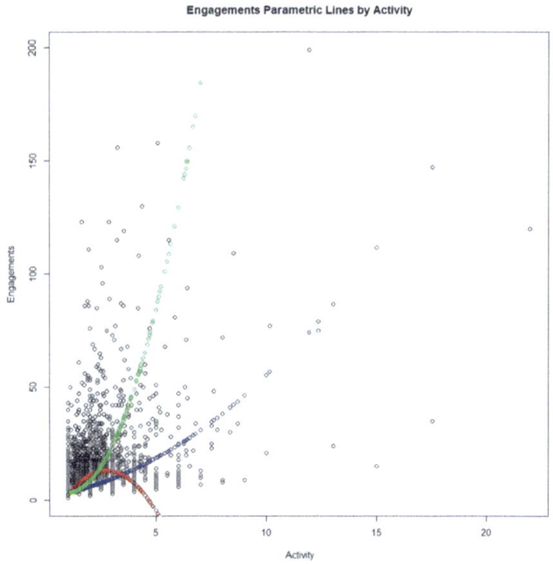

4 Discussion

The model provides an approach for estimating and predicting social media engagements, the initial intent of the study. Twitter was specifically used; however, other social media platforms could be analyzed with appropriate data sources. A limitation was that statistical inference only supports the less than10 degree centrality graph model. To illustrate the model framework in a business setting, it is assumed that business stakeholders require a guarantee of an outcome of no less than 500,000 engagements.

Using this scenario as an example, the model parameters are.

$$E(Engagement) = \alpha + 1 * Activity + 2 * Activity2$$

where,

$$E(Engagements) = 500,000 = 2.08 + 1.35 * 1,120.544 + 0.397 * (1,120,544)$$

The required activity equals 1,120,544 re-tweets per influencer.

Results demonstrate that network analysis, and in particular degree centrality, adequately functions for choosing the initial set if analysts iterate the choices. This result aligns with other influencer marketing frameworks. The model also shows that choosing a good initial set depends on testing its possible outcomes by considering engagement of activity plus the activity. Future research could adapt the model to account for other factors that influence engagement, such as content quality or audience demographics.

Regression analysis and parameter estimations shed light on the expected output and on the certainty and statistical significance of the findings for each of the initial sets chosen in the first step. Regression output, in turn, allows for an iteration over the first step on degree centrality ranges. Finally, once the model parameters are chosen, the expected reach can be calculated as the expected number of engagements while providing inference support via the statistical evidence run by the algorithm.

5 Conclusion

This study proposes an analytical framework and algorithm for optimal influencer marketing decision-making. The model recognizes identifiable problems in influencer network analysis, namely identifying the initial set, parameter estimation, and maximization of reach; thereby, articulating the models into a step-by-step framework for iterative decision making and choice. The outlined model enables approximate calculations for initial sets and further insight into a projected outcome that allows users to explore the limitations and possibilities in influencer marketing approaches. This algorithm offers a solution to reach a given marketing goal by choosing an initial set of influencers based on the graphs and the parameters observed from the empirical data. Through this dynamic model, influencer marketing expectations can be supported with empirical assessments in support of future investigations.

References

1. Albalawi, Y., Sixsmith, J.: Identifying Twitter influencer profiles for health promotion in Saudi Arabia. Health Promot. Int. **32**(3), 456–463 (2017). https://doi.org/10.1093/HEAPRO/DAV103
2. Arora, A., Bansal, S., Kandpal, C., Aswani, R., Dwivedi, Y.: Measuring social media influencer index- insights from Facebook, Twitter and Instagram. J. Retail. Consum. Serv. **49**, 86–101 (2019). https://doi.org/10.1016/J.JRETCONSER.2019.03.012
3. Boerman, S.C., Muller, C.M.: Understanding which cues people use to identify influencer marketing on Instagram: an eye tracking study and experiment. Int. J. Advert. **41**(1), 6–29 (2022). https://doi.org/10.1080/02650487.2021.1986256
4. Brown, D., Fiorella, S.: Influence Marketing: How to Create, Manage, and Measure Brand Influencers in Social Media Marketing. Que Publishing, Indianapolis (2013)
5. Campbell, C., Farrell, J.R.: More than meets the eye: the functional components underlying influencer marketing. Bus. Horiz. **63**(4), 469–479 (2020). https://doi.org/10.1016/J.BUSHOR.2020.03.003
6. Cao, L.: Decentralized AI: edge intelligence and smart blockchain, metaverse, Web3, and desci. IEEE Intell. Syst. **37**(3), 6–19 (2022)
7. DataReportal. Global digital insights. https://datareportal.com/. Last accessed 3 Oct 2022
8. Feng, J.: Network analysis method of multi-directional music influence based on graph theory. J. Front. Art **1**, 9–12 (2021). https://doi.org/10.23977/artpl.2021.020504
9. Friestad, M., Wright, P.: The persuasion knowledge model: how people cope with persuasion attempts. J. Consum. Res. **21**(1), 1–31 (1994)
10. Garon, J.M.: Legal implications of a ubiquitous metaverse and a Web3 future. SSRN Electron. J. (2022). https://doi.org/10.2139/SSRN.4002551
11. Hermanda, A., Sumarwan, U., Tinaprillia, N.: The effect of social media influencer on brand image, self-concept, and purchase intention. J. Consum. Sci. **4**(2), 76–89 (2019)
12. Huo, Z., Huang, X., Hu, X.: Link prediction with personalized social influence. Proc. AAAI Conf. Artif. Intell. **32**(1), 2289–2296 (2018). https://doi.org/10.1609/AAAI.V32I1.11892
13. Kovavova, M., Horak, J., Higgins, M.: Behavioral analytics, immersive technologies, and machine vision algorithms in the Web3-powered metaverse world. Linguist. Philos. Invest. **21**, 57–72 (2022)
14. Lee, P.-Y., Koseoglu, M.A., Qi, L., Liu, E.-C., King, B.: The sway of influencer marketing: Evidence from a restaurant group. Int. J. Hosp. Manag. **98**, 103022 (2021). https://doi.org/10.1016/j.ijhm.2021.103022
15. Leskovec, J.: SNAP: Network datasets: Higgs Twitter dataset (2015). https://snap.stanford.edu/data/higgs-twitter.html. Last accessed 24 April 2022
16. Leung, F.F., Gu, F.F., Palmatier, R.W.: Online influencer marketing. J. Acad. Mark. Sci. **50**(2), 226–251 (2022). https://doi.org/10.1007/S11747-021-00829-4
17. López, M., Sicilia, M., Verlegh, P.W.J.: How to motivate opinion leaders to spread e-WoM on social media: monetary vs non-monetary incentives. J. Res. Interact. Mark. **16**(1), 154–171 (2022). https://doi.org/10.1108/JRIM-03-2020-0059/FULL/HTML
18. Twitter ads glossary. https://business.twitter.com/en/help/overview/twitter-ads-glossary.html. Last accessed 01 May 2023
19. Valerio, C., William, L., Noémier, Q.: The impact of social media on E-commerce decision making process. Int. J. Technol. Bus. **1**(1), 1–9 (2019). https://doi.org/10.5281/zenodo.2591569

Metaverse: A New Frontier for Influencer Marketing

Amandeep Dhaliwal

Abstract This study delves into the emergence of the metaverse, a virtual realm mirroring the physical world, facilitating user interactions. With the metaverse swiftly gaining traction, influencer marketing emerges as a novel avenue for brands to expand their outreach, craft immersive brand encounters, and gather valuable data and insights. Nonetheless, influencer marketing in the metaverse presents challenges like data privacy concerns and ensuring influencer authenticity. The chapter endeavors to bring clarity on the basics of the metaverse and influencer marketing and further how their amalgamation can provide advantages to businesses. It also addresses the issues of how businesses can adeptly harness influencer marketing in the metaverse to fulfill their marketing objectives using the Case of Metafluence. By scrutinizing the metaverse's attributes, various forms of influencer marketing, and the hurdles and prospects encountered by brands, this study aims to furnish invaluable insights for companies eager to capitalize on this innovative milieu.

Keywords Metaverse · Influencer marketing · Virtual influencers · Virtual reality

1 Introduction

The Metaverse is anticipated to revolutionize various facets of life, education, and entertainment. It is poised to offer new experiences driven by artificial intelligence, blockchain, and augmented reality technologies, all within the infrastructure of Web 3.0. This futuristic realm is expected to catalyze a paradigm shift, which is evident in the substantial investments made by renowned brands in Metaverse platforms and technologies [1].

Recent developments, such as Facebook's rebranding to "Meta," Microsoft CEO Satya Nadella's commentary on the Metaverse's transformative potential, and

A. Dhaliwal (✉)
SLM-PG, Manav Rachna International Institute of Research & Studies, Faridabad, India
e-mail: amandeep.slm@mriu.edu.in

S. Dutta et al. (eds.), *Advances in Data Analytics for Influencer Marketing: An Interdisciplinary Approach*, Information Systems Engineering and Management 9,
https://doi.org/10.1007/978-3-031-65727-6_11

Microsoft's acquisition of Activision Blizzard for approximately $68.7 billion [2], underscore the Metaverse's capacity to reshape future communication channels.

Central to the Metaverse's evolution is the emergence of influencers, who wield significant power to sway their followers' opinions and behaviors. Influencer marketing, facilitated through collaborations with these authoritative figures on social media platforms, has become instrumental in enhancing brand visibility and credibility as influencers share their experiences with products and services [3].

As the Metaverse evolves, influencer marketing will too. This burgeoning field, already pivotal in reaching target audiences on social media, will adapt to the dynamics of this new digital frontier. Influencers will leverage the Metaverse's unique blend of physical and digital experiences to engage audiences beyond the confines of traditional social media platforms like Instagram and Twitter [4]. This transition presents a fertile ground for the development of innovative digital marketing strategies, further intertwining technology with influencer-driven brand promotion. Already demonstrating its impact across sectors ranging from user experience to social dynamics, economics, and advertising, the Metaverse prompts a comprehensive examination of influencer marketing's role within its evolving landscape [5].

Influencer marketing, hailed as one of the most potent strategies of the digital age, is poised to adapt to this virtual metamorphosis, reaching wider audiences and crafting more immersive experiences. As the Metaverse expands, terms like "meta-influencer" and "metafluencer" have emerged, though their definitions remain fluid [6, 7]. Examples include Beymen's introduction of an avatar influencer in Decentraland as a "meta-influencer" and Ilonsi's NFT project, Meta-Fluencer, which aims to decentralize e-commerce and streamline. brand-influencer collaborations [6].

In projects like Metafluence, physical world influencers are represented in the Metaverse through avatars, fostering unprecedented interaction between followers and influencers [7]. Platforms like Instagram, YouTube, Twitter, and TikTok are unified within the Metafluence universe, enabling seamless engagement in this virtual city. These influencers are also known as "Metahut owners." Additionally, the emergence of CGI product virtual influencers, such as Kyra from India, adds another dimension to the meta-influencer landscape, with AI-driven capabilities that transcend traditional CGI integration into Metaverses and 3D films [8].

As businesses navigate the virtual landscape, they ponder the relevance and potential actions for their brands within these immersive environments. These emerging and immersive spaces offer a fresh approach to engaging existing customers while attracting new ones [9]. Consequently, established and emerging enterprises must explore diverse engagement strategies to connect with audiences of varied demographics. At the same time, the strides made by numerous companies and brands in influencer marketing within the Metaverse reveal a need for more clarity and understanding of the apparent fusion of influencer and Metaverse concepts, resulting in the creation of novel terminology and a hazy landscape. Many of the terms need to be universally recognized. The emergence of these new terms, alongside the development of projects in this domain, highlights that burgeoning opportunities are afforded by influencer marketing within virtual platforms. Metaverse still needs to have an in-depth understanding of these concepts, how influencer marketing operates in the Metaverse, and its evolving dimensions and opportunities, which is crucial in navigating this new environment. This study thus focuses on creating a basic understanding of the terms Metaverse and influencer Marketing and the opportunities this fusion of these terms offers for future marketing trends. This trend signifies a notable brand-influencer and audience engagement shift toward these immersive platforms.

2 Metaverse and the Opportunities

The term "Metaverse" has gained significant traction in contemporary discourse. It represents a convergence of virtual reality, augmented reality, sensory technology, and spatial computing, offering humans novel experiences within an expansive virtual multiverse [9]. The Metaverse is expected to revolutionize various aspects of communication, representation, interaction, economic dynamics, knowledge dissemination, work culture, digital immersion, and privacy. It epitomizes a digital transformation era that merges physical reality with virtuality [10].

In today's rapidly advancing technological landscape, the Metaverse has become captivating and contentious. Substantial investments in its underlying technologies, including artificial intelligence, blockchain, and augmented reality, as well as its ecosystem comprising elements like the virtual economy, avatar usage, and content creation, underscore the Metaverse's growing significance [11].

Metaverse transcends mere technological innovation; it embodies a sci-fi fantasy turned reality, poised to evolve into an entirely new global economy [12]. Even corporate giants like McDonald's and Nike envision the Metaverse as a bustling hub where consumers can dine, shop, play games, forge friendships, attend concerts, and cultivate virtual identities. This transformative concept continues to blur platform boundaries, ushering disparate ecosystems into a new decentralized universe [11]. Despite lacking a universally agreed-upon definition, the Metaverse draws upon elements like social media, virtual reality, online gaming, and cryptocurrencies, which have long been intertwined with human life.

This innovative realm, characterized by integrating spatial technologies such as virtual and augmented reality alongside Web 3.0 advancements, offers immersive multisensory experiences, transforming the traditional internet interface into a more dynamic and engaging platform. Moreover, individuals can converge, collaborate, and communicate through social media platforms and virtual workspaces [13].

Defined as a collection of interconnected, experiential, and three-dimensional virtual worlds enabled by Web 3.0 technology, the Metaverse offers real-time socialization. It fosters a user-owned internet economy spanning digital and physical realms. Mark Zuckerberg, envisioning this as "the embodied internet," encapsulates its essence, highlighting its potential to transcend traditional online experiences [14]. Interactive platforms such as GTA, Second Life, Fortnite, Minecraft, and Roblox provide early glimpses into the possibilities offered by the Metaverse, where individuals can engage in a myriad of activities, blurring the lines between work, leisure, and social interaction [15].

Participating in events and trading individually crafted products and services for real-world currency [16] showcase the Metaverse's gradual integration into daily life. A prime example of this fusion of digital and physical realms is seen in games like Pokémon Go, which intertwines real-world locales and streets with digital elements, enabling players to hunt for Pokémon [17]. These instances underscore the Metaverse's immersive nature, where conventional interfaces like screens and keyboards are eliminated, facilitating direct interaction with the digital environment [18].

Figures like Tim Sweeney and Mark Zuckerberg emphasize that the Metaverse's current manifestations are just a fraction of its potential, akin to the early stages of the internet's development. Its evolution will extend far beyond any single company's influence [19]. Mark Zuckerberg's vision for Meta as a Metaverse company rather than merely a social media entity highlights the importance of a profound sense of presence within this digital space. It is about experiencing content as if physically present, transcending traditional two-dimensional screens to coexist with others truly [20].

The projected market size of the Metaverse for 2025 and 2030. Estimates forecast a market size of USD 476.4 billion and USD 1542.9 billion by 2025 and 2030, respectively [10].

Facebook's significant investment in Horizon, a new workplace platform and proto-Metaverse VR, underscores the growing interest in virtual reality technologies. Brands are also leveraging virtual reality for business purposes, with luxury brand Gucci offering virtual products surpassing the value of physical items. Notably, Nike has ventured into selling virtual Jordan shoes within Fortnite, while Coca-Cola has introduced virtual wearables in Decentraland [21]. According to Bloomberg, the Metaverse market is valued at $800 billion, signaling substantial potential for growth. Experienced crypto investors are positioning themselves to capitalize on the burgeoning meta-universe and token trading of high-growth startups, anticipating an acceleration in the virtual reality boom. The prospect of a virtual reality platform rivaling major social networks within the next five years is increasingly plausible [10].

3 Influencer Marketing

As consumers' trust in traditional marketing continues to erode, businesses face mounting challenges in capturing and influencing their target audience. To navigate this shifting landscape, many marketers are leveraging the power of social media platforms such as Instagram, Facebook, and Weibo by collaborating with online influencers. This practice has given rise to "influencer marketing," a strategic communication approach wherein companies enlist and compensate online influencers to engage with their followers on social media, effectively promoting the brand's offerings [22]. These online influencers, whether individuals, groups, or virtual avatars, boast extensive networks of followers and are selected and remunerated by companies through methods like pay-per-post [23].

Influencer marketing is a collaborative effort in which companies partner with online influencers to endorse their products or services, thereby enhancing brand visibility. Unlike traditional celebrities, influencers hail from diverse backgrounds, nationalities, and social statuses, encompassing a spectrum of expertise and influence. These influencers can range from renowned fashion photographers on Instagram to knowledgeable cybersecurity bloggers on Twitter and prominent marketing executives on LinkedIn [4]. While many influencers may appear as everyday individuals, some boast substantial followings, numbering in the hundreds or even millions, and establish themselves as authorities in their respective fields. Consumers often turn to these influencers for product-related queries drawn by their expertise and engaging content creation across various social media platforms [2].

Influencer marketing stands out as one of the most effective strategies in social media marketing, leveraging influencers' sway on popular platforms like TikTok, Facebook, and Instagram to sway consumer purchasing decisions and promote businesses' offerings. Companies seek to harness influencers' storytelling prowess to sway consumer perceptions, as these influencers hold sway over their followers due to their industry knowledge and experience in domains like food, film, and fashion [24]. The prevalence of social media usage further underscores the importance of influencer marketing, with around one-third of the global population now active on these platforms, significantly influencing consumer behavior [25].

Despite its origins dating back to 2006, influencer marketing has evolved significantly over the past 15 years, coinciding with the growth of social media platforms like Twitter, Facebook, and Instagram. Initially dominated by bloggers sharing personal experiences and endorsing everyday items, influencer marketing now encompasses a wide array of content creation across platforms like TikTok, Instagram, and Facebook Live, offering influencers diverse avenues to engage with their audiences [4].

Influencer marketing leverages the influence and audience of influential individuals, known as influencers, to endorse products, services, or brands across social media platforms and online channels. These influencers have cultivated dedicated and engaged followings, enabling brands to tap into their credibility and connect with potential consumers in an authentic and compelling manner [26]. Its rapid emergence in the digital age is attributed to the widespread adoption of social media and the consumer shift towards seeking genuine and reliable recommendations from trusted individuals.

While some influencers may amass sizable followings akin to celebrities, their path to fame distinguishes them from traditional stars. Influencers garner recognition through their content creation efforts, seamlessly integrating brand endorsements into their personal narratives to craft authentic and value-driven content for their audience [27].

At its core, influencer marketing revolves around establishing partnerships with individuals who possess credibility, authority, and a significant following within specific niches or industries. These influencers span diverse backgrounds, including content creators, bloggers, vloggers, celebrities, and subject matter experts, wielding influence over their followers' purchasing decisions [28]. Brands collaborate with influencers to harness their storytelling prowess and authenticity, fostering personalized and relatable connections with potential customers.

Influencer marketing has burgeoned into a thriving industry, with influencers carving out careers as professional content creators and marketers [22, 23, 29]. Brands allocate substantial budgets for influencer marketing campaigns, compensating influencers for their collaborations. It has become a pivotal component of brand marketing strategies across various sectors [30, 31]. By the end of 2023, spending on influencer marketing activities is projected to reach $23.4 billion, with over 75% of marketers planning to allocate resources to influencer marketing initiatives Hub IM [31].

Businesses can gain a competitive edge in the increasingly competitive influencer marketing landscape by effectively allocating resources and fostering trust among consumers [24]. Strategic selection of influencers and optimization of individual posts to enhance engagement can bolster their efficacy, enabling businesses to maximize their returns and establish enduring relationships with their target audience.

4 Amalgamation of Metaverse and Influencer Marketing

Influencer marketing has already become a significant industry other than being a prevalent strategy for reaching target audiences on social media. It is poised to further expand in the future as new social media platforms, content categories, and brand outreach channels emerge [4, 29]. It is poised to undergo significant transformation influenced by the dynamics of the Metaverse. This emerging virtual universe provides influencers with the opportunity to seamlessly integrate physical reality with digital experiences, surpassing the limitations of traditional social media platforms like Instagram, Twitter, and TikTok [32]. Influencer marketing is considered an ideal market for the Metaverse due to its heavy reliance on technology. As influencer marketing transitions from social media to Metaverse platforms, it is anticipated to evolve into a realm offering novel experiences across various domains such as life, education, and entertainment. This evolution will be driven by advancements in artificial intelligence, blockchain, and augmented reality technologies, unlocking diverse creative opportunities and ushering in new possibilities within the framework of Web 3.0 [33].

Top influencers continuously adapt their content to reflect current trends, providing brands with essential tools for success. With the social media landscape evolving rapidly, it is crucial for influencers to leverage current trends in influencer marketing to solidify their authority and appeal to companies seeking partnerships. Brands, in turn, seek to capitalize on these trends to expand their audience, generate leads, and boost revenue. Three key trends in influencer marketing have emerged in 2022 [3].

Firstly, there is a shift towards prioritizing audio and video content over traditional text and graphics. With improvements in global bandwidth and the proliferation of 5G technology and fiber internet, people are increasingly consuming more video and audio content. Platforms like TikTok and Instagram Reels have witnessed surges in popularity, with a 50% increase in Facebook live video viewership observed early in 2020 amidst global lockdowns [3].

While Millennials and older demographics lean towards platforms like YouTube, Instagram, and Facebook for product research, TikTok has captured the attention of 30% of Gen Z users. This demographic preference underscores the importance for brands targeting younger audiences to engage in social listening, provide valuable information, and collaborate with influencers to build trust and brand exposure [34].

Secondly, micro and nano influencers are gaining traction, with businesses increasingly focusing on influencers with more minor, highly engaged followings. Micro-influencers, typically with fewer than 25,000 followers, boast the highest engagement rates, making them attractive partners for brands seeking authentic connections with their target audience. By collaborating with a larger pool of influencers, brands can optimize their influencer marketing budgets while ensuring alignment with their target demographics [3].

Thirdly, the emergence of CGI influencers presents new opportunities for brands to tailor their messaging to exact specifications. CGI influencers, created by designers, programmers, and social media managers, offer brands complete control over their narratives, from captions to content planning [35]. Brands such as UGG and high-end labels like Diesel and Prada have already tapped into the potential of CGI influencers like Lil Miquela, whose collaborations and features in prominent fashion publications signal a shift in influencer marketing dynamics [36].

Influencer marketing, a cornerstone of digital marketing strategies and hailed as one of the most potent marketing tools of this generation, is expected to adapt to this virtual metamorphosis, reaching broader audiences and crafting more immersive experiences [33]. The proliferation of the Metaverse and the integration of the influencer concept into this realm have led to the emergence of terms like "meta-influencer" and "metafluencer," albeit without universally accepted definitions [21]. For instance, Beymen introduced its avatar influencer in Decentraland as a "meta-influencer," while Ilonsi launched an NFT project named "Meta-Fluencer" aimed at revolutionizing brand-influencer collaborations and decentralizing e-commerce [6].

In projects like Metafluence, real-world influencers are represented in the Metaverse universe through avatars, fostering unprecedented interaction between followers and influencers. Platforms such as Instagram, YouTube, Twitter, and TikTok, where influencers wield significant influence, are unified under the Metafluence universe. Influencers in this Metafluence city are also known as " Metahut owners," with Beymen's avatar in Decentraland dubbed a "meta-influencer," and Ilonsi's digital NFT characters providing advantages in its marketplace labeled as "Meta-Fluencer" [7].

4.1 Advantages of Metaverse-Based Influencer Marketing

While influencer marketing remains one of the most potent marketing strategies today, it is not without its limitations. The capabilities of these social media platforms often constrain traditional influencer marketing activities conducted on social media platforms. Issues such as payment challenges, centralized data control, and fake follower count, among others, have become increasingly prominent [22]. However, upon examining examples of influencer marketing implementation within the Metaverse, it becomes evident that Metaverse platforms, bolstered by Web 3.0 infrastructure and blockchain technology, offer opportunities for amplifying the efficacy of influencer marketing through decentralized, secure communication and interaction, supported by cutting-edge technological tools [15].

The Metaverse presents an array of capabilities to brands for real-time interaction in influencer marketing, facilitating diverse event organization and experiential opportunities [13]. As elucidated by [37], the Metaverse holds vast potential and evolving applications, including advanced blockchain integration, virtual work and learning environments, virtual commerce, and innovative marketing avenues. This transition to the Metaverse redefines user experiences in influencer marketing, enabling followers to engage in real-time interactions with their favorite influencers, participate in concurrent events regardless of location, and interact with influencer avatars [38].

Furthermore, the burgeoning NFT technology, underpinned by blockchain, introduces a novel dimension to influencer marketing within the Metaverse [39]. As the Metaverse continues to expand, NFTs are poised to become increasingly pivotal digital assets for validating digital ownership [40]. Influencers venturing into the NFT realm can collaborate with brands and market their digital creations, earning cryptocurrency revenue. The entry of notable figures like Paris Hilton into the NFT arena, coupled with their promotion via social media, underscores the significance of NFTs in the market [41].

In the context of the Metaverse, avatars serve as representations of users' virtual identities and facilitate interaction within virtual events. Within the realm of influencer marketing, followers have the opportunity to engage with the avatars of collaborating influencers, sharing immersive experiences within the Metaverse [38, 42]. Notably, the emergence of avatar influencers, such as Bella introduced by Beymen, exemplifies the integration of influencer marketing into the Metaverse. Unlike traditional CGI influencers, Bella exists as an avatar within Decentraland, bridging the gap between virtual influencers and the Metaverse universe [33].

As the Metaverse landscape evolves, integrating influencer marketing into this virtual realm heralds a new marketing innovation and engagement era. With its unique capabilities and technological foundations, the Metaverse offers boundless opportunities for influencer marketing, transcending the limitations of traditional social media platforms and fostering immersive brand-consumer interactions.

4.2 *The Case of Metafluence*

"Metafluence" stands out as the pioneering influencer platform within the Metaverse, conceived by the minds behind Keepface. Established in 2017, Keepface boasts a database of over 500,000 influencers and has forged partnerships with renowned brands [7]. With Metafluence, influencer marketing strategies employed on social media are being redefined and adapted to the virtual realm, capitalizing on the unique opportunities offered by the Metaverse. The primary objective of Metafluence is to unite influencers, brands, and audiences within the Metaverse, elevating influencer marketing to unprecedented heights. Embracing the ethos of "Influence to Earn," Metafluence aims to construct an ecosystem that addresses prevalent challenges in influencer marketing, including fraud, data scarcity, and payment issues, by establishing sustainable mechanisms for influencers to monetize their influence. This ecosystem introduces innovative features tailored for influencer marketing, such as Metahuts, Metaclans, Metarooms, and events. Currently, three distinct room concepts are being developed within Metahuts: the "event room," the "NFT room," and the "shopping room." These Metahuts function as living spaces and as functional hubs where various Metaverse events can be hosted [7]. The shopping rooms provide a platform for virtual and physical product collaborations with brands, the NFT rooms enable the sale, display, and trade of NFTs, and the event rooms serve as venues for diverse events, product launches, and brand collaborations within the Metaverse. The Metafluence ecosystem aims to unlock a myriad of current and future opportunities, including hosting events with NFT-based ticket sales, trading NFTs, earning $METO (the currency of the Metafluence ecosystem) through property rentals within Metafluence, participating in governance, forging brand collaborations, pioneering in the Web 3.0 realm, conducting MetaCommerce by vending virtual and physical products, forming social clans for networking and event organization, and orchestrating events [6].

Prominent social media platforms like Meta, Instagram, and TikTok each possess their own advantages and limitations. Often, influencers' fans find themselves compelled to follow them across multiple platforms. Metafluence seeks to address these challenges by consolidating these user bases onto a unified platform, thereby streamlining brand-audience-influencer interactions. Leveraging Web 3.0 technology and blockchain foundations, Metafluence aims to create an influencer-centric Metaverse, ultimately culminating in the convergence of influencer marketing and the Metaverse [6].

As a component of the operational Keepface influencer platform, Metafluence epitomizes the extension of conventional social media platforms into the virtual expanse of the Metaverse. As a Metaverse platform centered on influencer marketing, Metafluence provides influencers with a platform to showcase their influential prowess. The universe crafted by Metafluence serves as an exemplary model of how influencer marketing can thrive within the Metaverse [7].

5 Challenges

Influencer marketing in the metaverse presents unique challenges and limitations. Here are some key points to consider:

Accessibility: A significant limitation is that the majority of users may lack access to the necessary tools, gadgets, and technology required to engage with the metaverse fully [43]. It is evident that before people can even begin to benefit from the endless opportunities offered by the metaverse, they need to be able to access them. As an example, most VR headsets are currently designed for middle-aged adults, without considering the needs of people in the K-12 age range (4–18 years) or older individuals, meaning that young children or seniors wanting to access the metaverse may have a negative experience wearing headsets not designed with their age in mind [44]. The same applies to people with physical or sensory disabilities but also to those with low levels of digital literacy skills or little to no access to devices and technologies currently available due to high costs.

Cyberspace Limitations: While meta-influencers can simulate real-life situations, their actions are confined to the digital realm, which may limit their tangible impact [45]. Metaverse environments include realistic, unrealistic, and fused environments. The fused environment reflects some unrealistic elements based on a realistic environment. However, although sound and visual modalities are relatively realistic, there are limitations in the atmosphere, smell, and tactile sensations felt in the field [44].

Data Security: The collection of user data within the metaverse could raise security concerns as the protection of personal information becomes increasingly complex. In the metaverse, the user's alter ego, the avatar, creates various data such as intimate information (e.g., messages, voice, and video), corporate secrets used for work, and the personal information needed for services to continue. As a result, security threats to such data will continue [46]. Personal information and content stored in a virtual environment, metaverse platform, or service system can be forged and leaked. For example, an avatar's information, such as voice and video recording, could be hacked while the user is using the platform, or an attacker could forge the avatar and abuse it. Protection measures and management policies for the existing system are insufficient to protect the virtual environment of the metaverse from cybercrime and cyber-attacks [47].

Privacy-The metaverse involves a multisensory environment. Due to complex and sophisticated features such as more *graphic, 3D* design, and immersive visual and auditory experience, when *unwanted and privacy*-invasive contents proliferate in the metaverse, they may be felt as more intrusive and are likely to have a more significant negative impact on the users or victims. Alternatively, VR headsets can collect richer data about users than traditional screens. Companies thus have more incentives to collect user data and share such data with third parties, which can be used for profiling to deliver customized advertising [48].

Brand Positioning: Brands may face challenges in positioning themselves within the metaverse, as it is a new and evolving space with different dynamics compared to traditional media [49]. Entering the Metaverse is not risk-free. By its nature, brand presentation in the metaverse will diverge in some respects from traditional contexts. Well-known brands, in particular, might struggle with presenting an image in the metaverse consistent with deeply ingrained expectations among consumers in the real world. This could weaken brand distinctiveness if not done correctly [50].

Human Element: Influencers are human behind the scenes, and their marketability is contingent on maintaining a curated image that may not always remain popular or trending [45]. User avatars can collaborate and share experiences, creating new value. Unlike in the real world, this collaboration makes it possible to transcend time and space [42]. It also gives users a common purpose and allows the metaverse to continue as a society. However, because communication is based on sensor information that is limited relative to reality, it is possible to misunderstand or make erroneous judgments about hidden intentions [38].

These limitations highlight the need for careful consideration and strategic planning when engaging in influencer marketing within the metaverse. It is a new frontier with great potential and hurdles that must be navigated thoughtfully.

6 How Businesses Can Use Metifluencers?

Influencer marketing has become a part of the Metaverse, marking the beginning of a new chapter in marketing innovation and engagement as the Metaverse environment changes. Influencer marketing prospects are virtually endless because of the Metaverse's distinct features and scientific underpinnings. Thus, businesses and brands must implement cross-platform integration, integrating Metaverse marketing strategies with traditional social media platforms. This is important as this integration facilitates the creation of a cohesive brand narrative and enables brands to reach a wider audience effectively. Some other recommendations are:

Investment in Metaverse and User Experience Design: To thrive in the Metaverse, brands must prioritize investing in user experience design tailored explicitly for this digital realm. This entails crafting immersive and interactive virtual environments, events, and interactions that resonate with the unique characteristics of the Metaverse [45]. By focusing on user experience design, brands can create compelling and memorable interactions that captivate users and foster more profound engagement with their brands. Additionally, investing in Metaverse-specific user experience design enables brands to differentiate themselves in this increasingly competitive space, positioning themselves as leaders in innovation and creativity [49]

Curated Experiences in the Metaverse Brands should prioritize providing curated experiences in the Metaverse, tailoring interactions to each user's preferences and interests. These personalized experiences may include customized shopping journeys, access to exclusive events, and other initiatives that offer users a unique and memorable brand interaction [51]. By curating experiences this way, brands can deepen engagement with their audience, fostering stronger connections and loyalty.

Additionally, personalized experiences demonstrate a brand's commitment to understanding and meeting the individual needs of its customers, enhancing overall satisfaction and brand perception. As the Metaverse continues to evolve, offering curated experiences will be essential for brands seeking to differentiate themselves and stand out in the digital landscape [48]. Through thoughtful curation, brands can create meaningful moments that leave a lasting impression on users, driving long-term success and growth in the Metaverse.

Creation of Meta-Influencers: Brands are encouraged to delve into the creation of meta-influencers, akin to the example of Bella, tailored specifically for the Metaverse environment. These digital avatars serve as brand ambassadors, capable of resonating deeply with virtual audiences, thereby offering a novel and immersive brand experience [33]. Meta-influencers are crucial in bridging the gap between brands and consumers in the virtual realm, fostering more robust connections, and driving engagement. By developing their meta-influencers, brands can establish a unique presence in the Metaverse, enhancing brand recognition and loyalty among users. Additionally, meta-influencers offer brands greater flexibility and control over their messaging and brand image in the virtual space [7].

Exploration of NFTs and Digital Assets: Brands should explore NFTs and digital assets in their marketing strategies, offering unique digital products or experiences for trade or collection to enhance brand engagement and foster consumer loyalty [40, 52]. Integrating NFTs and digital assets into marketing initiatives adds value to brand interactions, attracting consumers with exclusive offerings and immersive experiences. This approach not only differentiates brands in the competitive landscape but also creates opportunities for deeper connections with audiences through the exchange and ownership of digital assets. Embracing NFTs and digital assets enables brands to tap into emerging trends and technologies, driving innovation and growth in the digital marketplace [40].

Leveraging Blockchain for Transparency: Incorporating blockchain technology into influencer marketing strategies offers a promising avenue for enhancing transparency and trustworthiness. By leveraging blockchain, brands can effectively track and verify transactions, ensuring transparency in influencer collaborations. Blockchain's decentralized nature enables immutable records of transactions, preventing data manipulation and fraud [53]. Additionally, blockchain-based smart contracts can automate payment processes, ensuring fair compensation for influencers and reducing payment disputes [54]. This transparency not only enhances trust among stakeholders but also mitigates issues such as fake follower counts and undisclosed partnerships. Moreover, blockchain enables real-time tracking of campaign performance, providing brands with valuable insights into the effectiveness of their influencer marketing efforts [55]. By embracing blockchain technology, brands can establish credibility and integrity in their influencer marketing endeavors, ultimately fostering more robust relationships with consumers and influencers [52].

Engagement through Gamification: Gamifying marketing initiatives within the Metaverse, exemplified by Beymen's "Find Bella" game, boosts user engagement and creates memorable experiences for consumers [52]. This approach aligns with the trend of gamification in marketing strategies, enhancing brand interaction and loyalty [53]. By incorporating game elements, brands can captivate audiences and foster deeper connections with their products or services [33].

Focus on Community Building and partnerships: Fostering a strong community within the Metaverse is imperative for driving increased engagement and loyalty. Brands should prioritize cultivating communities centered around their meta-influencers and virtual environments. Additionally, forming partnerships with established influencers and brands within the Metaverse is vital. By tapping into their existing audiences and expertise, brands can amplify the impact of their marketing initiatives, reaching a broader audience and achieving more significant results [56]. Collaborative efforts enable brands to leverage the influence and credibility of other entities in the Metaverse, enhancing the effectiveness of their campaigns and establishing deeper connections with their target audience. Thus, focusing on community building and forging partnerships within the Metaverse is critical to driving brand success and fostering meaningful engagement with consumers [57].

7 Conclusion

The future of the Metaverse is poised to instigate a revolutionary shift, as evidenced by the substantial investments made by established brands in Metaverse platforms and technologies. Already, the Metaverse has demonstrated its impact across various sectors, ranging from user experience to social relations, economics, and advertising, prompting a closer examination of influencer marketing within this evolving landscape. Understanding how influencer marketing unfolds in the Metaverse and the evolving dimensions and opportunities it presents is crucial amidst this transformative era. Notable developments such as Facebook's transformation into the parent company "Meta," Satya Nadella's remarks on the Metaverse's global impact, and Microsoft's monumental acquisition of Activision Blizzard underscore the Metaverse's potential to reshape future communications.

As companies and brands delve into influencer marketing within the Metaverse realm, blending influencer and Metaverse concepts to form new paradigms becomes apparent. However, these terms lack universally accepted definitions as of yet. The initiatives undertaken in this domain and the introduction of novel terminology highlight the burgeoning opportunities facilitated by influencer marketing on virtual platforms and within the Metaverse milieu. This trend signals a transition of brand-influencer and audience engagements towards these innovative platforms, marking a significant shift in the influencer marketing landscape.

Influencer marketing is a futuristic trend that continues to evolve, presenting new avenues for brands to connect with consumers in an authentic and engaging way. As the industry progresses, influencers and brands alike must adapt to these trends to stay relevant and capitalize on emerging opportunities.

References

1. Bhatia, S.: The Metaverse: Framework, Building Blocks and Market Map. Level Up Coding [Internet] (2021). Available from: https://levelup.gitconnected.com/the-metaverse-framework-building-blocks-and-market-map-3bb2ccf0241c
2. Bulut, B.: The Rise of Virtual Influencers. Digital Age Press (2022)
3. Santora, J.: influencer marketing trends to watch in 2022 [Internet] (2022). Available from: https://influencermarketinghub.com/influencer-marketing-trends
4. Corbitt, L.: Influencer marketing in 2022: Strategies + examples. The BigCommerce Blog [Internet] (2022). Available from: https://www.bigcommerce.com/blog/influencer-marketing/#what-is-influencer-marketing
5. Canavesi, B.: What is the Metaverse: Where we are and where we're headed: ATD [Internet] (2022). Available from: https://www.td.org/atd-blog/what-is-the-Metaverse-where-we-are-and-where-were-headed
6. Metafluencer. Metafluencer whitepaper [Internet] (2022). Available from: https://metafluence.com/
7. M.T.S. Meet Metafluence, An influencer-centric ecosystem in the Metaverse [Internet] (2021). Available from: https://martechseries.com/technology/meet-metafluence-an-influencer-centric-ecosystem-in-the-metaverse/
8. Trendsdesk. Meet Kyra, India's first virtual influencer (2022). Available at: https://indianexpress.com/article/trending/trending-in-india/meet-kyra-indias-first-virtual-influencer-with-nearly-100k-followers-on-instagram-7956632
9. Reaume, A.: What is the Metaverse? Meaning and what you should know [Internet] (2022). Available from: https://seekingalpha.com/article/4472812-what-is-Metaverse
10. Traders, V.: Opportunities in the metaverse. Varchev Finance [Internet] (2021). Available from: https://www.varchev.com/en/%D0%B2%D1%8A%D0%B7%D0%BC%D0%BE%D0%B6%D0%BD
11. Kilzi, M.: Council post: The New Virtual Economy of the Metaverse [Internet]. Forbes (2022). Available from: https://www.forbes.com/sites/forbesbusinesscouncil/2022/05/20/the-new-virtual-economy-of-theMetaverse/
12. Robertson, A., Peter, J.: What is the metaverse, and do ı have to care? [Internet] (2021). Available from: https://www.theverge.com/22701104/metaverse-explained-fortnite-roblox-facebook-horizon
13. Masterson, V.: We could be spending an hour a day in the metaverse by 2026. But what will we be doing there? World Economic Forum [Internet] (2022). Available from: https://www.weforum.org/agenda/2022/03/hour-a-day-in-metaverse-by-2026-says-gartner/
14. Dionisio, J.D.N., WG III, B.: G R. 3D virtual worlds and the metaverse: Current status and future possibilities. ACM Comput. Surv. (CSUR) **45**(3), 1–38 (2013)
15. Rijmenam, M.: Step into the Metaverse: How the Immersive Internet Will Unlock a trillion-Dollar Social Economy. Wiley, New York (2022)
16. Pladson, K.: Nedir bu metaverse? DW (2021)
17. Leorke, D., Leorke, D.: A Definition and Brief History of Location-Based Games (2001–2008), pp. 17–44. Location-Based Gaming: Play in Public Space (2019)
18. Yanık, C.: Metaverse Platforms and User Engagement: A Comprehensive Study. Future Tech Publications (2022)
19. Ball, M.: The Metaverse: What it is, where to find it, and who will build it [Internet] (2020). Available from: https://www.matthewball.vc/all/themetaverse
20. Zuckerberg, M.: Founder's Letter [Internet] (2021). Available from: https://about.fb.com/news/2021/10/founders-letter/
21. Barrera, K.G., Shah, D.: Marketing in the Metaverse: conceptual understanding, framework, and research agenda. J. Bus. Res. **155**, 113420 (2023)
22. Leung, F.F., Gu, F.F., Palmatier, R.W.: Online Influencer Marketing. J. Acad. Mark. Sci. **50**, 226–251 (2022)

23. Veirman, M.H.L.: Disclosing sponsored instagram posts: the role of material connection with the brand and message-sidedness when disclosing covert advertising. Int. J. Advert. **39**(1), 94–130 (2020)
24. Chen, J.: What is influencer marketing: How to develop your strategy [Internet] (2020). Available from: https://sproutsocial.com/insights/influencer-marketing/
25. Awobamise, A.: Increased consumerism in a networked-Nigeria: a study on the effect of emalls and social media on youth spending. J. Dig. Media Interact. **1**(2), 40–55 (2018)
26. Daramola, O., Abdullah, S.N.: Can i trust your word? investigating the impact of influencers' perceived credibility on consumer trust and purchase intention on TikTok. Telematics Inform. **54**, 101468 (2020)
27. Lou, C., Yuan, S.: Influencer marketing: how message value and credibility affect consumer trust of branded content on social media. J. Interact. Advert. **19**(1), 58–73 (2019)
28. Casaló, L.V., Flavián, C., Ibáñez-Sánchez, S.: Influencers on Instagram: antecedents and consequences of opinion leadership. J. Bus. Res. **117**, 510–519 (2020)
29. Hunichen J, Welte D. Navigating the New Era of Influencer Marketing: How to Be Successful on Instagram, TikTok, & Co. California Management Review. 2020;63(1):5–25.
30. Freberg, K., Graham, K., McGaughey, K., Freberg, L.A.: Who are the social media influencers? a study of public perceptions of personality. Public Relations Rev. **37**(1), 90–92 (2011)
31. Hub, I.M.: The State of Influencer Marketing 2023: Benchmark Report [Internet] (2023). Available from: https://influencermarketinghub.com/ebooks/Influencer_
32. Hutchinson, A.: New study looks at the rise of virtual influencers [infographic]. Social Media Today [Internet] (2022). Available from: https://www.socialmediatoday.com/news/new-study-looks-at-the-rise-of-virtualinfluencers-infographic/621544/
33. Başev, S.E.: Influencer marketing in the age of metaverse: beymenverse and meta-influencer bella. Int. J. Eurasian Educ. Culture **9**(25), 29–41 (2024)
34. Sheikh, M.: 30 TikTok stats marketers need to know in 2022. Sprout Social [Internet] (2022). Available from: https://sproutsocial.com/insights/tiktok-stats
35. Sobande, F.: CGI influencers: When the 'people' we follow on social media aren't human. The Conversation [Internet] (2021). Available from: https://theconversation.com/cgi-influencers-when-the-people-we-followon-social-media-arent-human-165767#:~:text=Created%20by%20tech%2Dsavvy%20teams,become%20increasingly%20attractive%20to%20brands
36. Barker, S.: Influencer marketing trends you should stay on top of. Medium [Internet] (2019). Available from: https://bettermarketing.pub/influencer-marketing-trends-you-should-stay-on-top-of-64affebf0087
37. Lee, J.K., Fortuny, E.J.: Influencer-generated reference groups. J. Consum. Res. **49**(1), 25–45 (2022)
38. Brito Silva, M.J., Oliveira Ramos Delfino, L., Alves Cerqueira, K., Oliveira Campos, P.: Avatar marketing: a study on the engagement and authenticity of virtual influencers on Instagram. Soc. Netw. Anal. Mining **12**(1), 130 (2022)
39. David, L.E.E., Won, L.S.: Nft of nft: Is our imagination the only limitation of the metaverse? J. Br. Blockchain Assoc. (2022)
40. Hwang, Y.: When makers meet the metaverse: effects of creating NFT metaverse exhibition in maker education. Comput. Educ. **194**, 104693 (2023)
41. Karabacak, Z.İ, Güngör, İ: The metaverse as influencer marketing platform: influencer-brand collaborations of Parıs Hilton with 'Superplastic', 'Bohoo', and 'Levi's.' Etkileşim **11**, 176–199 (2023)
42. Audrezet, A., Koles, B.: Virtual influencer as a brand avatar in interactive marketing. In: The Palgrave Handbook of Interactive Marketing, pp. 353–376. Springer International Publishing, Cham (2023)
43. Rathore, B.: Metaverse marketing: Novel challenges, opportunities, and strategic approaches. Eduzone: Int. Peer Rev./Refereed Multidisciplinary J. **7**(2), 72–82 (2018)
44. Chrimes, C., Boardman, R.: The opportunities and challenges of the metaverse for fashion brands. In: The Garment Economy: Understanding History, Developing Business Models, and Leveraging Digital Technologies, pp. 389–410 (2023)

45. Han, D.I.D., Bergs, Y., Moorhouse, N.: Virtual reality consumer experience escapes: preparing for the metaverse. Virtual Reality **26**(4), 1443–1458 (2022)
46. Polona, C., André, M.T., Maria, N.: Metaverse: Opportunities, Risks and Policy Implications (2022)
47. Chen, Z., Wu, J., Gan, W., Qi, Z.: Metaverse security and privacy: An overview. In: IEEE, pp. 2950–2959 (2022)
48. Fernandez, C.B., Hui, P.: Life, the metaverse and everything: an overview of privacy. In: Ethics, and Governance in Metaverse, pp. 272–277. IEEE (2022)
49. Prahara, G.A., Adler, J.: Study of digital branding as media strategic toward metaverse. In: Sustainable Development in Creative Industries: Embracing Digital Culture for Humanities, pp. 96–101. Routledge (2023)
50. Ismail AR. Virtual Shopping in The Metaverse: How Brand Engagement in Self-Concept Influence Brand Loyalty. Available at SSRN 4494222. 2023;
51. Balis, J.: Curated Experiences in the Metaverse should be provided (2022)
52. Balis, J.: How brands can enter the metaverse [Internet] (2022). Available from: https://hbr.org/2022/01/how-brands-can-enter-the-metaverse
53. O'Donoghue, O., Spurlock, A., Allen, P., Wilson, S.: A blockchain model for fair payment in influencer marketing. J. Advert. Res. **59**(3), 259–273 (2019)
54. Alam, S.S.: Smart contracts: The next big leap in the evolution of blockchain. In: 2021 IEEE 21st International Conference on Software Quality, Reliability, and Security (QRS), pp. 368–374. IEEE (2021)
55. Yuan, Y., Wang, F., Guo, Y., Qiao, Z.: Influencer selection in influencer marketing based on blockchain technology. In: 2021 IEEE 4th International Conference on Electronic Information Technology and Computer Engineering (EITCE), pp. 169–174. IEEE (2021)
56. Rogers, B.: Metaverse Marketing: The New Frontier of Digital Marketing. Forbes (2020)
57. Zhu, Z., Cui, G., Wang, G., Huang, L.: Influence of digital marketing on consumer purchase intention: a moderating role of brand credibility. J. Theor. Appl. Electron. Commer. Res. **16**(4), 1253–1272 (2021)

Influencer Marketing on New Platforms and the Metaverse for Male Grooming Brands

Nidhi Sharma, Anita Singh, and Anubha Vashisht

Abstract In today's dynamic digital landscape, influencer marketing has emerged as a pivotal strategy for brands to engage with their audience. Male grooming brands, in particular, are navigating this landscape by exploring new platforms and the burgeoning realm of the Metaverse. This abstract investigates the evolving trends, challenges, and opportunities pertaining to influencer marketing within the context of male grooming brands, specifically focusing on the integration of new platforms and the Metaverse. In exploring the Metaverse's implications, the abstract delves into the immersive and interactive nature of virtual environments. It examines how male grooming brands are leveraging the Metaverse to create innovative and experiential marketing campaigns. From virtual try-on experiences to branded virtual spaces, companies are harnessing this technology to foster deeper connections with their target audience, capitalizing on the immersive potential of this digital realm. Furthermore, the abstract outlines the challenges inherent in navigating these novel marketing landscapes. It elucidates issues related to authenticity, influencer credibility, and the complexities of integrating brand messaging within immersive digital spaces. The findings underscore the pivotal role of influencers in shaping consumer perceptions and purchasing decisions, particularly among younger demographics. Moreover, the research highlights the transformative impact of new platforms, such as TikTok, Clubhouse, and emerging social VR spaces, in redefining influencer-brand collaborations. The abstract culminates in a set of strategic recommendations for male grooming brands venturing into influencer marketing on new platforms and within the metaverse. Insights into selecting authentic influencers aligned with brand values, crafting immersive experiences, and fostering genuine engagement in virtual spaces serve as guiding principles for successful integration.

N. Sharma (✉) · A. Singh
School of Business Studies, Sharda University, Greater Noida, UP, India
e-mail: sharma.nidhi@imsuc.ac.in

A. Singh
e-mail: anita.singh2@sharda.ac.in
URL: https://scholar.google.com/citations?user=AcIc2ywAAAAJ&hl=en&oi=ao

A. Vashisht
School of Business Studies, JBS Noida, Noida, UP, India

S. Dutta et al. (eds.), *Advances in Data Analytics for Influencer Marketing: An Interdisciplinary Approach*, Information Systems Engineering and Management 9,
https://doi.org/10.1007/978-3-031-65727-6_12

Keywords Influencer marketing · Metaverse · Brand attitude · Purchase intention · Male grooming

1 Background Influencer Marketing

In social media, influencers are commonly labelled as 'opinion communities,' particularly those who wield persuasive influence over specific segments of society. Through the sharing of personal experiences and perspectives, influencers actively engage with their audience, fostering discussions and ensuring the prominence of their viewpoints. In the contemporary landscape of consumerism, influencers are perceived to embody the traits of opinion leaders on social media platforms, effectively supplanting traditional opinion leaders. This shift is a direct result of advancements in mass media, with opinion leaders now making their presence felt predominantly in online realms following the decline of traditional outlets such as newspapers, radio, and television. Influencers may encompass globally recognized celebrities with extensive fan bases or individuals who amass trust and a substantial following by imparting noteworthy information. Presently, there appears to be a distinction in the methods employed by renowned influencers and those who achieve prominence through their influencer status. Content creators are tasked with consistently generating fresh content, employing marketing strategies, and adjusting their posts to suit diverse social media platforms. This continual effort aims to elicit increased requests for posts from followers, thereby sustaining a strong bond between the influencers and their audience. Lou Yuan characterizes influencers as "content creators attaining celebrity status."

Li underscored the importance of certain attributes for internet celebrities, the influencers of today. These attributes include attaining a specific number of followers, ensuring substantial engagement, and demonstrating future business potential. Li proposes five distinct inferences regarding internet celebrities, encompassing the following aspects: CV (consistently producing creative and compelling content), technology value (keeping abreast of technological advancements parallel to those of social media luminaries), emotional value (fostering a sense of intimacy with followers without creating pressure or distance), and symbolic value (leading followers and enhancing the likelihood of being followed by projecting a distinctive image).In the historical evolution of marketing, the emergence of influencer marketing has dramatically reshaped the landscape, and nowhere is this transformation more evident than in the realm of grooming brands targeted towards males. The journey of influencer marketing within the male grooming industry is a captivating narrative that spans the transition from traditional advertising to the digital age, emphasizing relatability, authenticity, and the power of personal connection.

Historically, grooming brands for men relied heavily on traditional advertising channels, including television commercials and print media. Iconic figures such as Cary Grant and James Dean were often featured in campaigns, embodying a conventional image of masculinity. However, the rise of influencer marketing can be traced back to the advent of social media, particularly platforms like Instagram and YouTube, which provided a new avenue for content creators to connect with audiences in a more direct and personal way.

The male grooming industry witnessed a paradigm shift as influencers, often experts in grooming and skincare, began to gain prominence. These influencers, ranging from barbers and hairstylists to skincare enthusiasts, leveraged social media to share personal grooming routines, product reviews, and styling tips. The authenticity and relatability of these influencers struck a chord with their followers, fostering a sense of trust and connection that traditional advertising struggled to achieve.

One pivotal moment in the history of male grooming influencer marketing was the rise of grooming experts on YouTube. Influencers like Aaron Marino, whose channel Alpha M focuses on men's grooming and style, played a crucial role in bridging the gap between brands and consumers. Marino's down-to-earth approach and genuine passion for grooming resonated with a broad audience, positioning him as a trusted authority in the male grooming space.

As social media platforms continued to evolve, male influencers diversified across various grooming niches. Skincare experts like Hyram Yar bro became influential voices, providing valuable insights and recommendations to their audiences. Barbers and hairstylists like Chris Bossio showcased their skills and product preferences, contributing to the growing trend of authentic grooming content.

The shift towards authenticity and relatability in male grooming influencer marketing had a profound impact on societal perceptions of masculinity. Influencers challenged traditional norms by openly discussing grooming and self-care, creating a more inclusive narrative that encouraged men to embrace personal grooming as an essential aspect of well-being.

Brands quickly recognized the potential of collaborating with male influencers to reach their target audience more effectively. These partnerships often involved influencers creating content that seamlessly integrated grooming products into their daily routines, showcasing the practical benefits in a way that felt genuine and approachable.

The historical journey of male grooming influencer marketing is not without its challenges. The increasing demand for transparency and authenticity has required influencers and brands to navigate carefully, ensuring that collaborations align with the values and expectations of their audience. Influencers who maintain credibility and sincerity in their recommendations continue to thrive in an environment where trust is paramount.

In conclusion, the historical background of influencer marketing within the male grooming industry is a compelling narrative of evolution, from traditional advertising to the digital age of authenticity and relatability. Male influencers have played a pivotal role in reshaping societal attitudes towards grooming, fostering a positive and inclusive discourse. As the industry continues to evolve, the collaboration between grooming brands and male influencers remains a powerful force, shaping not only consumer choices but also contributing to a broader cultural shift in the perception of masculinity and self-care (Fig. 1).

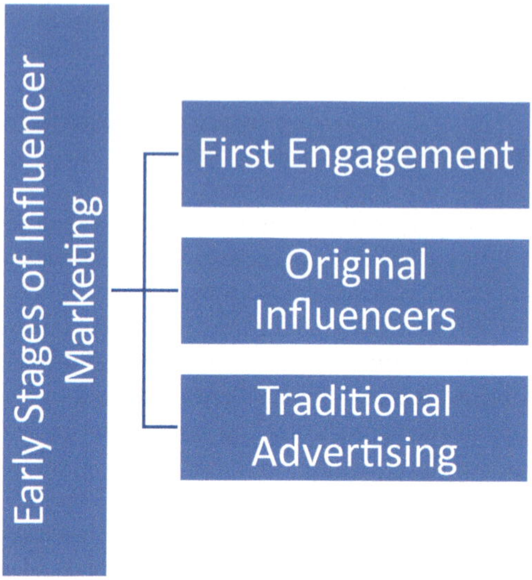

Fig. 1 Early stages of influencer marketing. *Source* Author compilation

Early Stages of Influencer marketing

First Engagement—In the early days, influencer marketing primarily consisted of brand endorsements by celebrities. Simple product placements and endorsements were the norm, lacking the personal touch that modern influencer partnerships bring.

Original Influencers—Early influencers were often industry experts or thought leaders, not the social media personalities they see today. They conveyed authority through their expertise, rather than personal appeal or relatability.

Traditional Advertising—Initial influencer collaborations were more akin to traditional advertising, with a focus on product features and sales pitches rather than authentic storytelling and genuine connections.

Evolution of Influencers Platform

Influencer marketing has evolved significantly over the years, transforming from a novel approach to a mainstream marketing strategy. It has paved the way for brands to reach their target audiences in a more authentic and engaging manner. As social media usage continues to soar, influencer marketing has become a pivotal tool for brands to connect with consumers on a personal level.

1. **Diverse Platforms**

   From niche forums to global platforms, influencer marketing expanded to include a variety of social media sites, catering to different demographics and interests.
2. **Advanced Analytics**

Platforms integrated comprehensive analytics, allowing brands to track campaign performance, determine ROI, and refine their influencer strategies based on data-driven insights.

3. **Regulatory Compliance**

   As the industry evolved, platforms enforced stricter rules and guidelines to ensure transparency, authenticity, and ethical practices in influencer partnerships.

Rise of Social Media Influencers

The rise of social media influencers has profoundly influenced the marketing landscape, and nowhere is this more evident than in the realm of male grooming brands. Over the past decade, the grooming industry has experienced a transformative shift as a new generation of male influencers took center stage on various social media platforms, revolutionizing the way grooming products are promoted and consumed.

The evolution of male grooming influencers can be traced back to the advent of platforms like Instagram, YouTube, and later, TikTok. These platforms provided a stage for individuals with a passion for grooming, skincare, and overall self-care to share their insights, tips, and product recommendations with a global audience. Unlike traditional advertising, these influencers offered a more personal and relatable approach, establishing a connection with their followers that transcended the conventional consumer-brand relationship.

One of the key factors driving the rise of male grooming influencers was the changing cultural landscape around masculinity. Influencers challenged stereotypical notions of what it means to be a man, fostering a more inclusive understanding of grooming and self-care. The stigma surrounding male grooming began to dissipate as influencers openly shared their grooming routines, skincare journeys, and experiences, creating a space for men to embrace and celebrate their personal care rituals.

As grooming influencers gained popularity, their impact on consumer behaviour became increasingly evident. These influencers were not just promoting products; they were building communities centered around self-improvement, confidence, and embracing one's individuality. The authenticity and transparency exhibited by male grooming influencers resonated with audiences, leading to a significant shift in how consumers discovered, evaluated, and purchased grooming products.

The rise of male grooming influencers also brought diversity to the forefront. Influencers from various ethnicities, backgrounds, and age groups shared their unique perspectives on grooming, ensuring that the content catered to a wide range of audiences. This diversity not only enriched the content landscape but also allowed brands to connect with a more comprehensive and representative consumer base.

Platform Emergence

- Social media platforms like Instagram, YouTube, and TikTok gained immense popularity, giving rise to a new breed of influencers who gained followers by sharing personal experiences and authentic content.

Engagement Revolution

- Engagement metrics became significant as brands realized the potential of influencers in driving conversations and interactions with their audience, leading to the birth of influencer marketing campaigns as we know them today.

YouTube played a pivotal role in the ascent of male grooming influencers. Channels dedicated to men's grooming routines, product reviews, and styling tips garnered millions of subscribers. Influencers like Alpha M (Aaron Marino), BluMaan (Joe Andrews), and Jose Zuniga (Teaching Men's Fashion) became authorities in the male grooming space, shaping trends and influencing purchasing decisions.

The advent of Instagram further intensified the influence of grooming influencers. The platform's visual nature allowed influencers to showcase the transformative power of grooming products through compelling images and short videos. Instagram also facilitated collaborations between influencers and brands, leading to the creation of sponsored content that seamlessly integrated grooming products into the influencers' narratives.

With the rise of TikTok, a new wave of male grooming influencers emerged, leveraging the platform's short-form, engaging content format. TikTok's algorithm-driven discoverability allowed influencers to rapidly gain followers, expanding their reach and impact. Bite-sized grooming tips, product demonstrations, and quick styling tutorials became viral sensations, propelling influencers to new heights of fame within the male grooming community.

However, the success of male grooming influencers has not been without challenges. Maintaining authenticity in a saturated market, navigating the disclosure of sponsored content, and addressing the increasing demand for transparency are ongoing considerations for both influencers and brands.

In conclusion, the rise of social media influencers in the male grooming industry represents a paradigm shift in marketing. These influencers have not only transformed how grooming products are promoted but also contributed to a broader cultural shift in attitudes towards self-care and masculinity. As male grooming influencers continue to evolve with the ever-changing landscape of social media, their impact on the grooming industry and societal perceptions is likely to endure, shaping the future of male grooming marketing for years to come.

2 Journey from Celebrity Endorsement to Influencer

Brands have used celebrities to sell products since long before athletes first appeared on cereal boxes. But influencer marketing and celebrity endorsements are not quite the same thing. Celebrity endorsements typically involve a company making a huge investment in someone, but it's hard to specify the exact return on that investment. With influencer marketing, it's easier to figure out the ROI because companies can closely monitor likes, shares, online conversations, and so forth.

For decades, the realm of personal care and beauty marketing existed firmly in the pink and purple domain. Ads focused on women, showcasing flawless faces and cascading locks. Men, meanwhile, were relegated to rugged aftershaves and deodorant commercials featuring stoic athletes. But like a perfectly sculpted beard, things have grown and evolved. Enter the revolution of influencer marketing, and with it, a seismic shift in how personal care products are marketed to men. Gone are the days of one-dimensional masculinity. Today's male consumer is multifaceted. He cares about his appearance, his well-being, and the message he projects. And who better to guide him through this newfound journey of self-care than relatable, authentic voices on social media? Influencers—the modern barbers, stylists, and confidantes—have emerged as trusted friends, sharing not just product recommendations, but personal experiences, vulnerabilities, and tips for navigating the often-confusing world of personal care. This journey hasn't been linear. From early stumbles with celebrity endorsements to the rise of niche micro-influencers, the evolution of influencer marketing for men's personal care has been fascinating. We've witnessed the normalization of daily skincare routines, the embrace of makeup, and the exploration of products far beyond the razor and shaving cream. And through it all, influencers have been the champions, fostering communities, dismantling outdated stereotypes, and building genuine connections with male audiences.

What is Micro Influencer?

You might assume that brands want to work only with macro influencers, but that's not always the smartest business decision. Micro influencers might produce more relatable content than their macro counterparts, who use professional photographers, stylists, makeup artists, and all kinds of consultants to put a post together. While macro influencers may advertise a lifestyle that seems unattainable to many people, micro influencers, on the other hand, can represent accessibility to their followers (Fig. 2).

Of course, there are obvious benefits to working with macro influencers: with their mass followings, they can help brands reach hundreds of thousands or even millions of people.

The Study talks about males grooming brands- For decades, the face of men's grooming was sculpted from Hollywood clay and projected onto cinema screens. Chiseled actors hawked cologne with smoldering gazes, athletically-blessed heroes flexed in deodorant ads, and rugged outdoorsmen battled mythical beasts (figuratively, of course) while lathering up with shaving cream. These larger-than-life figures were aspirational, yes, but also distant, their perfectly coifed hair and chiseled features seeming more attainable to gods than average mortals. Then, the 21st century rolled in, bringing with it a cultural upheaval that saw the dethroning of these celebrity titans and the rise of a new breed of heroes in the kingdom of male grooming: the influencers.

Influencers come in five different forms:

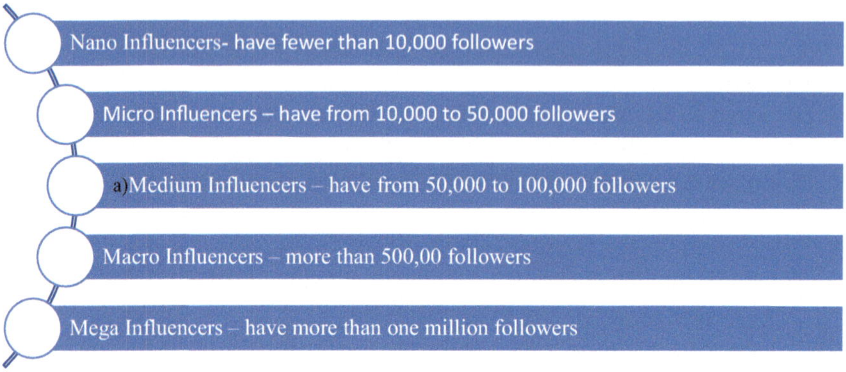

Fig. 2 Different forms of influencers *Source* Author compilation

The shift was subtle at first. Whispers of bearded bloggers brewing artisanal beard oils in their kitchens filled the internet, YouTube tutorials on skincare routines went viral, and Instagram feeds became showcases for impeccably sculpted eyebrows. These weren't Hollywood's golden gods, they were regular guys (and even some not-so-regular guys) sharing their personal journeys with grooming, demystifying the process, and making it relatable. Their voices were authentic, their tips grounded in everyday life, and their imperfections endearingly human. In a world saturated with glossy magazine perfection, these influencers offered a breath of fresh air, proving that good grooming wasn't reserved for sculpted superheroes.

Social media became the fuel that propelled this shift. Brands no longer needed the star power of a single A-lister to reach their target audience. They could strategically engage micro-influencers, men with dedicated communities who resonated with specific demographics or interests. The burly, bearded lumberjack influencer reached the outdoorsman, the tattooed barber appealed to the hipster crowd, and the fitness enthusiast inspired gym warriors to take care of their skin after pushing iron. This targeted approach allowed for nuanced messaging, building trust and loyalty through shared experiences and values.

Celebrity endorsements weren't entirely obsolete, though. They retained their allure for mass-market campaigns, instantly grabbing attention and boosting brand awareness. But their impact was no longer singular. Influencers, like a well-coordinated grooming squad, created a collective voice, amplifying messages and weaving them seamlessly into the fabric of everyday conversations. They weren't simply selling a product, they were sharing a lifestyle, inspiring a sense of belonging, and fostering a community where men could openly discuss their insecurities and aspirations around self-care.

As a result—a seismic shift in the male grooming landscape. Products once relegated to the "hygiene" aisle transformed into tools for self-expression. Skincare went beyond shaving cream, makeup became a way to enhance features, and even

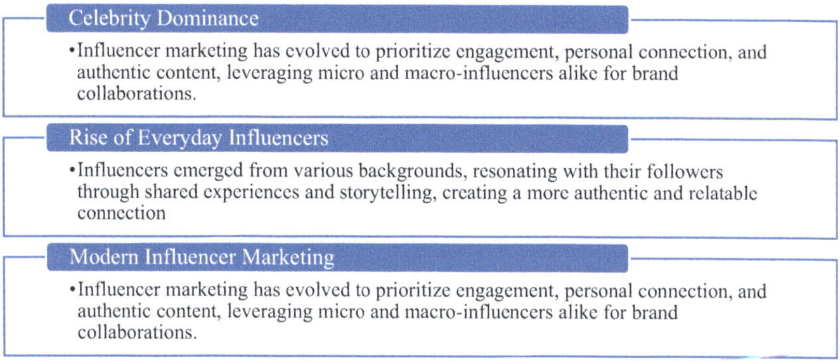

Fig. 3 From celebrity endorsement to influencer. *Source* Author compilation

hair colouring, once considered effeminate, became a badge of individuality. Influencers normalized these choices, showcasing their own routines and encouraging men to experiment without fear of judgement. But the journey is far from over. The influencer landscape is constantly evolving, with new platforms and personalities emerging every day. Brands must keep pace, diversifying their influencer choices to reflect the ever-changing demographics of their target audience. Authenticity remains paramount, and brands that partner with influencers who feel genuine and relatable will thrive.

From Celebrity Endorsement to Influencer

This shift is more than just a marketing trend; it's a reflection of a deeper cultural change. Men are increasingly comfortable exploring their self-image, taking ownership of their appearance, and prioritizing their well-being. Influencers, not unlike the barbers of old, have become confidantes and guides, helping men navigate this new terrain. The journey from celebrity endorsers to influencers has been one of democratization, authenticity, and self-discovery. It's a journey that's still unfolding, and one that promises to further blur the lines between masculinity and self-care, one perfectly sculpted eyebrow at a time (Fig. 3).

3 Maturation—Influencer Marketing Grows up from Hype to Hub (Authenticity to Algorithm)

Remember the early days of influencer marketing? Sponsored Instagram posts filled with #ad hashtags, unboxing videos screaming excitement, and product placements so blatant they felt like awkward blind dates. It was a Wild West of content creation, fueled by inflated follower counts and the promise of overnight fame. But just like a teenager sheds braces and discovers their unique style, influencer marketing has matured, evolving into a multifaceted tool capable of fostering genuine connections and driving measurable results.

This maturation journey is paved with several key milestones. Transparency took front stage, replacing shady endorsements with clear disclosures and honest reviews. Consumers, bombarded with marketing messages, craved authenticity. Influencers who shared their genuine experiences, vulnerabilities, and opinions, not just polished scripts, garnered trust and loyalty. They became friends, confidantes, and guides, navigating specific niches and building dedicated communities around shared interests.

- **Quality eclipsed quantity**: Micro-influencers with targeted audiences emerged as powerhouses, their deep connections and nuanced messaging proving far more effective than the megastar approach. Brands discovered the magic of partnering with experts, hobbyists, and passionate individuals who resonated with specific buyer personas. A fitness instructor promoting protein powder resonated far more than a Hollywood A-lister.
- **Metrics moved beyond vanity figures**: Likes and follower counts faded as brands embraced sophisticated analytics and engagement metrics. Measuring brand sentiment, website traffic, and actual sales provided a clearer picture of influencer effectiveness. It was no longer just about clicks and shares, but about understanding the impact of influencer recommendations on consumer behaviour.
- **Content diversified**: Ditching the one-size-fits-all approach. Interactive live streams, educational tutorials, insightful reviews, and collaborative co-creations became the norm. Influencer expertise was harnessed to develop products, craft compelling narratives, and tailor campaigns to specific platforms and trends. The focus shifted from simply promoting products to creating engaging experiences that resonated with audiences on a deeper level.

Impact of Influencer Marketing on Brands

Brand Awareness—Influencer endorsements greatly enhanced brand visibility, introducing products to a wider audience and generating buzz about new launches (Fig. 4).

Consumer Trust—Authentic and relatable content from influencers fostered trust among consumers, leading to increased brand loyalty and positive sentiment towards products.

Fig. 4 Impact of influencer marketing on brands. *Source* Author compilation

Sales Conversion—Influencers played a crucial role in driving purchasing decisions, with their recommendations and reviews directly impacting consumer buying behaviour.

This evolution is not without its challenges. **Navigating the ever-changing social media landscape** requires constant adaptation and nimbleness. Ensuring diversity and inclusivity remains crucial, as influencer communities need to reflect the multifaceted nature of our world. Combating influencer fraud and inauthenticity requires brands to be vigilant and prioritize genuine partnerships.

Despite these challenges, the future of influencer marketing is bright. Brands that embrace authenticity, cultivate trusted partnerships, and leverage influencer expertise will stand out in the crowd. It's no longer about quick wins and fleeting trends, but about building long-term relationships, fostering brand loyalty, and driving meaningful engagement. Influencer marketing, once a noisy teenager, has grown into a mature partner, ready to contribute meaningfully to the marketing landscape.

Remember, the influencer is not just a voice, but a hub, connecting brands with real people who are hungry for authentic connections and genuine experiences. In this era of information overload, it's not about shouting the loudest, but about whispering to the right ears, building trust, and sparking conversations that resonate. Influencer marketing, in its mature form, is poised to do just that, becoming a vital tool for brands to navigate the future, one compelling story at a time.

The study talks about Men's grooming used to be shrouded in secrecy, whispered tips traded amongst friends and fathers. The marketing? An awkward dance between hyper-masculine cologne ads and embarrassing "hygiene product" aisle placements. Then came the internet, and with it, a revolution: influencer marketing. Initially, the landscape was akin to a frat party bathroom; boisterous personalities spouting dubious product claims over shaky phone videos. But just like that messy party evolves into a sophisticated gathering with craft cocktails and artisanal cheese, influencer marketing for men's grooming has matured, becoming a nuanced and effective tool for reaching discerning consumers.

Gone are the days of generic "gym bro" influencers hawking everything from protein powder to hair wax. Modern brands understand the power of micro-influencers and niche communities. The burly powerlifter promotes functional shaving cream, the bearded barber demonstrates beard oil application techniques, and the tattooed artist showcases coloured hair products with artistic flair. Each influencer speaks to a specific tribe, building trust through shared experiences and genuine passion.

This shift reflects a deeper cultural change. Men are no longer beholden to outdated notions of masculinity. They're embracing self-care, experimenting with their appearance, and prioritizing well-being. Influencers have become trusted guides in this new terrain, normalizing these choices and fostering a sense of community. Think of them as modern-day barbers, swapping out scalpels and razors for tutorials and reviews, creating spaces where men can openly discuss skincare routines and product recommendations. The metrics have evolved too. It's no longer just about likes and follower counts. Brands are looking for engagement, authenticity, and brand alignment. They collaborate with influencers on co-created content, host interactive live streams, and leverage influencer expertise to develop new products. This deeper partnership cultivates genuine connections with consumers, driving brand loyalty and sales.

Let's take the example of Harry's, a razor brand that rose to prominence through influencer marketing. They eschewed celebrity endorsements, partnering with barbers, comedians, and everyday guys passionate about shaving. Harry's content wasn't just product demos; it was funny, relatable, and tackled topics like shaving anxiety and body image. This authenticity resonated with men, resulting in a loyal following and exponential growth. However, maturation doesn't mean perfection. Challenges remain. Transparency is paramount, and brands must ensure influencers disclose sponsored content clearly. Diversity and inclusion are crucial, as niche communities within male grooming deserve representation. Finally, navigating the ever-evolving social media landscape requires agility and adaptability.

The journey from bro-science to beard balm is still ongoing. But one thing is clear: influencer marketing for men's grooming has shed its awkward adolescence and evolved into a sophisticated tool for reaching modern consumers. By embracing trust, authenticity, and a nuanced understanding of men's evolving needs, brands can build genuine connections and cultivate communities that extend far beyond the product itself. The future of men's grooming, then, isn't just about flawless beards and sculpted abs; it's about fostering self-confidence, encouraging self-expression, and redefining what it means to be a man in a world where taking care of yourself is no longer a secret, but a celebration.

4 The Current Landscape of Influencer Marketing

Influencer marketing, once a fringe tactic of the early internet, has become a behemoth in the modern marketing landscape. Brands scramble to harness the power of social media personalities, hoping to tap into their engaged audiences and drive sales. But the influencer game is no longer a simple matter of throwing free products at celebrities and waiting for the magic to happen. Today's landscape is complex, nuanced, and constantly evolving, demanding from brands both strategic agility and a deep understanding of consumer psyche.

One of the defining trends is the rise of micro-influencers and niche communities. Gone are the days of chasing mega-star endorsements. Consumers crave authenticity and relatability, finding it in voices closer to their own. The tattooed barber resonates with the alternative crowd, the fitness enthusiast inspires gym-goers, and the skincare blogger guides beauty enthusiasts with targeted advice. This hyper-specific approach allows brands to reach highly engaged audiences and build genuine connections that translate into sales. Content reigns supreme. Forget the polished, scripted posts of the past. Today's consumers want raw, engaging experiences. Live streams, tutorials, reviews, and user-generated content are king. Consumers want to peek behind the curtain, see the real person behind the influencer persona, and experience the product in action. This shift demands from brands a willingness to collaborate, co-create, and cede some control to the influencer's creative vision.

Diverse Influencer Ecosystem—The current landscape encompasses a wide array of influencers, from micro-influencers with niche followings to macro-influencers with broad reach.

Brand Collaboration—Brands are adopting multi-faceted collaboration strategies, integrating influencers into various marketing campaigns, from product launches to large-scale events.

Transparency is king—Consumers are savvy, skep-tical, and unforgiving of inauthenticity. Brands and influencers who fail to disclose sponsored content risk facing backlash and losing trust. Clear labelling, honest reviews, and a willingness to engage in open dialogue are crucial for building long-term relationships with audiences. Data and analytics are driving the narrative. Influencer marketing is no longer just about gut feelings and follower counts. Sophisticated analytics tools track engagement, sentiment, and conversions, providing brands with invaluable insights into campaign effectiveness. Data helps determine which influencers resonate best with their target audience, what type of content performs best, and where to allocate resources for maximum impact.

Fake followers—One of the major challenges was the prevalence of fake followers and engagement, leading to misrepresentation of an influencer's reach and impact.

Transparency Issue—The lack of transparent disclosure about paid partnerships and sponsored content led to credibility issues, undermining the trust between influencers and their audience.

However, this dynamic landscape also presents challenges. Navigating the ever-changing algorithms **of social media platforms** can be a constant struggle. Staying ahead of emerging trends and adopting new platforms demands nimble adaptation. And of course, combating influencer fraud and inauthenticity remains a persistent issue, requiring brands to conduct thorough due diligence and prioritize partnerships built on trust and shared values.

The Males grooming brands—The rise of influencer marketing has revolutionized how brands connect with male consumers, forging a path towards self-expression, authenticity, and a redefined masculinity. The landscape is now segmented, a bustling bazaar where micro-influencers reign supreme. Barbers with their tips and tricks, fitness enthusiasts showcasing workout-proof skincare, and fashion bloggers sporting effortlessly sculpted beards—these are the voices resonating with specific tribes within the male audience. This targeted approach allows brands to speak directly to the passions and insecurities of their core demographics, building trust and fostering community beyond the fleeting appeal of a celebrity endorsement. Content has transcended the polished perfection of yore. Now, authenticity holds the crown. Tutorials bursting with real-life tips, live streams peppered with honest reviews, and user-generated content showcasing unfiltered product experiences—these are the narratives captivating consumers. Brands that collaborate, co-create, and cede some control to the influencer's creative vision unlock a treasure trove of genuine engagement. The audience wants to see the man behind the beard, the real person grappling with blemishes and experimenting with hair colours. This raw vulnerability creates a space for open dialogue and fosters genuine connections.

Transparency is not merely a buzzword, it's a lifeline. Today's consumers are savvy, skep-tical, and unforgiving of inauthenticity. Clear disclosure of sponsored content is a non-negotiable, while honest reviews and open conversations build trust, the currency of influence. The days of veiled product placements are over; transparency is the foundation upon which long-term relationships with audiences are built. But this dynamic playground also facing challenges. Algorithms shift like desert sands, demanding nimble adaptation and constant content reinvention. Emerging platforms blur the lines, requiring brands to explore beyond the familiar comfort of Instagram and TikTok. And of course, the area of fraudulent influencers lurks ever present, necessitating careful due diligence and partnerships built on shared values. Data and analytics hold the key to unlocking insights. Understanding engagement metrics, sentiment analysis, and conversion rates enables brands to fine-tune their strategies, identify the most impactful partnerships, and optimize campaigns for maximum reach and resonance. By leveraging data, brands can move beyond guesswork and intuition, crafting influencer campaigns with laser-like precision.

Despite these challenges, the future of influencer marketing is bright. Brands that embrace authenticity, leverage data, and prioritize long-term partnerships stand to thrive. It's about more than just pushing products; it's about building communities, fostering conversations, and creating engaging experiences that resonate with consumers on a deeper level. Influencer marketing, in its evolved form, is poised to become a cornerstone of modern marketing, connecting brands with real people in a way that resonates and drives sustainable growth.

Future Trends in Influencer Marketing

Here are some key trends that are poised to shape the way brands and influencers connect in the years to come:

1. **Hyper-personalization and Micro-influencers Take Center Stage**: The era of generic campaigns targeting broad audiences is fading. Consumers crave personalized experiences, and brands are turning to hyper-local and niche micro-influencers to reach these specific communities. Expect to see partnerships with experts in even the most specialized fields, from sustainable fashion advocates to financial literacy gurus (Fig. 5).

2. **Beyond Sponsored Posts: Embracing Co-creation and Authentic Storytelling**: Consumers are tired of formulaic content. Brands will increasingly work with influencers as co-creators, developing engaging narratives and product lines together. Think collaborative capsule collections, interactive live streams, and behind-the-scenes glimpses into the influencer's creative process. Authenticity will be paramount, with audiences seeking relatable stories and real experiences.

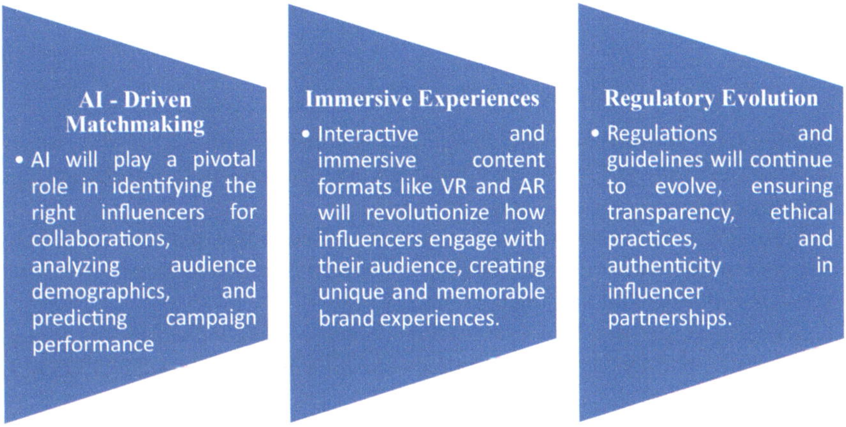

Fig. 5 Future trends in influencer marketing. *Source* Author compilation

3. **Data-driven Decisions and Performance Measurement**: Metrics will move beyond vanity likes and follower counts. Advanced analytics will track sentiment, brand awareness, and conversion rates, allowing brands to measure the true impact of influencer campaigns. Expect sophisticated attribution models and AI-powered insights to guide influencer selection and campaign optimization.
4. **Rise of Virtual Influencers and the Blurring of Reality and Digital**: Get ready for the era of impeccably curated digital personalities influencing trends and driving brand engagement. These AI-powered virtual influencers offer brands a new canvas for storytelling, allowing them to explore fantastical narratives and push creative boundaries. Expect ethical considerations and concerns about authenticity to be at the forefront of this emerging trend.
5. **Live Commerce and Interactive Experiences Take the Lead**: The lines between social media and e-commerce will continue to blur. Live-streaming platforms will become powerful sales tools, with influencers showcasing products in real-time and interacting directly with viewers. Augmented reality and virtual try-on technologies will further enhance the interactive shopping experience.
6. **Focus on Social Responsibility and Purpose-Driven Partnerships**: Consumers are increasingly conscious of brand values and seek connections with influencers who champion social causes and environmental sustainability. Expect to see more partnerships focused on promoting positive change, with brands and influencers collaborating on educational campaigns and supporting worthy initiatives.
7. **Trust and Transparency Remain Cornerstones**: In a world saturated with information, trust will be the ultimate currency. Transparency will be non-negotiable, with clear disclosure of sponsored content and authentic interactions being essential for building lasting connections with audiences. Brands and influencers who prioritize genuine relationships over quick wins will emerge as leaders in the future landscape.

5 Conclusion

The conclusion comes with three point as follows:

Authenticity Importance—Authentic and genuine content will be the cornerstone of successful influencer marketing, driving trust and long-term relationships between influencers, brands, and consumers.
Adaption and Innovation—Continuous adaptation to new platforms and technologies, along with innovative storytelling methods, will be crucial for staying relevant and impactful in influencer marketing.
Collaborative partnership—Successful influencer marketing will rely on strong, collaborative partnerships between brands and influencers, built on mutual respect, shared values, and aligned objectives.

The future of influencer marketing is dynamic and exciting. By embracing these trends and focusing on authenticity, data-driven insights, and meaningful experiences, brands can forge genuine connections with consumers and unlock the true potential of this powerful marketing tool. The future is about collaboration, creativity, and a shared pursuit of building a more transparent and impactful ecosystem for brands and influencers to thrive together.

Bibliography

1. Alldredge, K., et al.: How US consumers are feeling, shopping, and spending—and what it means for companies. In: The Global Economy. McKinsey & Company (2022)
2. Arora, Arun, et al. "Reimagining marketing in the next normal." McKinsey Article (2020).
3. Charumilind, S., et al.: When Will the COVID-19 Pandemic End? March 2022 update. McKinsey & Company (2022). https://www.mckinsey.com/industries/healthcare-systems-and-services/our-insights/when-will-the-covid-19-pandemic-end (Erişim Tarihi: 17 Mayıs 2022)
4. Crystal Gazing: Trends Shaping the Future of Influencer Marketing. https://adage.com/article/digital-marketing-ad-tech-news/ai-chatbots-predict-2024s-hottest-trends-social-media-and-ad-tech/2534701
5. Esber, D., Wong, J.: Discussions in digital: Influencer marketing is ready for its close-up. McKinsey Co. (2020). Available online at: https://www.mckinsey.com/~/media/McKinsey/BusinessFunctions/MarketingandSales/OurInsights/DiscussionsindigitalInfluencermarketingisreadyforitscloseup/Discussions-in-digital-Influencer-marketing-is-ready-for-its-close-up.pdf
6. Fowler, K., Thomas, V.L.: Influencer marketing: a scoping review and a look ahead. J. Mark. Manag.Manag. **39**(11–12), 933–964 (2023)
7. From Bro-science to Beard Balm: The Maturation of Influencer Marketing in Men's Grooming. https://www.researchgate.net/publication/273022774_From_Subservient_Chickens_to_Brawny_Men
8. Gunday, G., et al.: How European Shoppers Will Buy Groceries in the Next Normal. McKinsey & Company (2020)
9. Hazan, E., et al.: Marketing in the metaverse: an opportunity for innovation and experimentation. McKinsey Q. (2022)

10. How Influencer Marketing Will Evolve in 2024. https://impact.com/influencer/influencer-marketing-trends-to-watch-in-2023/
11. Influencer Marketing Grows Up: From Hype to Hub, Authenticity to Algorithm. https://digiday.com/media-buying/marketers-see-potential-in-googles-rumored-ai-infused-search-overhaul/
12. Influencer Marketing Playbook for the Men's Grooming Industry—Affable.ai. https://www.affable.ai/blog/influencer-marketing-for-mens-grooming-brands
13. Karabacak, Z.İ., Güngör, İ.: The metaverse as influencer marketing platform: influencer-brand collaborations of Parıs Hilton with 'Superplastic','Bohoo', and 'Levi's'. Etkileşim **11**, 176–199 (20230
14. Men's grooming brands that nailed their influencer marketing strategy—ThoughtLeaders. https://www.affable.ai/blog/influencer-marketing-for-mens-grooming-brands
15. The Current Landscape of Influencer Marketing. https://adage.com/article/digital-marketing-ad-tech-news/ai-chatbots-predict-2024s-hottest-trends-social-media-and-ad-tech/2534701
16. Vidani, J., Das, S.G.: A review on evolution of social media influencer marketing: reflection on consumer behaviour and consumer's decision-making process. Turkish Online J. Qual. Inquiry **12**(9) (2021)
17. Weekly Shorts: Building a robust men's grooming brand with influencer marketing. https://www.affable.ai/blog/influencer-marketing-for-mens-grooming-brands.

Impact of Influencer Marketing Through Instagram Among Youth: A Study

Pooja Verma and Monica Sainy

Abstract In the era of digital marketing Influencer marketing became a trending technique for the brands. From the use of a celebrity to an influencer, marketing techniques are getting more advanced and different every day. Nowadays Instagram is a very common and effective platform for influencers to do influencer marketing. Social media has developed into a widely used tool for businesses to establish their brands. Influencer marketing is an emerging field that has shown to be a successful means of engaging youth on social media. Here, businesses hire social media celebrities to leverage their sizable fan bases to advertise goods and services on their behalf. Although influencer marketing is associated with word-of-mouth marketing, it does not have a theoretical basis in the academic discussion. This study attempts to evaluate the impact of Influencer Marketing through Instagram among youth. This study also found the different factors which affects the Influencer marketing among youth and how it affects different genders through Instagram.

Keywords Influencer marketing · Instagram · Social media marketing · Marketing communications · Branding · Media strategy

1 Introduction

Influencer marketing is a strategy used by companies to promote their products on the internet by acquiring endorsements or recommendations from influencers and content creators. Most influencers may be found on social media and blogs, but other producers, such as podcast broadcasters are also explored as influencers. According to the definition of influencer marketing "A non-promotional strategy to marketing in which businesses focus their efforts on opinion leaders, rather than direct target

P. Verma
Shri Vaishnav Vidyapeeth Vishwavidyalaya, Indore, India

M. Sainy (✉)
Amity Business School, Amity University, Raipur, Chhattisgarh, India
e-mail: monica.sainy1@gmail.com; msainy@rpr.amity.edu

S. Dutta et al. (eds.), *Advances in Data Analytics for Influencer Marketing: An Interdisciplinary Approach*, Information Systems Engineering and Management 9,
https://doi.org/10.1007/978-3-031-65727-6_13

marketing touch points" [1]. Influencer marketing is typically used to raise brand recognition, but it may also lead to many conversions and purchases. Instagram has rapidly become the trending medium for business owners who are trying to reach new customers, and influencer marketing on the network has become a popular method of doing so. Instagram Influencers are Instagram users with a large following and established reputation, as well as the ability to influence others due to their sincerity and authenticity. The individuals who promote the brand hashtag and have the most followers are the brand's influencers. Social media influencers are a new breed of independent, third-party advocates that use their online personas to sway the opinions of their audience. Social media influencers use social media platforms to produce content that promotes specific brands in an effort to gain recognition and a following.

An influencer is a type of social micro-celebrity; they don't always need to be journalists, bloggers, or celebrities. Social media's ascent has given marketers a new avenue to interact with customers. Through social media, people follow the lives of influencers, and when done well, influencer marketing can appear more organic than commercial. The influencer marketing industry has evolved over the years into a recognized form that can be divided into four groups: mega-influencers, macro-influencers, micro-influencers, and nano-influencers which are based on the number of their followers. It would appear that the only difference between these four groups is the size of their fan bases. The differences are actually much more subtle than that, though.

According to Lin et al. [2] among the most well-known and widely used social networking sites available today are Facebook, Twitter, YouTube, and Instagram. More than ever, buyers are consulting other buyers to help them make decisions about what to buy. Influencer marketing today is usually managed from the console of just one online personal's digital device. Influencer marketing management firm Mavrk defined influencers as online personals that have the power to affect their followers with their real or perceived authority, knowledge, position, or relationship. Instagram is currently leading the market in terms of reach and scope among online photo sharing services because of the app's visual nature and high rate of user engagement. 98% of fashion brands had an Instagram account as of March 2016. As of December 2016, an average of 27.9 image brand posts was made on Instagram each month. Advertisers seeking to engage quality, niche audiences at scale are increasingly drawn to influencer marketing (Instagram 2018). Instagram serves as a platform for visual storytelling for a wide range of users, including musicians, celebrities, newsrooms, teenagers, and anyone with a passion for the arts. Some users gain a lot of followers over time by cultivating a strong relationship with their followers and offering a distinctive visual aesthetic for their content.

As the increase in sponsored content the differentiation between promotional adverts and content is becoming increasingly important as digital media consumption becomes the standard. Influencer marketing is becoming more commonplace as the marketing environment changes. As "Influence equals action" is now a reality, people buy influencer-endorsed goods and services as well as the brand narratives these individuals develop. As a result, these rules protect the interests of advertisers, marketers, influencers, and consumers. Therefore, the standards aim to introduce transparency to influencer marketing by letting consumers know what content has been paid for by brands. In 2021 a definitive criteria for influencer advertising on digital media were released by the Advertising Standards Council of India (ASCI). According to which Influencers are required by the standards to credit the brands on the promotional content they share [3].

Social media influencer marketing is similar to classic celebrity endorsements in traditional mass media, with the exception that the interactions are more content-driven in influencer marketing [4]. The traditional approach is primarily based on one-way broadcasting communication, in which followers are typically unable to respond to the messages of celebrities. Therefore the degree of audience engagement in influencer marketing is slightly higher than with traditional celebrity endorsements [5]. Social Media Influencers have become capable of altering information flows and playing an important role for spreading information about the brands and social causes because of their fan following amongst Instagram users.

2 Review of Literature

In the context of digital marketing, individuals favor social media influencers above traditional celebrities due to their perceived sociability. The concept of celebrity influence, meaning transfer, and celebrity legitimacy in social media marketing would be transformed by this new group of micro-celebrities [6]. According to Primasiwi et al. [7], the three most important key performance indicators, in accordance with digital marketers, are followers, like/comment ratios, and audience reachability. Niloy et al. [8], showed in contrast product source credibility to be an irrelevant factor impacting attitude toward influencer, the study discovered source attractiveness, product match up, and source familiarity to be major factors that affect attitude of audience towards influencer. Hudders et al. [9], found in their study that, when it comes to a subjective dislike of the endorsed product, a defensive answer hurts influencer credibility more than an accommodating one. As a result of this trustworthiness decreased, followers have a worse perception of product quality and have less positive brand perceptions. Also the commercial bias of influencer-created content decreases followers' trust in them and the content's perceived credibility. The mediation results show that followers' willingness to look up more information about the posted product is negatively impacted by the commercial orientation of influencer-created content due to its perceived credibility and dependability [10].

Instagram influencers with a large number of followers are thought to be more likable, owing to their popularity. Importantly, perceptions of popularity based on an influencer's number of followers can lead to perceptions of opinion leadership in some cases [11]. According to Veissi [12], If the audience suspects that an influencer is not authentically advertising things that they use or believe in as products, posts by Instagram influencers are seen as more trust worthy than traditional ads when it comes to purchase choices. Jin et al. [6] shows that consumers exposed to Instagram celebrity's brand posts view the source as more reliable, have a more favorable opinion of the endorsed product, feel more present on social media, and feel more envious of the source than consumers exposed to traditional celebrity's brand posts. In terms of the effectiveness of Instagram influencer promotional posts, the effects of disclosure types (explicit/implicit/no disclosure), influencer credibility (high/low), and brand credibility (high/low), had a good influence on message credibility, attitude toward the ad, buy intention, and eWOM intention [13].

According to Afifah et al. [14] E-WOM and Instagram Influencer had a good influence on brand image of Wardah and Maybelline cosmetics in part and con currently. As a result, marketers must take into account product feedback from consumers and influencers. Primasiwi et al. [7] demonstrated that the commu nity that an influencer has (Community), the experiences that exist on the influencer page (Interaction), the costs associated with interacting with an influencer (Rate Card), and the qualities held by the influencer (Value of Influencer) are the Key Performance Indicators that are taken into consideration when choosing influencers for digital marketing promotion. Niloy et al. [8] found that if a food brand wants to initiate influencer marketing strategy, the brand must consider whether the customers will have a positive attitude towards the influencer, and in order to generate a positive attitude towards the influencer, the brand must ensure that the influencer is attractive, compatible, and familiar, with familiarity as the prime concern.

Casaló et al. [15], stated that, rather than perceived quality or quantity, perceived originality and distinctiveness of an Instagram account's postings are the major characteristics that contribute to an account being seen as a thought leader. Schorn et al. [16] showed disclosures and benefits appeal have no direct effect on the desire for further information about the product, but they do affect the influencer's trustworthiness. However, emphasizing human rewards rather than environmental benefits improved the trustworthiness of the influencers.

The Instagram post's advertising impact is less than the traditional print advertisement. Where customers already follow influencers, the Instagram post has the biggest impact. When it comes to endorsements, influencer marketing is most effective when customers are already following the source. The follower status generally has a favorable impact on the attitude effect, while not being significant in this study. If the test subject already follows an influencer, there is a positive and significant influence on their inclination to buy [17]. According to Arora et al. [5], when generating the influencer index, there is no relative prominence of various social media sites. When modeling the structures, each platform is assigned the same weight age. Evans 'Phua' Lim and Jun [18] found in their study that language featuring "Paid Ad" increases ad recognition favorably, which then interacted with participants' recall of a disclosure and moderated the effect of disclosure language on brand opinion and sharing intention.

According to Hazari et al. [19] extroverted Instagram users preferred lifestyle commercials and brand influencer advertisements more than introverts do for both videos when they were combined as well as separately. There are no differences in attention or attitudes of audience based on gender towards the advertisements. Also there is no significant differences for males and females which shows males are more attentive and attitude towards the advertisements than females. Study shows that gender segmentation for the sort of Instagram advertising have little impact. Study discovered that men and extroverts paid more attention to and had a more positive attitude toward commercials. For brand influencer advertisements, no gender differences were identified. The segmentation of audience by their personality rather than gender shows more differences.

As social media platforms grow in popularity, more people will be able to become influencers or follow the influencers. Instagram has quickly become the go to platform for businesses looking to reach out to new consumers, and influencer marketing on the platform has become a popular way to do it. With the Conversion of Word of mouth into E-WOM the audiences started to believe in the social media reviews and references towards brands and products. Hence the influencers are getting more value and trust because of social media followers. There are many factors that can affect the image of any influencer and their popularity among social media users. Nowadays Instagram has become a more popular and trending social media platform, especially for youngsters. Instagram has a wide range of audiences that are affected by the influencers. Studies in the Indian context are needed as a concern for Influencer marketing with respect to the Instagram platform.

3 Objectives

To identify the Factors affecting Influencer Marketing on Instagram among Indian youth.

To Study the effect of gender on Influencer Marketing on Instagram among Indian youth.

4 Research Method

4.1 The Study

The study was exploratory in nature and explored as well examined the various factors that signify the impact of Instagram Influencer Marketing on consumer perception.

Sample and Area of the Study:

1. For the study, a sample of 135 respondents has been taken.
2. The data has been collected from social media users located in cities of Madhya Pradesh.

4.2 Tool for Data Collection

For the study, primary data was collected with the help of a self-designed questionnaire. The respondents were asked to respond on "Likert Scale" (Five Point Scale) ranging from "Strongly Disagree" being 1 to "Strongly Agree" being 5.

4.3 Tool for Data Analysis

The analysis of collected data was carried out using Statistical Package for Social Science (SPSS). Principle component method of factor analysis with varimax rotation was used. Z-test was applied to assess the effect of independent variables on the dependent variable.

5 Results and Analysis

In this study, it is desirable to consider the various factors affecting Influencer marketing on Instagram among youth. Thus, factor analysis was executed using Principal component Analysis (PCA). The appropriateness of any data for factor analysis depends upon Bartlett's test of sphericity. The significance value 0.000 was obtained, which indicates that Data do not produce an identity Matrix and thus could be used for factor Analysis (Table 1). Cronbach's Alpha Value is 0.972. The KMO measured the adequacy of the sample, and the reading was 0.931 as shown in the Table 1, which was considered acceptable for factor Analysis. For the factor extraction, only the factors with Eight values greater than 1 were considered. The statement whose factor load was more than 0.4 was considered for Factor Analysis.

The 45 variables are grouped under eight derived factors depending upon Eigen Values of each factor. The factor Load was rotated by varimax rotation as shown in the Table 2. The highest factor loading criterion was used to select the constituent factor. Eight factors identified by the present study explain 70.9% of the variance as shown in the Table 3.

Extraction Method: Principal Component Analysis.

Table 1 Showing KMO and Bartlett's test

KMO and Bartlett's test		
Kaiser–Meyer–Olkin measure of sampling adequacy		0.931
Bartlett's test of sphericity	Approx. Chi-Square	4994.500
	Df	990
	Sig	0.000

Table 2 Showing rotated component matrix

Rotated component matrix								
	Component							
	1	2	3	4	5	6	7	8
VAR 00,041	**0.818**	0.009	0.230	0.114	0.131	−0.015	0.000	0.140
VAR 00,044	**0.793**	0.147	0.137	0.254	−0.010	0.038	−0.048	0.140
VAR 00,043	**0.783**	0.181	0.207	0.172	−0.049	0.010	0.124	0.141
VAR 00,040	**0.762**	0.239	0.068	0.262	0.034	0.112	0.098	0.196
VAR 00,035	**0.756**	0.211	0.336	0.144	−0.064	0.196	−0.117	0.071
VAR 00,042	**0.746**	0.106	0.085	0.249	0.247	0.135	0.097	0.001
VAR 00,038	**0.703**	0.209	0.211	0.077	0.297	0.289	0.083	−0.065
VAR 00,036	**0.685**	0.283	0.326	−0.021	0.111	0.192	−0.108	0.080
VAR 00,025	**0.626**	0.425	0.220	0.180	−0.008	0.231	0.112	0.123
VAR 00,037	**0.624**	0.162	0.180	0.073	0.231	0.404	0.118	−0.128
VAR 00,045	**0.622**	0.315	0.223	0.075	0.359	−0.084	0.119	0.157
VAR 00,022	**0.607**	0.394	0.319	0.229	−0.023	0.129	−0.033	0.029
VAR 00,034	**0.605**	0.351	0.195	0.250	−0.026	0.237	−0.097	0.099
VAR 00,039	**0.601**	0.219	0.043	0.178	0.310	0.263	0.093	−0.062
VAR 00,031	**0.599**	0.493	0.009	0.178	0.193	−0.001	0.116	0.161
VAR 00,019	**0.575**	0.326	0.232	0.317	0.051	0.124	−0.374	−0.001
VAR 00,020	**0.541**	0.486	0.279	0.130	0.210	−0.042	−0.056	0.110
VAR 00,033	**0.518**	0.413	−0.007	0.004	0.439	0.106	0.297	−0.101
VAR 00,024	**0.512**	0.496	0.302	0.182	0.027	0.207	0.152	0.151
VAR 00,021	0.439	**0.616**	0.324	0.094	0.231	0.051	0.027	0.006
VAR 00,030	0.324	**0.600**	0.102	0.133	0.166	0.313	0.238	0.084
VAR 00,029	0.420	**0.564**	0.173	0.315	0.105	0.294	−0.069	0.106
VAR 00,023	0.522	**0.523**	0.410	0.128	0.075	0.140	−0.011	0.095
VAR 00,032	0.460	**0.506**	0.005	0.223	0.267	0.120	0.222	−0.059
VAR 00,028	0.465	**0.478**	0.329	0.137	0.058	0.351	0.167	0.071
VAR 00,015	0.280	**0.462**	0.403	0.379	0.039	0.070	−0.241	0.055
VAR 00,018	0.369	**0.454**	0.240	0.167	0.413	0.123	−0.256	0.227
VAR 00,012	0.282	**0.092**	0.706	0.195	0.336	0.185	0.128	0.016
VAR 00,013	0.353	0.236	**0.679**	0.249	0.071	0.134	0.132	−0.053
VAR 00,011	0.194	0.310	**0.629**	0.051	0.395	0.038	0.311	0.104
VAR 00,006	0.218	0.080	**0.567**	0.529	−0.014	0.037	0.160	0.120
VAR 00,010	0.321	0.155	**0.541**	0.308	−0.052	0.155	0.065	0.206
VAR 00,014	0.267	0.372	**0.450**	0.436	0.097	0.077	−0.193	0.128
VAR 00,005	0.071	0.295	−0.007	**0.713**	−0.022	0.144	0.107	0.118
VAR 00,008	0.242	0.023	0.238	**0.700**	0.317	0.052	0.028	0.081

(continued)

Table 2 (continued)

Rotated component matrix								
	Component							
	1	2	3	4	5	6	7	8
VAR 00,007	0.286	0.006	0.398	**0.689**	0.185	0.070	−0.086	0.093
VAR 00,009	0.335	0.233	0.196	**0.649**	0.134	0.077	0.219	0.005
VAR 00,016	0.076	0.114	0.143	0.184	**0.729**	0.212	0.145	0.200
VAR 00,017	0.100	0.424	0.206	0.308	**0.486**	0.209	−0.031	0.058
VAR 00,026	0.191	0.142	0.187	0.123	0.161	**0.773**	0.020	0.119
VAR 00,027	0.430	0.240	0.065	0.155	0.184	**0.612**	0.008	0.097
VAR 00,004	−0.032	0.107	0.215	−0.048	0.048	0.303	**0.650**	0.088
VAR 00,001	0.138	0.004	0.076	0.252	0.117	−0.166	**0.578**	0.082
VAR 00,002	0.116	0.014	0.051	0.071	0.096	0.230	0.177	**0.828**
VAR 00,003	0.256	0.244	0.152	0.300	0.165	−0.123	−0.043	**0.596**

Table 3 Showing total variance

Component	Extraction sums of squared loadings			Rotation sums of squared loadings		
	Total	% of Variance	Cumulative %	Total	% of Variance	Cumulative %
1	21.127	46.949	46.949	10.806	24.013	24.013
2	2.506	5.570	52.518	4.973	11.050	35.064
3	2.087	4.638	57.157	4.132	9.183	44.246
4	1.458	3.241	60.398	3.907	8.683	52.929
5	1.344	2.986	63.384	2.407	5.350	58.279
6	1.221	2.713	66.097	2.360	5.244	63.523
7	1.129	2.510	68.606	1.737	3.860	67.383
8	1.033	2.296	70.902	1.584	3.519	70.902

Extraction Method: Principal Component Analysis. Rotation Method: Varimax with Kaiser Normalization.

Rotation converged in 8 iterations.

There are 8 factors identifies after conducting factor analysis that impacts moment marketing among the youth. The 8 factors were given appropriate names based on the variables. Factor loads and names of the factors are shown in the above table (Table 4).

Table 4 Showing identified factors

Factors	Name of the factors	Items	Factor load
F1	Influencer affinity	I feel influencers are expert I feel influencers are experienced	12.476
		I trust influencers who is referred by someone I feel that the information shared by influencers is supported by strong arguments I feel the influencers shares important information I follow the influencers who interacts frequently I follow the influencers who provide new and unique information to me I like the product if I like the influencer I purchase the product if i like the influence I trust the brand if I trust the Influencer I like the influencer marketing on Instagram because it is easy to handle I like the influencer marketing on Instagram because it is more reliable I like influencer marketing on Instagram because it is more informative I feel personal with the Instagram influencer I get motivated by the Instagram influencer to become an influencer I like to share the posts of	
		Instagram influencers because it is knowledgeable I like to share the posts of Instagram influencers because it is trending I like to share the posts of Instagram influencers because I feel personal with the influencers I feel updated when I get connected with the Instagram influencers	
F2	Influencer as a content poster	Influencers attracts me towards the product or brand I follow influencers who look smart I trust the content shared by the influencers I feel influencers are Knowledgeable about the product I feel that the information shared by the influencers is convincing I feel that the influencers provide credible information I feel the influencers provide trustworthy information I feel the influencers get interactive with the audience I follow the influencers who interacts with the strong arguments	4.295

(continued)

Table 4 (continued)

Factors	Name of the factors	Items	Factor load
F3	Influencer as a catalyst	I like to listen to the Instagram influencers I like to find the new influencers Influencers give me knowledge about the product I take interest in product because of the influencer I follow influencers who have good followers on their social media accounts	2.866
F4	Instagram as a media diet	I like to follow the celebrities and influencers on Instagram I like to listen to the Instagram influencers on reels I like to listen to the Instagram influencers on reels because it is less time consuming I like to listen to the Instagram influencers on reels because it is attractive	2.751
F5	Influencer as an expert	I follow influencers on the basis of content they post I follow influencers who are popular or known to me	1.215
F6	Content sponsorship	I check if the post is sponsored or not I get more trust on unsponsored posts by the influencers	1.385
F7	Instagram as a marketing tool	I use the Instagram for knowledge I feel Instagram useful for marketing purpose	1.228
F8	Instagram as a social media platform	I use Instagram for social connectivity I use Instagram to follow the trend	1.424

6 Hypotheses

1. **H01**: There is no significant impact of Influencer Affinity on Influencer marketing on Instagram towards Indian youth based on gender.
2. **H02**: There is no significant impact of Influencer as a Content Poster on Influencer marketing on Instagram towards Indian youth based on gender.
3. **H03**: There is no significant impact of Influencer as a Catalyst on Influencer marketing on Instagram towards Indian youth based on gender.
4. **H04**: There is no significant impact of Instagram as a Media Diet on Influencer marketing on Instagram towards Indian youth based on gender.
5. **H05**: There is no significant impact of Influencer as an Expert on Influencer marketing on Instagram towards Indian youth based on gender.
6. **H06**: There is no significant impact of Content Sponsorship on Influencer marketing on Instagram towards Indian youth based on gender.
7. **H07**: There is no significant impact of Instagram as a marketing tool on Influencer marketing on Instagram towards Indian youth based on gender.
8. **H08**: There is no significant impact of Instagram as a social networking tool on Influencer marketing on Instagram towards Indian youth based on gender.

7 Hypothesis Testing

The presented results are based on the result of Z-test (One tailed) which was used to assess the effect of independent variables on the dependent variable. The result of Z-test is presented in Table 5.

Table 5 shows the Z test results of demographic variable Gender with the factors Influencer Affinity, Influencer as a Content Poster, Influencer as a Catalyst, Instagram as a Media diet, Influencer as an Expert, Content Sponsorship, Instagram as a Marketing Tool and Instagram as a Social Media Platform. The test was conducted with the aim of investigating whether there is a significant impact of these eight factors on influencer marketing through Instagram among youth on the basis of their Gender. The sample in Table 5 is divided into two categories which are male and female. From the Table 5 it can be seen that all the z values are less than 1.96. Hence these factors are found to be not significant. Therefore, the null hypotheses H01, H02, H03, H04, H05, H06, H07 and H08 are accepted. Hence there is no significant impact of Influencer Affinity, Influencer as a Content Poster, Influencer as a Catalyst, Instagram as a Media diet, Influencer as an Expert, Content Sponsorship, and Instagram as a Marketing Tool and Instagram as a Social Media Platform on influencer marketing through Instagram among youth on the basis of their Gender.

Table 5 Showing Z-test for gender and factors affecting moment marketing

Motivational factor	Male sample		Female Sample		Calculated value (Z)	Critical value	S/NS
	Mean	Stand dev.	Mean	Stand dev.			
Influencer affinity	55.345	19.179	57.433	17.585	0.648	1.96	NS
Influ encer as a content poster	26.172	8.859	27.622	7.283	1.032	1.96	NS
Influencer as a catalyst	13.913	5.399	14.641	4.633	0.832	1.96	NS
Instagram as a media diet	11.604	4.423	11.490	4.534	0.144	1.96	NS
Influencer as an expert	6.827	2.295	7	2.328	0.422	1.96	NS
Content sponsorship	6.444	2.302	6.396	2.078	0.125	1.96	NS
Instagram as a marketing	7.407	1.708	7.018	2.107	1.122	1.96	NS
Tool							
Instagram as a social media platform	7.024	2.109	7.018	2.080	0.015	1.96	NS

Note NS—Not Significant, S—Significant

8 Discussion

It is very clear from the factor analysis that eight factors contribute majorly in the impact of Influencer Marketing on Instagram among Youth. In which Influencer affinity is the most important factor. In a study conducted in conjunction with Twitter and the analytics company Annalect, 56% of users stated they rely on friend recommendations, while 49% said they rely on influencers [20]. Also Influencer plays a very important role as a content poster, as a catalyst and also as an expert which attracts the young audience. In the support of this (Pew Research Centre 2015) showed Instagram as a good fit for influencer marketing because of its visually appealing features and billion active users. Instagram targets Millennials, a younger demographic. Lee and Kim [13] and Belanche et al. [21] also stated that the type of the content and presentation affects the audience towards Influencer marketing. According to Belanche et al. [21] the influencer gives the information matters for the audience and Influencer works as an Expert for them. Veissi [12] said in a study that the trust towards the Influencer is also an important factor. Evans 'Phua' Lim and Jun [18] mentioned the Content Sponsorship as a factor for influencer marketing and paid ads influence more but [22] found that there is no discernible main effects for product placement or sponsorship disclosure. The positive outcomes that marketers are witnessing are the reason behind the surge in popularity of influencer marketing. According to a collaborative research study, influencer marketing can generate an annual return on investment that is eleven times higher than that of other traditional forms of advertising [23].

9 Conclusion and Implications

The present study attempted to study the impact of influencer marketing on Instagram among youth. The study found a positive impact of influencer marketing on Instagram for both the genders among youngsters. The study determined that the affinity towards an Influencer enhances the involvement of youngsters. They see the Influencer as a content poster and attracts with the influencer which positively affect the Influencer marketing. There are some more factors were found are Influencer as a Catalyst, Instagram as a Media diet, Influencer as an Expert, Content Sponsorship, Instagram as a Marketing Tool and Instagram as a Social Media Platform which also affect influencer marketing among youth. Also through the statistical analysis the it was examined that there is no significant impact of Influencer Affinity, Influencer as a Content Poster, Influencer as a Catalyst, Instagram as a Media diet, Influencer as an Expert, Content Sponsorship, and Instagram as a Marketing Tool and Instagram as a Social Media Platform on influencer marketing through Instagram among youth on the basis of their Gender. Hence influencer marketing can be used by the brands and companies for the youngsters on Instagram. The involvement of the youth towards social media is growing rapidly and Instagram became a popular among the youth which discovered as an effective platform for influencer marketing.

References

1. Hall, J.: 8 Influencer Marketing Trends That Will Lead You to Success. Forbes (2016)
2. Lin, R.H., Jan, C., Chuang, C.L.: Influencer marketing on Instagram. Int. J. Innovat. Manag. **7**(1), 33–41 (2019)
3. ASCI issues final Guidelines for Influencer Advertising on Digital media, launches ASCI.Social platform (2021). [Press release]. https://www.ascionline.in/wp-content/uploads/2022/09/press-release-influencer-guidelines-2021.pdf
4. Lou, C., Yuan, S.: Influencer marketing: how message value and credibility affect consumer trust of branded content on social media. J. Interact. Advert. **19**(1), 58–73 (2019). https://doi.org/10.1080/15252019.2018.1533501
5. Arora, A., Bansal, S., Kandpal, C., Aswani, R., Dwivedi, Y.: Measuring social media influencer index-insights from Facebook, Twitter and Instagram. J. Retail. Consum. Serv. **49**, 86–101 (2019)
6. Jin, S.V., Muqaddam, A., Ryu, E.: Instafamous and social media influencer marketing. Mark. Intell. Plan. **37**(5), 567–579 (2019)
7. Primasiwi, C., Irawan, M.I., Ambarwati, R.: Key performance indicators for influencer marketing on Instagram. In: 2nd International Conference on Business and Management of Technology (iconbmt 2020), pp. 154–163. Atlantis Press (2021)
8. Niloy, A.C., Alam, J.B., Alom, M.S.: Influencer marketing: factors influencing a customer's purchase intention. Asian J. Bus. Environ. **13**(1), 23–32 (2023)
9. Hudders, L., De Jans, S., De Veirman, M.: The commercialization of social media stars: a literature review and conceptual framework on the strategic use of social media infuencers. In: Borchers, N.S. (ed.) Social Media Infuencers in Strategic Communication. Routledge, New York (2021)
10. Gamage, T.C., Ashill, N.J.: # Sponsored-influencer marketing: effects of the commercial orientation of influencer-created content on followers willingness to search for information. J. Prod. Brand. Manag. **32**(2), 316–329 (2023)
11. De Veirman, M., Cauberghe, V., Hudders, L.: Marketing through Instagram influencers: the impact of number of followers and product divergence on brand attitude. Int. J. Advert. **36**(5), 798–828 (2017)
12. Veissi, I.: Influencer Marketing on Instagram (2017)
13. Lee, S., Kim, E.: Influencer marketing on Instagram: How sponsorship disclosure, influencer credibility, and brand credibility impact the effectiveness of Instagram promotional post. J. Glob. Fash. Market. **11**(3), 232–249 (2020)
14. Afifah, L.I., Worang, F.G., Arie, F.V.: The influences of e-wom of reference group and instagram influencer toward wardah and maybelline brand image. Jurnal EMBA: Jurnal Riset Ekonomi, Manajemen, Bisnis dan Akuntansi **10**(1), 143–151 (2022)
15. Casaló, L. V., Flavián, C., Ibáñez-Sánchez, S.: Influencers on Instagram: antecedents and consequences of opinion leadership. J. Bus. Res. **117**, 510–519 (2021)
16. Schorn, A., Vinzenz, F., Wirth, W.: Promoting sustainability on Instagram: how sponsorship disclosures and benefit appeals affect the credibility of influencers. Young Consumers (2022)
17. Riedl, J., Luckwald, L.V.: Effects of influencer marketing on Instagram. In: Effects of influencer marketing on Instagram: Riedl, Joachiml uLuckwald, Lisa von. [Wei- denberg]: Access Marketing Management eV (2019)
18. Evans, N.J., Phua, J., Lim, J., Jun, H.: Disclosing Instagram influencer advertising: the effects of disclosure language on advertising recognition, attitudes, and behavioral intent. J. Interact. Advert. **17**(2), 138–149 (2017)

19. Hazari, S., Sethna, B.N.: A comparison of lifestyle marketing and brand influencer advertising for generation Z Instagram users. J. Promot. Manag. **29**(4), 491–534 (2023)
20. Swant, M.: Twitter says users now trust influencers nearly as much as their friends (2016). Accessed from the Adweek website: http://www.adweek.com/digital/twitter-saysusers-now-trust-influencers-nearly-much-their-friends171367
21. Belanche, D., Flavián, M., Ibáñez-Sánchez, S.: Followers Reactions to Influencers Instagram Posts (2020)
22. Ewers, N.L.: # sponsored–Influencer Marketing on Instagram: An Analysis of the Effects of Sponsorship Disclosure, Product Placement, Type of Influencer and their Interplay on Consumer Responses (Master's thesis, University of Twente) (2017)
23. Kirkpatrick, D.: Influencer marketing spurs 11 times the ROI over traditional tactics: study. Market. Dive **6** (2016)
24. Hudders, L., Lou, C., de Brabandere, M.: Understanding the impact of influencers responses to negative follower comments on the persuasiveness of sponsored Instagram posts. Int. J. Advert. **41**(1), 178–204 (2022)
25. Instagram.: Instagram (2018). Accessed 07 21, 2018, from Instagram: http://www.Instagram.com
26. Perrin, A.: Social Media Usage: 2005–2015. Pew Research Center. Pew Research Center, 08 Oct. 2015. Web. https://www.pewresearch.org/internet/2015/10/08/social-networking-usage-2005-2015/

E-Commerce Platforms Extend Beyond Selling. A Case Study on India

Sulagna Das

Abstract E-commerce is the buying and selling of goods and services via the Internet. People shop online. E-commerce websites benefit both shoppers and merchants. So, they benefit from being able to buy goods and services cheaper than local businesses or shopping malls. Nowadays, everything is bought online, saving time, money, and energy, and the products are customized, i.e. manufactured to the customer's specifications. All you need is a Smartphone to order. The best aspect is that these places are open 24/7. Anyone can shop at any time. Our research shows that perceived utility positively impacts online shopping intentions. Convenience is considered to boost internet shopping motivation. Reliability influences online shopping intent positively. Affordability influences people's propensity to shop online. Product availability influences the urge to shop online.

Keywords Online shopping · Convenience · Trust · Intention to buy

1 Introduction

E-commerce, which means electronic commerce, deals with the online buying and selling of goods and services. People shop online, through the internet. E-commerce websites offer several benefits to buyers and sellers. They get the benefit of purchasing goods and services at much lower prices than local shops or shopping malls. The common names of such online platforms are Amazon, and Flipkart for all types of products starting from electronic goods, electrical goods, Laptops, and computer-related gadgets like printers, keyboards, mouse, hard disks, etc., kitchen tools, utensils, furniture, FMCG, home decoration items, etc., Myntra, Meesho, Faballey, Shopperstop, Ajio, etc. are for dresses and Apparels, Nykaa, Purple for cosmetics, Amazon fresh/Pantry, Jio-Mart, Grofer, and Bigbasket for Groceries and ration, vegetables, fish, meat, and eggs, etc., Zomato and Swiggy for cooked food, Ola and Uber for taxi

S. Das (✉)
Department of Management Studies, JIS University, Kolkata, India
e-mail: sulagna.das@jisuniversity.ac.in

S. Dutta et al. (eds.), *Advances in Data Analytics for Influencer Marketing: An Interdisciplinary Approach*, Information Systems Engineering and Management 9,
https://doi.org/10.1007/978-3-031-65727-6_14

services, Redbus and IRCTC for bus and train services, different banking applications for managing banking and financial services online, Netflix, Hoichoi, Hotstar, etc. for OTT, Over the top media services. Today everything of requirement is ordered online, which saves time, money, and energy, and the best part of it is, the products are customized, i.e. according to the wish of the customer. The only thing that is required to order is a smartphone. The best thing is there is no opening or closing time, these shops are opened for 24 × 7. If someone wishes, he can shop at midnight. There are many payment options, somewhere if a specific debit card or credit card is used, a customer gets an instant offer of a discount, pays less, or cash back, EMI facilities through Bajaj Finance, SBI Cards, ICICI, HDFC Cards are always there. If the price of a mobile phone is Rs. 25,000, customers get it for three to four thousand rupees less on e-commerce websites. During Flipkart's Big Billion Day or Amazon's Great Indian Festival Days, the customers get offer of free shipping, low price, lower than any other day on all the products and services provided by these two websites. Other websites also give offers during the festival season, which serves two purposes, first one is it makes a customer happy, satisfied, and loyal, on the other hand, the sale of products and services also increases. This is another reason today e-commerce has become famous, as it offers affordability to its customers. By 2026, the e-commerce market is projected to grow to a value of $200 billion USD. Social Media these days are also contributing to e-commerce, Facebook, and WhatsApp, have provided a big platform for many, who buy and sell online. There are public groups and virtual marketplaces where the members display their products and others order and buy the same. The payment is done Online, and the products without problem reach the customer through courier services like Bluedart, DTDC, or Delhivery. The popularity of e-commerce has tremendously increased regionally, nationally, and internationally. An Indian who stays in the USA or anywhere in the World can easily buy the customized products of his own country through electronic channels.

E-commerce originated in 1994 when a guy used his website, net market, an American retail platform, to sell a CD to a friend. This was the first instance of an online purchase of a product. From the year 1994 till the year 2021, the e-commerce business has gone through so many changes. Traditional sellers and buyers have been forced to change and embrace new technology to cope with the present situation. The year 2020–2021 has shown us a big revolution, the pandemic has taught us a lot. During lockdown days consumers shifted from local shops to online shops for ordering their groceries, vegetables, fish, egg, meat, and other necessities. Now it has become a habit for us, we don't want to wait for any requirement to be fulfilled later on, just a click on the smartphone, and the website is ready to show us innumerable product varieties, of different companies. We have so many options while buying something or the other of our own choice. This new technology has become so friendly and easier to use, people nowadays download applications on their mobile and initiate purchase transactions.

E-Commerce is of four types, Business to Business, where plagiarism software is sold to Universities for checking Plagiarism in their work related to research, Business to Consumer, where a consumer buys kitchen appliances from Amazon, Consumer to Consumer, where a customer sells his old car to another customer through OLX or a reseller group where a customer buys a product from the manufacturer and resell the same to another customer and Consumer to Business, is when a photographer licenses a photo, clicked by him, for a business to use.

2 Literature Review

There are several pieces of literature on E-Commerce and Online Shopping intentions of customers. This is the need of the hour that consumer shifts to online shopping or e-commerce platform from going physical. E-Commerce has booming growth in India as well as globally [29] it offers advantages, and opportunities to different partners, businesses, and academics though it has some restrictions [17, 38]. Herekar [15] in his paper observed that e-commerce offers several benefits to the stakeholders of a company. E-Commerce uses different instruments to conduct online business, [20, 27]. The factors that influence [18] the E-Commerce are adoption of the Internet, buying and selling online, payment, delivery, service operations, cost operation reductions and profit levels [3, 25]. E-Commerce is changing the functional areas of each and every business starting from advertising to paying bills, creating huge competition between existing and new companies, new business models and products and services [35]. The companies get an opportunity to expand worldwide at a minimum cost and increase their share in the market. Customers regard this as convenient since it allows them to have easy access to a large range of goods and services while also saving money, time, and energy [33].

2.1 Literature on Customers' Buying Intention

Ayo et al. [2] focused on the risk part of e-commerce. They have found the interrelationship between consumer trust, perceived risk, and behavioral intention. While trust has a less substantial impact on a customer's behavioral intention than risk, risk has a highly significant effect on trust. On the other hand, task technology fit, perceived usefulness and easy to use, and buying intention are significant. Peiris et al. [26], and Ha et al. [14] according to their research, consumers are hesitant to use online platforms due to security risks; yet, trust, perceived utility, convenience of use, and readiness to reveal social and personal influence all have beneficial effects. Rahman et al. [28] found in their study that, in contrast to local stores, customers prefer to purchase online because it saves them time and gives them access to a wider selection of goods and services. They purchase after researching online. However, they have mixed feelings about their online buying experiences since they are worried about the security of the payment system.

2.2 Literature on Perceived Usefulness

The concept of Perceived Usefulness has come from TAM (Technology Acceptance Model) of [8]. Usefulness is the utility of any products or services that a consumer is looking for, and which he gets from e-commerce websites. Singh et al. [32], in their study found that perceived usefulness is a non-significant predictor for online shopping. Ahmed et al. [1] observed that consumers' intentions to purchase through online shopping platforms are positively correlated with perceived utility and enjoyment.

2.3 Literature on Convenience

The term 'convenience' was first used by [7]. These goods are easily accessible. Bhagyasree et al. [4] in their paper observed that customers move towards online shopping as is it convenient for them. Dost et al. [10] observed that online shopping was convenient for consumers who preferred to stay at home and shop. Sultan et al. [34] and Yaraş et al. [39] in their paper observed that convenience mostly has a positive influence on customer shopping online,

2.4 Literature on Reliability

The term Reliability means reliability, the shopping platform on which a customer can rely. Yaraş et al. [39], and Roudposhti et al. [30] observed that consumers' intentions to make online purchases are significantly positively impacted by trust. A customer will trust a reliable website that has good product and service information. They will always return to that site from where they have bought earlier and has a good experience. Zhao et al. [40] in their paper observed that trust increases the goodwill of a consumer toward a business, and this as a result reduces the risk.

2.5 Literature on Affordability

The term affordability means inexpensive and customers can afford it. The products and services available online are cheaper than the local store, so they are affordable. Affordability is a factor why people go for online shopping these days. Yaraş et al. [39] observed Price awareness and clients' intention to shop online are positively correlated.

2.6 Literature on Availability

The term availability means another reason why consumers these days gravitate toward online shopping platforms is the vast array of goods and services that are accessible online. Sharma [31] in her study found that a variety of products available online influence consumers' attitudes and buying behavior towards online shopping. Yaraş et al. [39] observed Variety of merchandise and clients' intention to shop online are positively correlated.

2.7 Literature on Benefits

The term benefit talks about gains and profits. Any advantage received from shopping online. Sharma [31] in her paper talked about the benefit of online purchases as easy returns and replacement items this factor has a favorable impact on customers' hassle-free internet buying. Deng et al. [9] discovered that one of the key elements influencing consumers' intentions to make online purchases is their perception of the benefits. Perceived benefits increase the willingness to search and the willingness to purchase online among customers.

3 Objectives for the Study

1. To comprehend the elements influencing consumers' intention to purchase goods and services through e-commerce platforms.
2. To understand the issue Indian customers encounter when making online purchases.
3. To understand India's potential for e-commerce.

4 Proposed Hypotheses

H1 = Customers' inclination to buy online is positively impacted by perceived utility.

H2 = Customers' intention to make an online purchase is positively impacted by convenience.

H3 = Customers' intention to make an online purchase is positively impacted by reliability.

H4 = Customers' intention to buy online is positively impacted by affordability.

H5 = Customer intention to buy online is positively impacted by availability.

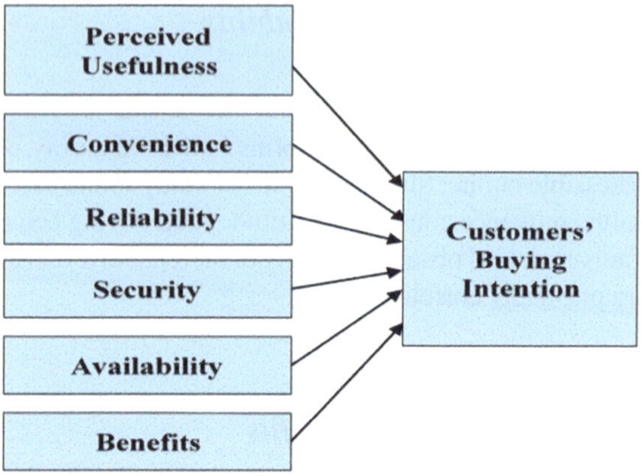

Fig. 1 Study model (By Author)

H6 = Customers' inclination to purchase online is positively impacted by benefits.

5 Proposed Model of the Study

See Fig. 1.

Dependent Variable = Customer Buying Intention

Independent Variable = Perceived Usefulness, Convenience, Reliability, Affordability, Availability, and Benefits.

6 Research Methodology

The responses from Indian citizens gathered via social media platforms form the basis of this paper. The sample size is 380. The period of the study is from 2021 to 2023.

6.1 Data Collection and Sampling

The proposed hypothesis model was evaluated in this study utilizing a seven-point scale (Likert) with a score ranging from strongly agree (7) to strongly disagree (1). Kim et al. used eight items to assess perceived trust. Kim et al. [21], as well as McKnight et al. [22]. The item includes phrases such as "Based on my previous interactions with the E-Commerce merchant, I am confident that it is reliable." Four items from Chan et al. (Ngai, E. W., Poon, J. K. L., and Chan, Y. H.) were used to assess perceived ease of use (2007). The item contains phrases such as "It would be simple for me to develop proficiency in the use of my cell phone for E-Commerce transactions." The item includes phrases such as "The product/service purchased through E-Commerce is reasonably priced." Attitude toward use is likewise examined using the same ratio of ten items as in [19]. The reliability was accessed with the help of four items adopted from [16]. The item includes phrases such as "The product/service purchased through E-Commerce is reasonably priced." Finally, five items adapted from Fernie, J., and McKinnon, A. C. were used to assess purchasing behavior (2004). The Item contains comments such as "In the future, I will use E-Commerce Apps to conduct online purchasing."

6.2 Data Analysis

This section provides a detailed breakdown of the demographic elements that influence e-commerce portal use, as well as the most popular categories of use. Attitudes toward use, subjective norms, and perceived behavioral control over the intention to use are tested in the regression analysis.

7 Analysis Using Descriptive Statistics

Male participants are more likely to use e-commerce than female participants, and those under the age of 20 are more likely to do so. It is well-known that e-commerce use is directly linked to one's socioeconomic status. Finally, people who have a background in technology are more likely to use e-commerce than those who don't (Table 1).

Table 1 Demographic profile

	Use		Do not use	
Socio-economic variable	Frequency	Gender (%)	Frequency	Gender (%)
Male	206	58.90	144	41.10
Female	174	47.80	190	52.20
Age				
18–25 years	208	51.50	196	48.50
25–40 years	130	59.10	90	40.90
40–55 years	28	46.70	30	53.30
55 and more	14	46.66	16	53.33

Source Author's Calculation

Table 2 Percentage of E-commerce usage by category of use

E-commerce usages		
Use categories	Frequency	Percentage (%)
Health and beauty	170	44.73
Electronic equipment	156	41.05
Online ticketing	130	34.20
Fashion	124	32.60
Entertainment	110	28.90
Tourism	86	22.60
Feeding	80	21.10
Home	54	14.20

Source Author's Calculation

Table 2 illustrates the results based on the categories that participants indicated as most in demand. The most common use of digital banking for e-commerce transactions is for health and beauty, electronics, online ticketing, and fashion. Health and beauty products can be purchased online by a wide range of participants; a significant number also claimed that they have purchased electronic equipment (mobile phones and digital accessories) via e-commerce portals. Young people are increasingly embracing online ticketing because it is more convenient and time-saving.

Because each participant's response may be numerous, the sum of the frequencies may equal or exceed the sample size (Table 3).

It can be observed from the above table that factors are reliable and have an influence on consumer buying intentions (Table 4).

Table 3 Descriptive statistics, with reliability & validity test of the dimensions of the models e-commerce user

Dimension	No. of items	Mean	Deviation	Cronbach α	AVE
Perceived Usefulness (4)	4	4.362	1.654	0.899	0.687
Convenience (4)	4	4.82	1.628	0.858	0.807
Reliability (4)	4	5.281	1.531	0.848	0.799
Security (3)	3	5.642	1.479	0.795	0.567
Availability (4)	4	5.82	1.415	0.889	0.687
Benefits (3)	3	4.985	1.721	0.798	0.758

Source Author's Calculation

Table 4 Model summary

Model	R	R Square	Adjusted R Square	Std. error of the estimate	Durbin-Watson
1	0.715	0.515	0.593	0.4190	2.083

Source Author's Calculation

It can be easily predicted from the above table that consumer buying intention has a direct and positive influence on the Perceived Usefulness, Convenience, Reliability, Security, Availability, and Benefits of the e-commerce product and platform. So, it can be stated that all framed hypotheses were accepted.

8 Conclusion

To assess the model's validity, the authors used two approaches: one based on TPB recommendations and the other on disaggregating the DTPB into dimensions rather than independent variables. Perceived usefulncss has a positive effect on the intention to shop online (H1). It is also stated that Convenience has a beneficial effect on the intention to shop online (H2). Reliability has a positive effect on the intention to shop online (H3). The Affordability of products and services has a beneficial effect on the propensity to shop online (H4). Product availability has a positive effect on the intention to shop online (H5). A consumer will use a digital platform for shopping if three conditions are met: (1) they like e-commerce; (2) they are subjected to social pressure in favor of e-commerce transactions; and (3) they have the resources and knowledge to carry out the transaction. According to Fishbein and Azjen, this occurs when people make sensible choices (1975). Thoughts about a given action influence a person's intention and subsequent performance. Chen et al. [6] found that these three criteria influenced the propensity to adopt digital transactions. The test result for hypothesis H1 suggested that positive sentiments about e-commerce enhance usage. People like e-commerce because it is convenient (perceived value), easy to use (perceived ease of use), and suitable for their work style. People are more likely to utilize something if it fits into their life (compatibility) and poses no security or privacy concerns (trust) (intention to use).

9 Limitation

According to the first criterion, the results indicate that the TAM may explain a larger percentage of the variation, in purchase intention on the Internet, implying that its explanatory power is greater. Additionally, this approach enables the analysis of factors affecting attitudes, which is particularly useful when this variable is critical to the inquiry. From what a person understands, this refers to the significance of the information supplied. For the models, the TAM contains simply their attitude toward online buying and technological aspects. On the contrary, the TPB enables the determination of the effect that regulatory influence from third parties has on the acceptance process, as well as perceived control over behavior. Thus, the TPB provides a broader picture of the variables affecting the adoption of electronic commerce by a segment of the ultimate customer population. Finally, both models meet the third requirement in terms of application complexity, as they both include the same independent variable and incorporate the same number of explanatory factors. In summary, the Technology Acceptance Model outperforms the Theory of Planned Behavior in terms of its prediction capacity for Internet shopping uptake. However, because the TPB provides a broader understanding of the elements affecting this behavior, it is more appropriate when the investigation's purpose is to investigate the causes of acceptance rather than to forecast whether it will occur or not.

Notwithstanding the methodical approach employed, there are certain limits with this work. First of all, the research's dependent variable—the intention to make an online purchase—was assessed subjectively based on the participants' impressions of their own actions [5, 37]. While many researches have employed this methodology, some authors recommend using objective measurements instead, like actual behavior [36]. Chen et al. [6] contend that utilizing both forms of measurement and evaluating their correlation would be practical.

Self-generated validity is linked to the second potential drawback, which is typical of this kind of research [11]. In this approach, there is a propensity for the responses to the initial questions to be the basis for the answers to following questions when the respondents are questioned about topics or actions on which they have not previously reflected, leading to exaggerated causal links. This problem may be especially relevant in the case of subjects with no previous experience with the behavior analyzed, as in the sample of non-buyers. However, in the present investigation, this effect seems to be attenuated, since the results obtained in the Study are highly consistent with those observed in the sample of subjects who have previously carried out an online transaction. The results of this study, in turn, raise new questions for the development of future research. In the first place, following the proposals of [37], it would be necessary to deepen the development of models for accepting purchases on the Internet that incorporate the approaches of the two models analyzed. This integrative approach would allow obtaining models with a high predictive capacity of the adoption of electronic commerce and that at the same time provide a comprehensive explanation of the factors that affect behavior. On the other hand, it would be interesting to incorporate other psychological variables into the models analyzed to determine if they improve their ability to predict and understand the purchase on the Internet. In particular, the inclusion of psychological variables traditionally associated with e-commerce, such as risk aversion or propensity for innovation, can provide a more complete view of the acceptance process. Finally, it would be interesting to investigate whether the perceived utility and ease of use are indeed the characteristics of Internet shopping that conform to the attitudes of the subjects, or it is possible to identify other more adequate and precise variables such as convenience, savings in prices or access to a wider offer.

References

1. Ahmed, E., Akhlaq, A.: Digital commerce in emerging economies: factors associated with online shopping intentions in Pakistan. Int. J. Emerg. Markets **10**, 634–647 (2015). https://doi.org/10.1108/IJoEM-01-2014-0051
2. Ayo, C.K., et al.: Business-to-consumer e-commerce in Nigeria: prospects and challenges. Retrieved African J. Bus. Manag. 5(13), 5109–5117 (2011)
3. Ajao, B.F., et al.: Factors influencing the implementation of e-commerce innovations: the case of the Nigerian informal sector. African J. Sci. Technol. Innovat. Dev. 10(4), 473–481. Informa UK Limited (2018). https://doi.org/10.1080/20421338.2018.1475541
4. Bhagyasree, J.D., Venugopal, J.S.: A study on the impact of online shopping on consumer's buying behavior. Retrieved Turkish J. Comput. Math. Educ. **12**(12), 2843–2851 (2021)
5. Chau, P.Y.K.: An empirical assessment of a modified technology acceptance model. J. Manag. Inf. Syst.: JMIS 13(2), 185–204. Informa UK Limited (1996). https://doi.org/10.1080/07421222.1996.11518128
6. Chen, L.-D., Nath, R.: Determinants of mobile payments: an empirical analysis. J. Int. Technol. Inf. Manag. 17(1), Article 2. https://doi.org/10.58729/1941-6679.1105. Accessed 5 May 2024
7. Copeland, M.T.: Relation of consumers' buying habits to marketing methods. Harv. Bus. Rev. **1**(2), 282–289 (1923)
8. Davis, F.D.: Perceived usefulness, perceived ease of use, and user acceptance of information technology. MIS Q. **13**(3), 319–340 (1989). https://doi.org/10.2307/249008
9. Deng, W., Ting, S., Zhang, Y., Chunli, T.: Factors Affecting Consumers Online Choice Intention: A Study Based on Bayesian Network, vol. 12, pp. 1–10. https://doi.org/10.3389/fpsyg.2021.731850. Accessed 5 May 2024
10. Dost, M.K.B., et al.: Online shopping trends and its effects on consumer buying behavior: a case study of young generation of Pakistan. J. Soc. Dev. **5**(1), 1–22 (2015)
11. Feldman, J.M., Lynch, J.G.: Self-generated validity and other effects of measurement on belief, attitude, intention, and behavior. J. Appl. Psychol. 73(3), 421–435. American Psychological Association (APA) (1988). https://doi.org/10.1037/0021-9010.73.3.421
12. Fernie, J., Mckinnon, A.: The development of retail logistics. In: Fernie, J., Kogan, L.S. (eds.) Logistics and Retail Management, Emerging Issues and New Challenges in the Retail Supply Chain, pp. 207–232 (2009)
13. Fishbein, M., Ajzen, I.: Belief, Attitude, Intention, and Behaviour: An Introduction to Theory and Research. Addison-Wesley Publishing Co, Inc. (1975)
14. Ha, N.T., Nguyen, T.L., Pham, T.V., Nguyen, T.: Factors influencing online shopping intention: an empirical study in Vietnam. J. Asian Finance Econ. Bus. **8**, 1257–1266 (2021)
15. Herekar, P.M.: E-Commerce in India—problems and prospects. Retrieved from NCON13-Fin-014 (2013)
16. Ijaz, M., Rhee, J.: Constituents and consequences of online-shopping in sustainable e-business: an experimental study of online-shopping malls. Sustainability 10(10), 3756. MDPI AG (2018). https://doi.org/10.3390/su10103756
17. Iqbal, J., Quazi, F.A.: Challenges and prospects of e-commerce in India. Retrieved J. Eng. Res. Appl. **7**, 14–16 (2017)
18. Kareem, T.S., et al.: Electronic commerce and business performance: an empirical investigation of business organizations in Nigeria. Retrieved from International. J. Acad. Res. Bus. Soc. Sci. **4**(8), 215–223 (2014)
19. Karjaluoto, H., et al.: Factors underlying attitude formation towards online banking in Finland. Int. J. Bank Market. 20(6), 261–272. Emerald (2002). https://doi.org/10.1108/02652320210446724
20. Kaur, H., Kaur, D.: E-commerce in India—challenges and prospects. Int. J Eng. Tech. **1**(2), 36–40 (2015)
21. Kim, D.J., et al.: Trust and satisfaction, two stepping stones for successful e-commerce relationships: a longitudinal exploration. Inf. Syst. Res.: ISR **20**(2), 237–257. Institute for Operations Research and the Management Sciences (INFORMS) (2009). https://doi.org/10.1287/isre.1080.0188

22. McKnight, D.H., et al. developing and validating trust measures for e-commerce: an integrative typology. Inf. Syst. Res.: ISR **13**(3), 334–359. Institute for Operations Research and the Management Sciences (INFORMS) (2002). https://doi.org/10.1287/isre.13.3.334.81
23. Nachar, M.: Factors That Predict the Adoption of Online Shopping in Saudi Arabia. Walden University (2019). https://scholarworks.waldenu.edu/dissertations/6630
24. Ngai, E.W.T., et al.: Empirical examination of the adoption of WebCT using TAM. Comput. Educ. 48(2), 250–267. Elsevier BV (2007). https://doi.org/10.1016/j.compedu.2004.11.007
25. Panigrahi, A.K., et al.: E-commerce services in India: prospects and problems. SSRN Electron. J. (2016) Elsevier BV. https://doi.org/10.2139/ssrn.2736822
26. Peiris, P.M., et al.: An empirical study of customer adoption of e-commerce: a customer trust model to support the adoption of e-commerce among small- and medium-sized enterprises in Sri Lanka. Retrieved Int. J. Bus. Inf. **10**(4), 491–518 (2015)
27. Raghunath, A., Panga, M.D.: Problem and prospects of E-commerce. Retrieved Int. J. Res. Dev. Manag. Rev. (IJRDMR) (2), 59–68 (2013)
28. Rahman, M.A., et al.: Consumer buying behavior towards online shopping: an empirical study on Dhaka City, Bangladesh'. Cogent Bus. Manag. 5(1), 1514940 (2018). Informa UK Limited. https://doi.org/10.1080/23311975.2018.1514940
29. Ravaliya, D.M.: Recent prospects and challenges In E-commerce. IJCRT **8**, 1283–1286 (2021). ISSN: 2320-2882
30. Roudposhti, V.M., et al.: A new model for customer purchase intention in e-commerce recommendation agents. J. Int. Stud. 11(4), 237–253 (2018). https://doi.org/10.14254/2071-8330.2018/11-4/17
31. Sharma, K.B.: A study of consumer attitude towards online shopping in india and its impact. Retrieved Global J. Soc. Sci. 35–54 (2020)
32. Singh, R.R, Mittal, S., Kukreti, A.: Effects of Online Shopping On Consumer Buying Behavior (2018). https://www.researchgate.net/publication/324694092. Accessed 5 May 2024
33. Sujana, Y.V., Narayanappa, G.L.: Impact of E-Commerce on brand loyalty and brand switching behaviour of consumers of electronic products: a study. Retrieved EPRA Int. J. Res. Dev. (IJRD) 5(5), 489–493 (2021)
34. Sultan, M.U., Uddin, M.N.: Consumers' Attitude Towards Online Shopping Factors Influencing Gotland Consumers to Shop Online (2011)
35. Sweety, A., Sriram, S.: Prospects of E-commerce In Indian Business context. Retrieved IJARIIE **2**, 333–337 (2016)
36. Szajna, B.: Empirical evaluation of the revised technology acceptance model. Manag. Sci. **42**(1), 85–92 (1996). Institute for Operations Research and the Management Sciences (INFORMS). https://doi.org/10.1287/mnsc.42.1.85
37. Taylor, S., Todd, P.: Decomposition and crossover effects in the theory of planned behavior: a study of consumer adoption intentions. Int. J. Res. Market. **12**(2), 137–155. Elsevier BV (1995). https://doi.org/10.1016/0167-8116(94)00019-k.
38. Tripathy, J.P., Khan, F.: Problems and Prospects of E-Commerce in India. Retrieved IOSR J. Bus. Manag. (IOSR-JBM) **18**, 54–58 (2016)
39. Yaraş, E., Özbük, M.Y., Ünal, D.A.: Factors affecting consumers' intention to purchase online. Retrieved IUYD J. Internet Appl. Manag. **8**(2), 63–74 (2017). https://doi.org/10.5505/iuyd.2017.91885
40. Zhao, X., et al.: The impact of reference effects on online purchase intention of agricultural products: the moderating role of consumers' food safety consciousness'. Internet Res. **27**, 233–255 (2017). https://doi.org/10.1108/IntR-03-2016-0082

Analyzing the Effectiveness of Influencer Campaigns in Driving Customer Acquisition and Retention: Perspectives from Unilever Pakistan

Arman Khan, Zahid Hussain, Muhammad Ussama Majeed, and Amarnath Padhi

Abstract This study explores the revolutionary potential of influencer marketing initiatives inside Unilever Pakistan with a concentrate on customer acquisition (CA) and retention (CR). Using a large sample size of 386 participants, the research uses a quantitative methodology to carefully examine the effects of independent variables. The study confirms the importance of influencer marketing (IM) using the F test. Unfiltered coverage throughout the population is ensured by the basic random sampling technique that was selected. The results shed light on the strategic value of influencers in the retail goods sector, especially in the fast-paced environment of Unilever Pakistan. Influencer campaigns become powerful instruments for improving brand-consumer communication and fostering loyalty. This study emphasizes how important it is for companies to leverage the impact of digital characters in order to build genuine relationships that appeal to today's savvy customers. Leading this revolutionary movement, Unilever Pakistan uses influencers to create lasting connections and mold brand identities. Recognizing the subtleties of influencer interactions is becoming more and more important as marketing terrain changes, serving as a guide to help organizations achieve long-term success.

A. Khan
Department of Business Administration, Shaheed Benazir Bhutto University, Nawabshah, Pakistan
e-mail: armankhan@sbbusba.com

Z. Hussain (✉)
Department of Business Administration, Khadim Hussain Shah Institute of Technology, Karachi, Pakistan
e-mail: Zahidhussain9341@gmail.com

M. U. Majeed
Department of Management Sciences, National University of Modern Languages, Lahore, Pakistan
e-mail: ussama.majeed@numl.edu.pk

A. Padhi
School of Management Studies, NIST University, Berhampur, India
e-mail: amarnath.padhi@nist.edu

S. Dutta et al. (eds.), *Advances in Data Analytics for Influencer Marketing: An Interdisciplinary Approach*, Information Systems Engineering and Management 9,
https://doi.org/10.1007/978-3-031-65727-6_15

Keywords Influencers marketing · Customer behaviour · Customer acquisition · Retention · Unilever Pakistan

1 Introduction

Within the ever-changing fast-moving consumer goods (FMCG) industry, Unilever Pakistan is a significant participant. Known with its iconic brands, Unilever Pakistan has left its imprint on the business with lifebuoy, lux, Surf, and Walls [16]. In 1952, Unilever Pakistan opened for business in Rahim Yar Khan [12]. Vegetable oil manufacturing was among of its innovative ventures, which demonstrates its dedication to innovation. Unilever Pakistan has changed throughout the previous 52 years, moving its Rahim Yar Khan location to Karachi as a result of manpower specialization and availability [14]. In the 1890s, Unilever launched its groundbreaking Sunlight soap idea. The people of Victorian England were drawn for this washing and sanitation product [12]. Sunlight soap gave customers an enjoyable experience in addition to improving their beauty. Over three millennia, Unilever has left a history of adaptation and resiliency. Its wide range of products meets different demands, making life easier at home and providing nourishment for bodies. Besides just its merchandise, the brand influences stories, builds bonds, and creates long-lasting connections [13]. Influencer marketing efforts become effective instruments in this situation. They promote expansion and engagement by establishing a bridge among businesses and customers [11]. As a pioneer, Unilever Pakistan uses influencer marketing to spread its purpose. These virtual personas appeal to astute customers who mold brand opinions [10]. Influencer intricacies need to be understood because advertising dynamics are changing. Because of its dedication to innovation and customer welfare, Unilever Pakistan is positioned for longevity [8]. The history of Unilever Pakistan combines effect, inventiveness, and legacies. Influencer marketing initiatives link Unilever with its target demographic like strands woven across time, guaranteeing an increasingly promising and captivating future [7]. Instant messaging (IM) has become a powerful tool in recent times, connecting customers with businesses via social media channels. Unilever Pakistan aims to investigate how influencer campaigns affect CA and CR in this particular setting. This research explores the revolutionary function of influencers, confirming their importance in modern marketing tactics. Unilever Pakistan hopes to create enduring relationships and mold brand impressions by utilizing digital characters. Recognizing influencer relationships is becoming more and more important as marketing terrain changes, serving as an orientation for organizations seeking long-term success. Current studies highlight the effectiveness of instant messaging (IM), linking it to increased customer involvement and steadfast loyalty. Bogner et al. show that customers are more likely to believe suggestions from influencers than from traditional forms of promotion. Additionally, [9] stress how important it is for influencer material to be relevant and relevant in order to generate strong client engagement and loyalty to a company. Even though the benefits of instant messaging are well known, empirical research is still necessary to determine how effective it is in different situations, especially the Pakistani consumer products sector. In order to fill the hole in research, our research surveyed 386 participants.

2 Literature Review

2.1 Influencer Marketing

At its core, IM is a strategic partnership between brands and influencers. It leverages the Influencer's credibility and reach to create and distribute authentic content promoting the brand's products or services. IM is a collaboration between popular social media users and brands to promote brands' products or services. These partnerships have been going on informally since the dawn of social media. The IM economy was valued at $21.1billion in 2023 [41]. Influencers are people with sizable social media followings who can influence their followers' behaviour as consumers [40]. IM is a hybrid of old and new marketing tools combining celebrity endorsement with modern content-driven campaigns [25]. The results of IM campaigns stem from collaborations between brands and influencers. IM involves promoting and selling products or services through Individuals capable of driving action from the target audience [21]. There are three main types of influencers, celebrities, macro-influencers and macro-influencers [19]. IM works due to its positive ROI for brands and the abundance of potential influencers [18]. IM is a process where businesses partner with individuals ranging from celebrities to macro-influencers who have a significant social media presence [36]. It leverages the influencer's credibility and reach to create and distribute authentic content promoting the brand's products or services [30]. IM bridges the gap between brands and consumers utilizing the influence of social media personalities to drive engagement, trust and sales [28]. Influencer characteristics identity that popularity and attractiveness as heuristic cues that influence judgements of influencer efficacy. Additionally category involvement and altruistic motives for collaboration moderates follower's reliance on these cues [39]. Factors like audience, brand, content and technology play pivotal roles in consumer reactions to IM [21]. IM has grown significantly in recent years. The Influencer market increased form $1.7 billion in 2016 to $9.7 billion in 2020 [27]. By 2021, it reached $23.8 billion and projections estimate it to be a massive $16.4 Billion business by 2022 [20]. Influencers resonate with their audience due to shared characterstics. Influencer's knowledge and authority impact their effectiveness [24]. Trust plays a crucial role in IM [33]. Influencers' appeal influences followers' perceptions [23]. Authenticity and credibility enhance influencer efficacy [26]. Influencer content should provide value and engage followers [20]. Alignment between influencer persona and brand image matters [17]. Social media influencer marketing began to flourish around 2018 approximately 10 years after the concept was introduced in 2008 [16]. IM involves partnering with individuals who have specialised knowledge, authority and social reach [14]. By 2022, marketing managers are projected to spend

$15 billion on IM leveraging both offline and electronic word-of-mouth [12]. Social media influencers are everyday people who have organically built an online presence [11]. They use their social media platforms as opposed to more conventional ones to interact with fans and exercise influence. Socoal influencers, data release, origin trustworthiness, prosocial communication and reliability & engagement are main topics of recent studies. The impact of influencers on customer behavior is exmained in these themes. Influencers are powerful conduits for infleunce, influencing the views, tastes and choices of consmuers [12]. Consumer decisions are influenced by influencer marketing, which is motivated by genuineness and trust [16]. The worldwide influencer market is estimated to be worth billions of dollars. Influencers bridge the divide between companies and their customers by influencing choices and impressions [10]. Influencer marketing is a constantly evolving sector that combines connections, reliability, and coordinated strategy. It's an intriguing field for research and application since scholars are still delving into its processes, effects, and changing patterns.

2.2 *Customer Acquisition (CA)*

The procedure of drawing in holding the interest and then turning potential clients into purchasing ones in order to grow the client base is known as CA [43]. It covers the whole customer experience from the time a potential customer first learns about the business website until the point at which they make a purchase and might be represented as a CA funnel. Finding a methodical and long-lasting approach to draw customer to their organisation and influence them to take action like buying a product or subscribing to a paid service is the primary objective of CA [5]. Acquiring clients entails persuading prospective buyers to purchase the goods [8]. A successful CA approach draws in leads, develops into ready-to-sell prospects, nurturing and implementing prospects in selling channels in order to seal the transaction is known as CA. The procedure of drawing in and winning over new clients to the company is know as CA. It is essential to the expansion and sustainability of a company [3]. The goal of CA is to identify potential clients and convince them to make a purchase form company in a reliable, measurable manner rather than at random [9]. Using a CRM, email advertising and offering top-notch client service are a few ideas gaining more customers is crucial for growing clientele, building brand recognition and proving the viability to outside parties [12]. For long-term development, it entails methodically bringing in and winning over new clients while simultaneously putting a strong emphasis on CR [20]. The online gaming trends and regional growth of the casino industry create a competitive operating environment. Even it it costs money expanding a firm still requires acquiring new clients [31]. CA methods are still important for long-term success even though techniques for retention are more important [26]. CA is still required for long-term company development, yet CR is crucial. The cost of obtaining new customers is Higher than that of focused retention and devotion programs [32]. Gaining new clients is crucial to the expansion of any organisation. In order to pay expenses and engage in the business, it can broaden the client base, raise reputation and create income [13]. A crucial indicator of the entire cost of gaining a new client comprising all sales and marketing outlays is the Customer Acquisition Cost (CAC) [12]. It's essential for comprehending the worth of clients and the effectiveness of acquisition tactics. A number of procedures and tactics are involved in turning prospective into actual consumers throughout the CA process, which is a crucial corporate activity that promotes revenue growth and profitabilty [15]. Because IM can take advantage of the connection and confidence that influencers have among others, it has emerged as a key tactic in CA [16]. With their target market, influencers already have a degree of trust that can be extended to the businesses they support [8]. An influencer's audience tend to view a brand more reputable when they endorse a product [1]. Influencers frequently specialise in a certain market giving brand success to a niche market that's more inclined to be engaged in the goods or services. An influencer's endorsements can result in immediate revenue and sales particularly when networks such Facebook have shoppable content that let users buy products without existing the app [7]. Relationship advertising is another term for IM. It's key goals are establishing and preserving long-term connections and clienr engagement [9]. Companies may create a following about their brand by choosing influencers that communicate well with their target market [6]. Higher participation rates might result from content created by influencers that connects to their target market [5]. Connected consumers are more inclined to get brand evangelists and return customers. Companies use influencers to foster sustainable and lasting connections with clients as they strive to go from just acquiring new clients to retaining existing ones and committing to them [34]. IM has a substantial effect on acquiring new customers since it cultivates long-lasting relationships, drives sales, target certain groups and sustains a high degree of participation [28]. It's a tactic that fits into the need of contemporary customers for real, intimate brand experiences [33].

2.3 *Customer Behaviour (CB)*

CB is an investigation of the way consumers select, utilize and discard goods and services as well as the variables that affect these choices. It entails the interaction interplay of behaviour, mental processes, affect and external events [13]. It's the societal, psychological and bodily actions people take when they purchase, utilize and discard goods, services, concepts and customs [12]. It includes the methods by which customers choose what to buy as well as the elements that encourage or impact these choices [2]. It is a psychological analysis of the various CB trends that affect their choice-makinh process prior to completing any purchases [1]. The purchasing habits of customers is influenced by their choices, views and feelings. Analysis of CB relates to the purchasing habits of an individual taking into account background variables, social developments and frequent trends that may have an impact on their choice to make a purchase [12]. Businesses need to comprehend CB since it has a direct impact on the creation of products, advertising tactics and the customer encounter in general [7]. Businesses may make the buying process less useful and interesting for everybody by personalising it based on their comprehension of CB [6]. A greater level of fulfilment and rate of conversion result form this. By locating and removing obstacles in the purchasing process, CB analysis facilitates the buying procedure and raises conversion rates. By providing more specialised goods and services that cater to customers' unique requirements and preferences, firms can raise the value in dollars per client through comprehending behaviour patterns [30]. Understanding CB helps to ensure that company services, marketing initiatives and product designs are appealing to the intended market [20]. Behavioural psychology may be used to enhance relations with clients, which will enhance the client experience and boost loyalty [33]. Businesses have the power to shape nascent habits and reinforce good new attitudes in consumers, which may contribute to long-term behavioural shifts [37]. Greater profitability may come from recognizing and addressing CB since it makes targeting and managing client relationships more efficient [42]. Analyzing CB is crucial to developing plans that meet the wants and demands of customers, which will eventually end up in a more prosperous and successful firm [35]. Building a rapport of respect and confidence with customers is key to fostering brand loyalty and maintaining a competitive edge [43]. Because influencers have such as powerful effect on the choices and behaviours of their followers, there is a strong correlation between IM and CB [35]. Influencers have the power to change CB by using their audience's trust and thier reputation. Purchase decisions may directly alter as a result of their recommendations [31]. Today's customers are less open to direct marketing and more interested in genuine experiences [32]. Influencers have the ability to successfully influence customer habits due to their alleged authenticity and actual life experiences [39]. Influencers frequently specialise in a certain industry allowing for extremely focused marketing campaigns [34]. IM allows brands to target particular markets resulting in more effective and memorable advertising campaigns [38]. A brand earns credibility by association as it cooperates with an influencr. Customer sentimens towards a brand can be greatly influenced by the influencers trustworthiness, past experiencs and mindset [33]. Influencers have an impact on customer's for a long-time attitudes and opinions about businesses and goods in addition to their immediate purchasing behaviour [23]. IM has an important effect on CB influencing anything form instant buying choices to loyalty to a company overtime [24]. This can result in enduring shifts in CB and brand devotion. Brands can use it as a very effective tool in their advertising efforts particularly in online environments where customers appreciate suggestions from friends and family and authenticity [20].

2.4 Customer Satisfaction (CS)

CS a gauge of how satisfied customers are using a business's goods, services and capabilities indicatinv how effectively these fulfil or exceed their requirements [29].

CS score, ranking and quality measure of how well an item or service corresponds to or exceeds client demands frequently employed to assess a business's performance through the viewpoint of its clients [32]. The assessment of a business's customer's interactions with its goods and services as well as the degree to which it's offerings either match or exceed expectations [31]. It is a gauge of how successfully a firm meets customer expectations through its goods, services and overall client experience [40]. This serves as a barometer for both CS and the business's overall health [34]. A crucial indicator for firms is CS, which shows how well goods and services correspond to or exceed what consumers want [26]. It's significance are not emphasized because it has a direct impact on staff retention, loyalty and the general health of a business [31]. Customers that are happy with a brand were more inclined to stick with it, which lowers customer turnover and develop ls a steady client base [38]. When customers are satisfied, they are more likely to come back and make more purchases [20]. Overtime, happy customers prefer to spend more increasing the lifetime value to the business [22]. Word-of-mouth and internet reviews from satisfied customers can result in recommendations and the acquisition of new clients [25]. Ratings on CS can reveal information about how well a company's assistance and assistance teams work [23]. CS may be a crucial distinction in an overcrowded marketplace enticing buyers to pick the company over rivals [36]. Profitability and CS are directly correlated. Increased net revenue may be attributed to greater amounts of CS since happy consumers are inclined to recommend the brand to others, which boosts it's reputation and draws in new business [41]. CS is critical to the well-being and expansion of a firm. Satisfaction surveys may identify opportunities for enhancement assisting companies in fine-tuning their offers and customer service strategies [20]. It affects every facet of a business's activities including customer service, manufacturing, sales and advertising. Businesses may increase customer loyalty enhance financial success and fortify their standing as market leaders by putting a high priority on CS. CS and IM are closely related with IM has a big effect on CS. By giving a brand legitimacy, influencers can increase CS by giving consumers greater confidence to make purchases [10]. Influencers facilitate more intimate brand-customer interactions, which can increase CS by making them feel as though their needs and desires have been taken into consideration [14]. Influencers frequently provide an immediate means to communicate between the business and it's clients enabling immediate feedback and swift correction of any problems, this enhancing CS [13]. When influencers positively impact CS, it may result to higher intent to buy within the audience. Businesses may improve CS by using IM to change their attention to just CA and CR and dedication and this in consequently encourages brand loyalty [6]. Good experiencs encouraged by influencers may boost CS that in consequently encourages brand loyalty. By fostering a loop of positive feedback, increasing engagement and fostering trust, IM significantly contributes to the molding of CS [25]. If companies want to succeed in the long run and keep their consumer base happy the connection is essential [22].

2.5 Customer Retention (CR)

The ability of a business to maintain its clientele over decades and promote returning customers is known as Customer retention (CR). It gauges consumer satisfaction alongside a business's goods and services in addition to their propensity to stick with it in the long-run [35]. Connecting current consumers to sustain their devotion and promote repeat business is known as CR. It is a crucial performance metric for businesses indicating the calibre among their goods and services as well as the general satisfaction with their clients [32]. The percentage of returning consumers that show ongoing appreciation within the client based was a good indicator of CR [39]. Rate of turnover usually shows the proportion of consumers who keep utilizing a business's goods or services as opposed to the number whose discontinue doing so [33]. It is sometimes stated either an amount or percentage. It entails tactics meant to establish enduring bonds with clients and guarantee that they will keep picking a business's products over rivals [31]. Techniques for retaining customers that work may improve profits and make a company's structure stronger. Connecting alongside current consumers to preserve their devotion and lower the risk of attrition is known as CR [18]. It entails establishing enduring bonds with clients by giving consumers a satisfying experience via efficient communication, first-rate client care and customised offerings [34]. An evelanted level of CR were a sign of happy clients and a profitable business plan. Improved profitability may result from preserved client's tendency for making numerous purchases offer helpful suggestions and recommed new clients [5]. IM and acquiring clients combined constitute an effective approach in contemporary business procedures [21]. Any firm that wants to succeed needs to focus on acquiring clients, which involves finding ways to draw in as well as win over new clients. It is essential to an organisation's development, marketplace expansion and permanent success [3]. In order to encourage a company, IM makes use to the reputation and reach of people who have an extensive social media following. Influential Individuals may introduce an organisation to more consumers considerably enhancing the company's awareness and supporting it's CA activities [24]. It's a successful method to swiftly create trust with targeted audiences. The Influencers recommendations may give a business immediate trust and increase the receptivity of prospective consumers for converting attempts [25]. Influencers frequently serve particular markets, which makes it possible for CA efforts to choose audiences with greater accuracy [4]. Because of this targeted strategy and influencers' natural reach, IM may be more affordable than conventional marketing when it comes to bringing in new clients [1]. IM was a useful tactic for companies since it may be quantified immediately by looking at response rates, sales conversations and return on investment [43]. IM and acquiring clients working together may result in an additional flexibile and successful client growth and market share approach [2]. Companies may enhance their CA operations and achieve higher enthusiastic and committed CR by utilizing their confidence and accessibility of influencers[6].

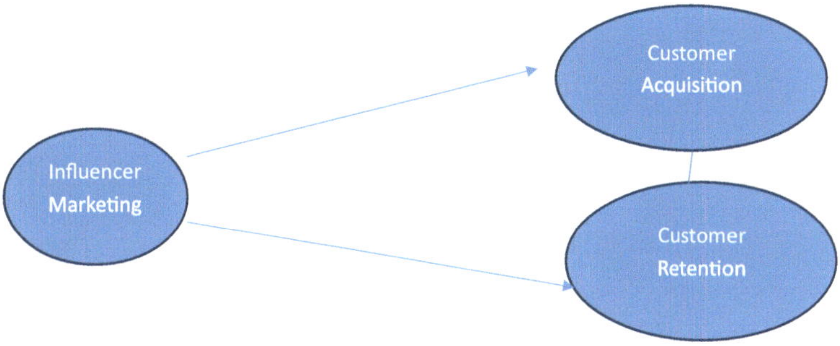

Fig. 1 Conceptual Framework of the study

3 Hypothesis Development

We have generated the following hypotheses, which are linked visually as shown in Fig. 1, based on our results from previous research.

H1: Influencers marketing has a positive effect on customer acquisition.

H2: Influencers marketing has a positive effect on customer retention.

H3: Customer Acquisition has a positive influence on Customer retention.

4 Methodology

The straightforward random sample procedure, a probability sampling approach typically chooses participants among the overall population at randomly and ignores demographic classification, is used within the quantitative approach employed in this research. The research was done at Unilever Pakistan. A questionnaire was used to collect information about an intake of 386 individuals, while the SMART-PLS program was then used for analysis as given in Table 1.

5 Data Analysis

The sample's gender composition appears to be almost equal according to the demographic information making it helpful for a fair examination of consumer behaviour through genders. The age breakdown reveals a notable prevalence in the 25–34 age range indicating that young people make up the bulk of Unilever Pakistan's clientele. This is consistent with the general pattern with these ages becoming very engaged customers. According to the educational background information a significant proportion of those surveyed hold a bachelor's or higher degree suggesting

Table 1 Demographic profile

Variables	Number of respondents	Percentage
Gender		
Male	195	50.5%
Female	191	49.5%
Age		
18–24 years	77	20%
25–34 years	154	40%
35–44 years	92	24%
45–54 years	41	10.6%
55+ years	22	5.7%
Education		
High school	58	15%
Bachelor's	174	45%
Master's	123	32%
PhD or Higher	31	8%
Employment status		
Full-time employed	289	74.9%
Part-time employed	58	15%
Unemployed	23	6%
Student	16	4.1%
Monthly income (PKR)		
Under 30,000	58	15%
30,000–60,000	116	30%
60,001–90,000	102	26.4%
Above 90,000	110	28.5%

indicating the client population is educated plus might posses higher standards of preferences and higher demands for consumer products. The bulk of the participants work full-time that may indicate a steady income and therefore more money available for consumption of consumer items. WihtA sizeable percentage of the participants earning over 90,000 PKR the median income category demonstrates a decent distribution across different income brackets and may indicate a stronger purchasing capacity and accessibility to Unilever Pakistan's luxury items. With the assistance of a fictitious demographic data Unilever Pakistan's clientele may be thoroughly viewed. This information can then be utlized to customise product offers and advertising strategies towards the particular needs and tastes of various client categories. We first create an indicator rating in order to assess the efficacy of IM, CA and CR. The most desirable scores could be compared below considering an adequate number of 386 participants and a rating system with five choices (derived strongly disagreeing to strongly agree). The optimal score for IM is 9.650 which is obtained by dividing the biggest response rating with the total amount of people surveyed and reaction alternatives. The formula used to determine the ideal score for CA yields a comparable result 9.650. with just four response alternatives the optimal rating overall CR is 7.720 which is calculated as the good of the most favourable response value a single fewer reaction choice plus the total amount of responders. In order to evaluate the comparative efficacy of the various tactics under consideration actual survey responses may be contrasted with these highest achievable ratings as shown in Table 2.

Utilizing a Likert scale 386 respondents were asked to rate their opinions of IM, CA and CR. The results are shown in the following table as follows strongly disagree (SD), disagree (D), neutral (NA), agree (A) and strongly agree (SA). IM has a cumulative score of 8.210 with IM3 having the greatest specific item rating of 1.711. Rhis implies that participants have a broadly positive opinion of instant messagings efficacy. The overall rating for CA is 8.180 Significantly less then IM with the greatest single component rating reaching 1.669 for CA1. Like IM the replies are biased toward A and SA showing that almost all respondents agree that IM has a good impact of CA. Compared to IM and CA, the overall rating for CR is 6.499 making it less. CR3 has the largest single item rating of 1.658. Even if the overall score was less the pattern for acceptance is still present having the most frequent

Table 2 Respondents assessment

Influencer Marketing (IM)	SD	D	NA	A	SA	Total
IM1	7	19	184	583	814	1605
IM2	5	26	194	652	701	1576
IM3	1	10	79	600	1021	1711
IM4		14	153	646	835	1648
IM5	2	14	121	602	930	1668
Total						8210
Customer Acquisition (CA)						
CA1	2	14	128	576	953	1669
CA2		22	153	625	855	1641
CA3		9	159	587	911	1662
CA4	4	36	233	613	668	1548
CA5		17	137	665	835	1648
Total						8180
Customer Retention (CR)						
CR1	1	13	207	682	697	1599
CR2		18	203	665	715	1601
CR3		15	141	613	889	1658
CR4		20	169	557	894	1641
Total						6499

Table 3 Path coefficients

Path	Coefficient
IM to CA	0.814
IM to CR	0.766
CA to CR	0.759

Table 4 R square results

Relationship	R2 value
IM to CA	0.664
IM to CR	0.779

answers including A as well as SA indicating that IM is still thought to have been helpful for CR albeit maybe to a somewhat lower level that acquire as promoted in the study of Table 3.

The path coefficients from the initial collection that indicate the trajectory and magnitude of the link among IM, CA and CR are displayed in the following table. The IM to CA coefficient of 0.814 indicates a highly favorable connection among IM and CA. The IM to CR 0.766 coefficients, which is marginally less than the IM to CR path but still shows a strong positive association suggests that CA is rather successful in promoting CR.

The values of R2 using the initial collection indicating how much the overall variation in the dependent variables such as CA along with CR might be accounted for by the independent Variable IM are displayed in the Table 4 above. This R2 value indicates that IM may clarify for about 66.4% of the variability in CA. An even larger proportion of the variation in CR may be explained by IM as evidenced by this higher R2 amount of 77.9%.

6 Discussion

The results of this investigation suggest that there exists a substantial positive connection among IM and CA (path coefficients of 0.814), highlighting the effectiveness of IM in bringing unfamiliar consumers to Unilever Pakistan's products. IM also demonstrates a strongly beneficial association with CR (path coefficients of 0.766), suggesting that IM is important for both bringing prospective business and keeping current consumers. The connection among CA and CR additionally displays a favorable path coefficients of 0.759, suggesting that efficient CA techniques are going to result in increased CR prevalence.According by the R2 coefficients for IM's effect on CA 0.664 & CR 0.779, IM offers a strong ability to forecast all results, which helps clarify a sizable amount from the variation in those dependent variables. For Unilever Pakistan, the elevated levels paths coefficients & R2 coefficients offer statistical proof using the efficacy of IM in generating CA as well as CR. From a practical point of view, this implies that Unilever Pakistan's marketing plan ought to prioritize instant messaging. The investigation's general findings support the theories and highlight how important IM is to improving CA and CR.These findings might inform Unilever Pakistan's strategy choices, spending decisions, and marketing strategies to maximize the advantages of IM for long-term, profitable company expansion.

7 Implications

This study has important practical ramfications for Unilever Pakistan on the role IM plays in increasing CA and CR. These consequences also extend to related sectors and markets. Given the compelling data demonstrating the effect of IM on consumer behaviour, Unilever Pakistan ought to think about increasing the amount of money dedicated to influencer campaigns as a component of its tactical advertising spends. Selecting the right influencers is essential. Unilever Pakistan ought to concentrate on collaborating with influencers whose ideals mesh perfectly with the company's image and who's characteristics match their intended audience. IM must be combined with additional advertising channels to create a unified approach that makes the most of each platform's advantages and improves both acquisition and retention. Statistical instruments including as path Coefficients and R2 values ought to be routinely employed in the evaluation of ad efficacy. This data-driven method can assist in optimising marketing tactics for increasef return on investment. The study's conclusions can be applied to improve customer pleasure along with at every touchpoint throughout the buying process, which will increase retention rates. Since CA and CR are positively correlated, beginning advertising campaigns must focus on developing long-term connections with consumers in addition to sales. Gaining frequently seen awareness of buyer habits and tastes through the measurement of influencer campaign efficacy with regard to of CA and CR may result to improved marketing strategies. This study emphasizes how crucial it is to keep up with evolving customer tastes and actions, which are greatly impacted by influencers on online platforms. Unilever Pakistan can use the investigation's practical recommendations to improve its IM tactics. Through comprehenson and implementation of these results, the organisation can enhance its CA and CR strategies culminating in long-term business expansion and a more robust competitive advantage within the industry.

8 Conclusion

The study's findings, which looked at how well IM works for Unilever Pakistan in terms of attracting new customers and keeping existing ones. The article unequivocally shows that IM is an effective strategy for bringing in prospective customers and keeping hold of existing clients. The strong path coefficients show that IM has a considerable predictive ability on CA and CR as seen by the R2 values, which account for a significant amount of the variation in these results. The acquisition and retention of customers are mutually beneficial. Creating the conditions for enduring client loyalty is just as important as growing the amount of customers through effective acquisition techniques. The results indicate that Unilever Pakistan can reap significant benefits frim intentional expenditures in IM including the expansion and sustainability of their client base. The study emphasizes how cirital it is for marketers to adopt a data-driven strategy in which statistical proof not but feeling, guides decisions. In order to provide a cohesive customer experience, it promotes a comprehensive marketing strategy that incorporates IM with other channels. The results emphasize the importance of having a customer-centric approach since improved satisfaction and loyalty can result from knowing and meeting client preferences. Potential areas of study are made possible by this study, especially when it comes to examining the long-term impacts of IM on client lifetime value and reputation. The Study offers thorough and useful information that can have a big impact on Unilever Pakistan's marketing plans. Effective use of IM can help the firm improve its attempts to attract and keep customers strengthening its standing in the marketplace and fostering continued success.

References

1. Al-aziz, H.Y., Monalisa, S.: Comparison of facebook and instagram to assess the efectiveness of advertising channels in customer acquisition. Jurnal Teknologi Informasi dan Pendidikan **15**(3) (2023)
2. Ascarza, E., Neslin, S.A., Netzer, O., Anderson, Z., Fader, P.S., Gupta, S., ... Schrift, R.: In pursuit of enhanced customer retention management: review, key issues, and future directions. Cust. Needs Solut. **5**, 65–81 (2018)
3. Abbas, M., Ibrahim, I., Hasanuddin, R., Fitri, F., Umar, R.: The effect of affiliate programs and consumer behavior on profit margins: the mediating role of Customer Acquisition Cost (CAC) and customer loyalty. Atestasi: Jurnal Ilmiah Akuntansi **6**(2), 453–468
4. Bakri, Z.F.: Analyzing the influence of digital marketing strategies on business performance in the beauty industry: a comprehensive analysis of social media engagement and influencer collaborations. J. Econ. Manag. Bus. Technol. **2**(1), 37–48 (2023)
5. Belanche, D., Casaló, L.V., Flavián, M., Ibáñez-Sánchez, S.: Understanding influencer marketing: the role of congruence between influencers, products and consumers. J. Bus. Res. **132**, 186–195 (2021)
6. Bakker, D.: Conceptualising influencer marketing. J. Emerging Trends Mark. Manag. **1**(1), 79–87 (2018)
7. Berne-Manero, C., Marzo-Navarro, M.: Exploring how influencer and relationship marketing serve corporate sustainability. Sustainability **12**(11), 4392 (2020)
8. Campbell, C., Farrell, J.R.: More than meets the eye: The functional components underlying influencer marketing. Bus. Horiz. **63**(4), 469–479 (2020)
9. Cordova-Buiza, F., Urteaga-Arias, P.E., Coral-Morante, J.A.: Relationship between social networks and customer acquisition in the field of IT solutions. IBIMA Bus. Rev. **10**, 1–10 (2022)
10. De Veirman, M., Hudders, L., Nelson, M.R.: What is influencer marketing and how does it target children? A review and direction for future research. Front. Psychol. **10**, 498106 (2019)
11. Dimitrieska, S., Efremova, T.: The effectiveness of the influencer marketing. Econ. Manag. **18**(1), 109–118 (2021)

12. Feng, Y., Chen, H., Kong, Q.: An expert with whom i can identify: the role of narratives in influencer marketing. Int. J. Advertising **40**(7), 972–993 (2020). https://doi.org/10.1080/02650487.2020.1824751
13. Gachassin, E., Prentice, C., Thaichon, P. (2023, April). The influence of brand digital storytelling using the hero's customer journey communication technique on customer acquisition and retention. In: Digital marketing & eCommerce conference, pp. 129–136. Springer Nature Switzerland, Cham
14. Ginting, Y., Chandra, T., Miran, I., Yusriadi, Y.: Repurchase intention of e-commerce customers in Indonesia: an overview of the effect of e-service quality, e-word of mouth, customer trust, and customer satisfaction mediation. Int. J. Data Netw. Sci. **7**(1), 329–340 (2023)
15. Halvadia, N. B., Menon, S.: A study on the impact of existing and emerging trends in digital marketing on consumer buying behavior. SKIPS Anveshan **2**(1)
16. Hudders, L., Lou, C.: A new era of influencer marketing: lessons from recent inquires and thoughts on future directions. Int. J. Advert. **41**(1), 1–5 (2022)
17. Haenlein, M., Anadol, E., Farnsworth, T., Hugo, H., Hunichen, J., Welte, D.: Navigating the new era of influencer marketing: how to be successful on Instagram, TikTok, & Co. California Manag. Rev. **63**(1), 5–25 (2020)
18. Hamzah, A.A., Shamsudin, M.F.: Why customer satisfaction is important to business?. J. Undergrad. Soc. Sci. Technol. 1(1) (2020)
19. Irawan, C., Susanto, B., Juliadi, R.: Implementation of digital marketing at printing company to increase customer acquisition and customer retention (Case Study: Citra Kreasindo Mandiri Company). In: Conference Series, vol. 3, no. 1, pp. 619–639 (2020)
20. Jin, S.V., Muqaddam, A., Ryu, E.: Instafamous and social media influencer marketing. Mark. Intell. Plan. **37**(5), 567–579 (2019)
21. Kumar, V., Reinartz, W., Kumar, V., & Reinartz, W. (2018). CRM in social media. Customer Relationship Management: Concept, Strategy, and Tools, pp. 311–326
22. Kusuma, A.: Unveiling the impact digital marketing strategies and business performance in the beauty industry. J. Econ. Manag. Bus. Technol. **2**(2), 65–72 (2024)
23. Kadekova, Z., & Holienčinová, M.: Influencer marketing as a modern phenomenon creating a new frontier of virtual opportunities. Commun. Today **9**(2) (2018)
24. Khattal, M.H.D., John William, A.: Assessment on enhancing business customer acquisition for Autonomous Mobile Robots (AMR) in industrial automation and robotics industry. In: International Conference on Economics, Business and Sustainability, pp. 257–264. Springer Nature Singapore, Singapore, Jan. 2023
25. Leung, F.F., Gu, F.F., Palmatier, R.W.: Online influencer marketing. J. Acad. Mark. Sci. **50**(2), 226–251 (2022)
26. Lendvai, E., Tóth, G.: Customer acquisition activities of web stores. Analecta Technica Szegedinensia **16**(1), 90–96 (2022)
27. Martínez-López, F.J., Anaya-Sánchez, R., Fernández Giordano, M., Lopez-Lopez, D.: Behind influencer marketing: key marketing decisions and their effects on followers' responses. J. Mark. Manag. **36**(7–8), 579–607 (2020)
28. Masuda, H., Han, S.H., Lee, J.: Impacts of influencer attributes on purchase intentions in social media influencer marketing: mediating roles of characterizations. Technol. Forecast. Soc. Chang. **174**, 121246 (2022)
29. Pick, M.: Psychological ownership in social media influencer marketing. European Bus. Rev. **33**(1) (2021)
30. Patel, M.R.: Customer acquisition and retention in E-commerce using AI & machine learning techniques. J. Harb. Eng. Univ. **44**(8) (2023)
31. Rathore, B.: Exploring the impact of digital transformation on marketing management strategies. Eduzone: Int. Peer Rev./Ref. Multidiscip. J. **8**(2), 39–48 (2019)
32. Sandra, J., Indrayanti, S., Suryana, A.: Influencer marketing strategies and brand image in boosting consumer purchase intent: the role of customer support intervention. Jurnal Informatika Ekonomi Bisnis 12–16 (2024)

33. Sabbeh, S.F.: Machine-learning techniques for customer retention: a comparative study. Int. J. Adv. Comput. Sci. Appl. **9**(2) (2018)
34. Sharma, A., Harsora, H., Menon, S.: A study on the impact of existing and emerging trends in digital marketing on consumer buying behavior. Int. J. Creative Res. Thoughts (IJCRT) **10**(2), 652–662 (2022)
35. Taylor, C.R.: The urgent need for more research on influencer marketing. Int. J. Advert. **39**(7), 889–891 (2020)
36. Vinerean, S., Opreana, A.: Social media marketing efforts of luxury brands on Instagram. Exp. J. Mark. **7**(2), 144–152 (2019)
37. Vrontis, D., Makrides, A., Christofi, M., Thrassou, A.: Social media influencer marketing: a systematic review, integrative framework and future research agenda. Int. J. Consum. Stud. **45**(4), 617–644 (2021)
38. Visser, M., Hoving, M., van Kempen, C., Jansen, M., Kloos, M., van Ommen, R.: Customer Acquisition: recruiting visitors through owned or earned channels. In: Digital Marketing Fundamentals, pp. 210–274. Routledge (2019)
39. Ye, G., Hudders, L., De Jans, S., De Veirman, M.: The value of influencer marketing for business: a bibliometric analysis and managerial implications. J. Advert. **50**(2), 160–178 (2021)
40. You, Y., Joshi, A.M.: The impact of user-generated content and traditional media on customer acquisition and retention. J. Advert. **49**(3), 213–233 (2020)
41. Yalamati, S.: Revolutionizing digital banking: unleashing the power of artificial intelligence for enhanced customer acquisition, retention, and engagement. Int. J. Manag. Edu. Sustain. Dev. **6**(6), 1–20 (2023)
42. Zhang, L., Erturk, E.: Potential lessons from Chinese businesses and platforms for online networking and marketing: an exploratory study. Soc. Sci. Hum. Open **6**(1), 100274 (2022)
43. Zheng, R., Li, Z., Na, S.: How customer engagement in the live-streaming affects purchase intention and customer acquisition, E-tailer's perspective. J. Retail. Consum. Serv. **68**, 103015 (2022)

Analysing the Impact of Hosting Influencer Giveaways on Purchase Intention of Luxury Items by Generation Z

Nibir Khawash, Prasenjit Chakravarty, and Sudeshna Pahari

Abstract The dynamic changes in purchase intentions for Gen Z consumers within the specific Indian context have been potentially impacted by influencer marketing. With the increasing accessibility to internet connectivity and the use of mobile phones in India among the Gen Z population, social media influencers have grabbed the attention of their followers. It leverages the cost-effectiveness of marketing efforts for different luxury brands in India. Since self-concepts, self-esteem, uniqueness, and product values drive young consumers' purchase intention for luxury goods, the brands get the opportunity to embrace the same in the Indian luxury market. The research investigates the potential impact of hosting giveaway campaigns through influencers in different platforms on the luxury product purchase intention of the Gen Z population in India. A quantitative study is conducted to gather diverse perspectives for influencer giveaways. The potential impact is analysed by adopting regression coefficients, and ANOVA tests through SPSS software for the identified variables. The findings have highlighted the potential implications of giveaway platforms, give-away mechanisms, giveaway items, and influencer types for the purchase intention of Indian Gen Z consumers.

Keywords Influencer giveaways · Gen Z consumer behaviour · Luxury goods purchase intention · Influencer marketing in India · Gen Z population in India

N. Khawash (✉) · P. Chakravarty · S. Pahari
Amity Business School Major Arterial Road, AA II, Newtown, Kadampukur 700135, WB, India
e-mail: khwashnibir@gmail.com

P. Chakravarty · S. Pahari
Amity Business School, Amity University Kolkata, Kolkata, India

S. Dutta et al. (eds.), *Advances in Data Analytics for Influencer Marketing: An Interdisciplinary Approach*, Information Systems Engineering and Management 9,
https://doi.org/10.1007/978-3-031-65727-6_16

1 Introduction and Background

In the dynamic social media influencer space, the influence on Generation Z population is more likely to be in the fashion industry. The rising popularity and brand endorsement by various social media influencers (SMIs) enable a significant change in the purchase behaviour of the Gen Z population. The specific population considered is the individuals born between 1997 and 2010 in India. Average Indians have started social media accounts after the revolution named digital India and many people feel lost without using their social media handles. In this study, a detailed evaluation of the purchase behaviour of Gen Z is done while considering the impact of hosting influencers through different online platforms. The potential impacts of hosting influencers and their giveaways through different online platforms are going to be analyzed by focusing on the Gen Z population and their intention to luxury goods in India. The study considers the quantitative survey of a significant number of individuals in the Indian luxury fashion market to conclude the findings of the study.

According to a report, Indians are used to engaging in their mobile phones for an average of 4.9 h per day which is ranked 8th in most often mobiles used around the world. In 2024, the fashion industry in India is estimated to reach a revenue of $105.50 billion and this is due to the rising influence of the influencers for fashion apparel and luxury items. Another study has suggested that 76% of total customers in the fashion industry find potential fashion deals through different social media and other online platforms. The Instagram reels are the key content segment that has influenced the purchasing of fashion apparel among Indians because of its popularity among the Gen Z population. For the specific fashion market of India, people often believe in Indian influencers rather than global celebrities and influencers. The shared information and brand awareness about the potential luxury brands are significantly trusted by Indians [1, 2]. The digital population in India has grown significantly over the decade by embracing internet services effectively and India holds the record number of active internet user base with the number of around 900 million. Emerging social platforms like YouTube, Instagram, Facebook, and Snapchat have potentially contributed to the rise of the numbers in India. In 2022, the number of active Facebook users has surpassed the other nations with the evolving rate of cheaper data packs and internet services. Due to the potential internet accessibility and better network connectivity, Indians are attracted to the digital lifestyle. Among the overall Indian population, the Gen Z population has significantly captured the potential opportunities for active networking through online accessibility. Influencers have also utilized the digital landscape of India to increase their reach among the young population in India. As most of the population in India consists of the young generation, there is great potentiality for digital marketing of different products [3]. Social Media Influencers (SMIs) refer to the micro-celebrities or opinion leaders in online media platforms and they have established themselves as credible experts for specific products or topics. These influencers are not only regarded as the reflection of a human being but also as an embodied digital agent by linking the designs based on their narratives and beliefs. As consumers are more likely to show significant trust in the digital influencer, luxury brands are partnering with the influencers for an effective brand communication strategy. Digital influencers are often regarded as opinion leaders in online platforms as they can effectively mediate brand message sharing to the digital communities in India. In recent years, brands have recognized the increasing influence of celebrities on Gen Z and they have included influencer marketing strategies as a key part of their digital marketing. Effective communication on branding is carried out by the influencers by highlighting their potential characteristics in digital marketing including their content quality, expertise, trustworthiness, and social interactions through the promotional brands. Luxury brands in India have adopted the practice of hiring digital influencers as online opinion leaders so that they can disseminate brand messaging and influence the digital communities [4].

To run a successful giveaway program, influencers are required to gain significant social media followers and website traffic to boost overall brand awareness among the followers. The significant presence of influencers in the niche marketplace makes it quite effective in launching luxury products for specific consumer segments. The rise of internet-based media platforms has enabled the influences of different types of celebrities including athletes, artists, pop stars, thought leaders, content creators, bloggers, and other micro-influencers. In the marketing strategies, the content creators, micro-influencers, and bloggers are engaged actively as they are a cost-effective option for the growing luxury brands in a new marketplace because of their credibility in the market. An influencer giveaway campaign enables the brand to utilize the authority and credibility of the reputation of various influencers. When the influencers are engaged in promoting the giveaways of specific luxury brands, the followers of the influencers are more likely to participate as per the recommendations of the influencers. The partnership with influencers also provides the scope for increasing awareness and learning about the brand. Increasing brand awareness can target the particular audience who are likely to get interested in the market niche and levels that are mentioned in Fig. 1. Through the strategy, the brands can effectively gather the followers to the platforms improving the engagement rates while generating more sales. The accessibility and efficiency of the giveaway virtually make it more effective in running the brand awareness program and generating sales leads. The influencer giveaway is significantly recommended to the growing luxury brands in the Indian sub-continent as it provides total control or the promotional aspects, programmer designs and customized content. It has the potential to gather direct website traffic rather than depending on different third-party platforms [5, 6].

India's luxury market is expected to show potential growth of up to 3.5 times its current market size and it is expected that it will reach the mark of $200 billion by the year 2030. Different studies have also suggested that more than 50% of luxury item buyers are using social media platforms significantly in India to discover trending luxury trends. 85% of the buyers have also reported that the virtual giveaways have shaped their purchase decisions in the market. A significant dichotomy is present in the Indian market between traditional luxury retail and modern luxury retail. Although traditional luxury products are considered more scarce and exclusive in the Indian market, modern luxury items are accessible to the young population. The

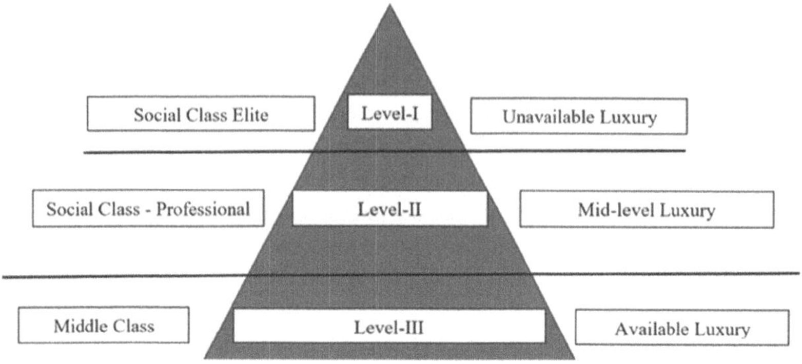

Fig. 1 Different levels of luxury products [5]

evolution in the luxury retail industry effectively focused on innovative strategies like the effectiveness of digital marketing, virtual showrooms, online influencers, live streaming shopping, and potential imports of new product categories into the marketplace. The ongoing trend is supported by the rising disposable income of Indian communities along with the significant exposure to worldwide trends of luxury goods. The exposure is highly obtained from the rise of digital platforms and people prioritize high-quality and uniquely designed products. The preferences of consumers are continuing to evolve with the increasing income and consumer mindset. It adds potential value to the storytelling and virtual experiences for luxury fashion brands. After the rising trend of e-commerce in the rural and small cities in India, people have realized the potential aspiration for a better lifestyle [7, 8]. Endorsing brands by influencers has generated positive sales outcomes for the brands as they are expected to influence their follower communities. However, potential strategies for marketing through key influencers are not developed in the present case scenario for the specific population in India. Although some studies have addressed the rise of digital influencers in the Indian market, potential topics focusing on luxury goods through giveaway programs are not covered in the studies. The following study is intended to bridge the gap by drawing the relationship between luxury brands with the consumers' purchase goals. The key characteristics of digital influencers along with their strategies to endorse the brands including giveaway programs have been addressed to explore efficient relationships with the brand and consumers. The purchasing behaviour of Gen Z and their luxury product preferences are also relevant to the study. Moreover, the relevance of the potential impact of sponsored blog posts and digital content creation has a relative influence on consumers. The popularity of the influencers can be endorsed with the marketing strategies to make aware of the product's potential consumer bases [9, 10].

In studying the above topic, the key objectives for the study include the analysis of the potential impacts of various influencer giveaways on the purchase decisions for luxury goods in India. It is going to identify the particular characteristics of the give-aways of influencers including the giveaway platforms, influencer types, giveaway mechanisms, and luxury items. By considering the most efficient characteristics of the campaign, the potential influences for the underlying mechanisms of Gen Z's luxury goods purchasing are going to be investigated through the study.

2 Discussion on Previous Literature

Vijaya et al. have conducted a similar study on the topic of Gen Z purchase intentions by comparing it with the purchasing attitude of millennials in India. They have considered the variable of purchase intention as the centre of their analysis based on the potential marketing strategies. The 4Ps including price, product, promotion, and place have been included as effective marketing strategies for evaluating the intention of purchasing in India. They have adopted the Hayes process-macro and regression analysis by using the SPSS software to find the potential strategies for the 4Ps in marketing while evaluating the impacts on purchase intention. The study has concluded that 4Ps strategies have a potential impact on online purchasing depending upon the income and age of the population. The evaluation of the moderators including the gender, age, and income of the considered population has provided the neutral influence of gender in consumer purchasing. Although the study has focused on the overall impact of purchasing decisions based on the 4Ps among Millennials and Gen Z, it did not evaluate the impacts of specific marketing strategies of influencer giveaways to Indian Gen Z. Moreover, the study has considered the participants from mainly the South Indian continent particularly from the state of Tamil Nadu limiting the potential findings for the overall Indian population [11]. Hemantha has significantly emphasized the strategies of celebrity endorsement for luxury brands in his book. The author focuses on evaluating the purchasing patterns among Gen Z customers by identifying the luxury brand's business goals for targeting the minds of consumers. He has cited that celebrity endorsement is considered one of the regular processes for promoting luxury items and raising awareness within any industry. In his study, the exploration of the mixed responses of Gen Z's consumer patterns has been made with the advancement of technologies in the market. The conclusion of the study has brought the moderate impact of celebrity endorsing on the purchasing patterns of Gen Z customers as consumers are inclined to social media influencers regarding learning about lifestyles and luxury fashions. The expertise and credibility of the celebrity significantly contributed to the strategic evolution to influence consumers. The study has directed to focus on influencer giveaways leveraging the study about the considered topic. From the insights, potential strategies for luxury brands can be effectively analyzed depending upon the changing preferences and attitudes towards influencer giveaways [12].

Singh et al. have studied the potential factors contributing to the interest and purchase intention for fashion clothing among Gen Z consumers in India. The study has provided a significant framework involving the potential factors which determine the motivation of purchasing clothing items in the specific Indian context. Authors have identified the potential predictors for analyzing the interest of Gen Z customers including the self-concept, uniqueness of products, brand image, perceived brand quality, and word-of-mouth promotions. By employing structural equation modelling, the authors have analyzed the data from the questionnaire survey of 211 customers across India. The key aspects of the findings are established based on the customer's role of self-concept, uniqueness, and brand image. Although the perceived clothing quality within the industry along with the word-of-mouth promotions have a low impact on clothing fashions among young consumers, other three factors have influenced consumers' purchasing decisions. The findings from the study have also supported the role of self-concept, uniqueness, and brand image in increasing interest in the clothing styles of Gen Z people. This study has provided significant opportunities to continue the study on influences on Gen Z's purchasing intention from other potential luxury retailers in the Indian market [13].

3 Consumer Behaviour of the Gen Z Population Towards Luxury Items in India

In 2022, the luxury industry has outlined the potential factors for its growth due to the rising aspirations of the population of Gen Z. They account for the entire growth of the luxury market in India and this generational factor is the key contributor to the market development. As India holds the highest population for Gen Z covering up to 51% of the population, the willingness of the generation to spend more money in enhancing their lifestyle has brought the opportunities for the luxury brands. More than 700 million young populations have contributed to the growth of the industry and this 1/3rd of urban India is significantly interested in spending more in their progress and success in the society. In today's market landscape, Gen Z is shifting their attitude to check for the brand's social media presence. In the online and e-commerce market, the increasing trends of transforming brick-and-mortar stores into online e-commerce are relevant to understanding the shopping pattern changes. As the population has preferred to shift their focus to the online website of the brand, the brands are required to strengthen their presence in the digital market. Due to the potential value of social media for the brand to increase sales, the brands try to improve the trust, reputation, and value of their product categories. By creating such dimensions, the brands are effectively communicating with potential customers. In India, the lack of potential awareness about luxury items is relevant and thus influencer marketing has emerged as the suitable option to develop their marketing criteria [14].

After the pandemic, the increasing trend of digital platforms necessitates the online presence of luxury brands creating potential value to consumer preferences. Authentic and reliable advertisements through online platforms have gained significant popularity among the population. Potential differences between artificial and original ads are distinguishable as consumers prefer the original ads over any artificial aspects of marketing. The brands are required to avoid artificial ads as well as sponsored ads while emphasizing influencer marketing. As the population significantly trusts the influencers in their unique segments, influencer giveaways can be a great opportunity for the brand to connect with consumers. Now, the Gen Z population is the mainstream customer for luxury items and this category can be divided into four key segments including convenience shoppers, economic quality seekers, deal-hunting convenience seekers, and brand or quality-conscious purchasers. The segmentation of these customers can help in making potential decisions in the marketplace to anticipate the needs of each customer segment. The potential understanding of the internet natives provides the importance of visually appealing content for the brands while emphasizing creating hype for these customers. The potential aspirations for luxury items of these customers are not valued for impressing friends or parents but rather for influencing their peers. They are more likely to be influenced by what they are watching and reading on social platforms while ignoring brand loyalty [15].

4 Influential Factors for Purchasing Luxury Items to Gen Z Consumers

The significant numbers of young consumers of luxury brands are influenced by the global trends from digital platforms. Indian Gen Z population counts for more than 600 million and they are regarded as the digital natives for significant access to social media and other digital platforms. This generation has grown up with the incident of rural-to-urban urban migration, increasing disposable income, urbanization, and enhanced proximity to Western culture. Moreover, digital literacy and the increasingly privatized education system in the country have contributed to the proliferation of the Internet. Therefore, consumers have typical influences for internet sensations while growing up and their behavioural pattern is also impacted by their Western counterparts for their greater influence in the digital industry. The potential attitudes and beliefs of consumers related to luxury goods refer to the purchasing motivations of consumers. The significant attention, curiosity, and concern about luxury lifestyles are also regarded as the internal influencers for luxury goods among customers. Since the retail reforms in 2014 in India, different international and Indian luxury fashion brands have strengthened their presence in the country by effectively anticipating the lifestyle needs of young consumers. As they have grown up in the digital landscape, they are harmonized with the latest trends in luxury lifestyles by celebrities and influencers [16]. Despite the influences of potential digital influencers, the generation of customers is neither loyal to the brands nor conscious about pricing for luxury goods as per their parents. This particular customer segment potentially seeks instant gratification from the brand while searching for the uniqueness and quality of the brand. The engagement of this customer might impact the production and innovative brand strategies for different luxury item manufacturing companies. The brand's potential recreational strategies aligning with the expectations of the customers have

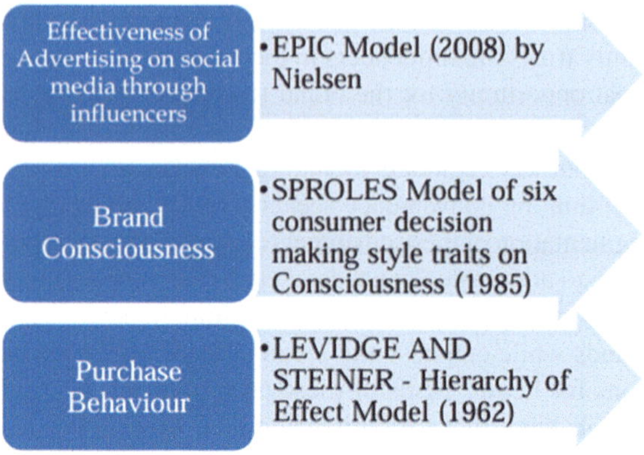

Fig. 2 Different theories for influencer marketing [18]

significantly changed for differentiating products in the current market landscape [17].

To potentially adopt significant lifestyle and fashion trends, Sproles's theory has provided significant strategies for identifying the key variables predicting the possible factors for adopting particular fashion trends. In product development strategies, it has become crucial to identify consumer expectations and brand differentiation strategies. Sproles has indicated the prediction of key hedonic and symbolic interests for consumers as per their emotional responses, personal beliefs, attitudes, curiosity, and attention to existing lifestyles. Hedonic values are regarded as the potential motivation factors for purchasing luxury products by young customers as compared to utilitarian approaches. It is associated with the enhancement of self-esteem, self-concepts, and social relationships. Although the demographic attributes for luxury fashion have become the potential indicators for customers' motivation and preferences, it becomes pertinent to acknowledge the potential psychographic factors for luxury interests and purchase intention [18] (Fig. 2).

The scalability of India's Gen Z population is significantly increasing with the increasing urbanization, middle-class population, and disposable income. Because of the potential income disparities within India along with the societal status in Indian societies, young customers are significantly attached to international and branded luxury items to improve their self-esteem as well as self-concept. The expectation for enhanced status and improved self-esteem is reflected in their luxury choices for luxury retail. Although the pricing of different luxury goods can be considered the key criterion for influencing the purchase decisions of consumers, it does not matter in the case of selecting luxury products for enhanced self-esteem. The understanding of the potential psychographic factors is crucial for luxury brands to anticipate the mindset of the Gen Z population in the country. The brands might be required to anticipate the needs of different customized luxury items along with effective customer services for potential customers. The young customer base often demands unique products because of their individual self-concept. They expect a separate identity from the original products building the self-different9iation concepts which would not be outdated for their fashion choices. Through their luxury choices, they are expecting to pursue symbolical arguments with their social standing, self-image, and self-concept [19].

4.1 Research Hypothesis

By considering the previous studies, a significant hypothesis can be made based on the key variables in the study.

H1: There is a potential relationship between the purchasing intention and influencer giveaway platforms.

H2: There is a significant relationship between potential characteristics of influencers on the giveaway process and purchasing intentions among the Gen Z population.

H3: There is a potential relation between customer's purchasing intention with various giveaway mechanisms.

H4: There is a potential relationship between the purchasing intention of Gen Z customers and giveaway products.

4.2 Proposed Framework

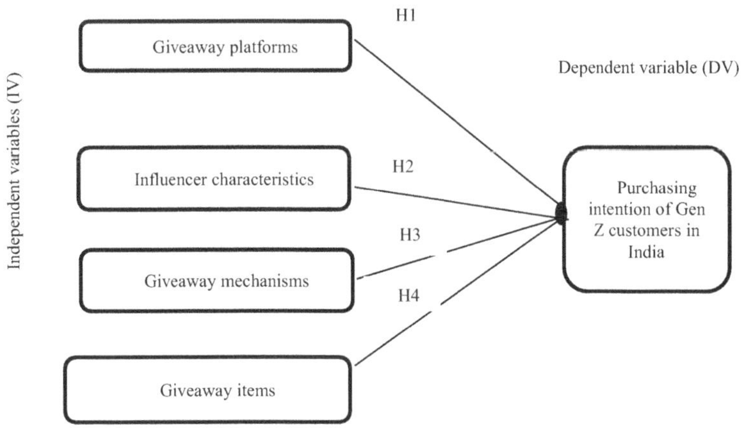

5 Methodology

The study is based on the quantitative method of data collection by deploying the deductive research approach. The methodology of a research study provides a systematic procedure to conduct a research process while entailing the specific methods of data collection, research designs, research approaches, and data analysis process. In analyzing the potential impacts of influencer marketing on Gen Z consumers, the positivism philosophy is considered to design the effective process of data collection and analysis of gathered information in the current marketplace. Based on the factual knowledge about luxury goods preferences among customers, the study has included quantifiable information in the analysis. The deductive approach also enables the researchers to analyse the information about the Gen Z population of India and their potential choices for luxury goods. The approach has also facilitated the analysis process of identifying the specific consumer patterns from the research project about the luxury product market in India. The hypothesis of the research is potentially selected based on the deductive strategies for the collected information about the luxury industry and evaluating the potential impacts on young customers [20]. The explorative design of the study enables potential implications for the identified research problem for the luxury market in India. The investigation of potential insights related to the problems of potential changes in consumer patterns is conducted to establish the foundation of the investigation. Based on the highlighted independent variable of influencer giveaways, the process of influencer marketing, and luxury products, the analysis of purchase intention is included in the study. By facilitating the quantitative survey of Indian consumers, the exploratory design can effectively elaborate the potential implications for the associated variables within Indian luxury retail. To investigate the potential impacts, the sample size of 500 consumers representing a total 472 million population of Indian Gen Z is taken into consideration [21].

5.1 Calculation of Sample Size

The sample size table of Krejcie Moragan is considered in this research to evaluate the impact on to purchase intention of Gen Z. In the survey process, the participants are chosen effectively within the specific age of under 27 years old and are specifically from the Gen Z category of the Indian population. The targeted population for the survey process is India's Gen Z population and the responses from the participants with more than 27 years old are not considered in the investigation. In the sampling process, the desired confidence level is assumed as 95% which is generally considered as the standard confidence level within a research study. The margin of 5% errors is kept for the underlying issues in the true value of the population. As per the significant model developed by Krejcie Morgan, the minimum sample size can be significantly obtained from the estimated total population while considering the standard level

Table 1 Considered sample size calculation from the table of Krejcie Morgan

Factor	Description	Value
Targeted population	Estimated population of Gen Z in India	472 million
Confidence level	Expected level of certainty in ensuring the appropriateness of the results	95%
Margin of error	Acceptable error percentage from the population value	5%
Recommended sample size from the table of Krejcie Morgan	Minimum sample size as mentioned in the table while depending upon chosen confidence level and margin of error	384
Targeted sample size	The estimated sample for the research	500
Selected sample size	Actual number of participants involved in the questionnaire survey	487

of confidence. For determining the sample size for this research, the total Gen Z population has been considered as 472 million and thus the minimum sample can be estimated as 384 from the table with a large population size. By considering a 95% confidence level, the actual size of participants in the survey is considered 487 from the assumed sample size of 500 due to potential dropouts and challenges in ensuring actual participation from the Gen Z population. The following table demonstrates the effective sampling determination process of the targeted Gen Z population [22–24] (Table 1).

5.2 Data Collection and Data Analysis Procedures

The data-gathering process for this specific research is conducted as the primary method of data collection which involves the gathering of consumer perspectives from quantitative surveys. For the addressed variables regarding the topic, the potential closed-ended questions have been structured with multiple choices for their preferences. By gathering the perspectives of different consumers from the Indian population, significant insights can be gained about the market from the analysis. Depending upon the relationship of different independent variables with the purchasing intention of young consumers, the analysis has drawn potential findings from the survey results. SPSS software is used to evaluate the results from the gathered responses of the participants. Significant results and findings have been obtained from the ANOVA test, coefficient test, and regression analysis test [25]. While continuing the data analysis and collection process, various ethical considerations have been adopted to validate the findings from this study. The study has effectively considered the ethical code and conduct of the data protection and regulation acts of India while adhering to the GDPR laws for consumer data protection. It has also been mandated that the gathered survey information will not be utilized for commercial purposes. Informed consent is taken from the participants for utilizing their responses for academic purposes. The research has also focused on emphasizing the anonymity and confidentiality of information during the survey. However, the research has faced some potential limitations during the study due to the potential requirement of financial resources and support. The time constraint is another limitation for the research to gather 500 authentic responses from potential consumers [26].

5.3 Demographics and Multiple Regression Analysis

The key demographics in the study enable the reliability and authenticity of survey responses. To conduct the research about Gen Z's evolving luxury preferences, the age of respondents is chosen below 27 as they are the potential representative of India's huge young population. Moreover, the potential education and occupation background of the participants contribute to the validation of the research. The regression analysis is going to estimate the possible value for the variables from the survey responses. ANOVA test is also significant in drawing the significance of the study from the quantitative statistical perspective. The functionalities between the variables can be effectively estimated by adopting the regression coefficient analysis and establishing the equation for the analysis. The equation can be made for the coefficients of the variables simplifying the quantitative research [27].

$$\begin{aligned} \mathrm{DV} &= \mathrm{C} + \mathrm{Beta\_1} * \mathrm{IV\_1.1} + \mathrm{Beta\_2} * \mathrm{IV\_1.2} \\ &\quad + \mathrm{Beta\_3} * \mathrm{IV\_1.3} + \mathrm{epsilon} \end{aligned}$$

where, DV = Purchase intention of Gen Z consumers in India, C = constant, IVs represent the corresponding independent variables, while Beta values denote the corresponding coefficients. The epsilon provides the corresponding errors from the regression coefficient table.

6 Results

6.1 Demographic Data

Gender

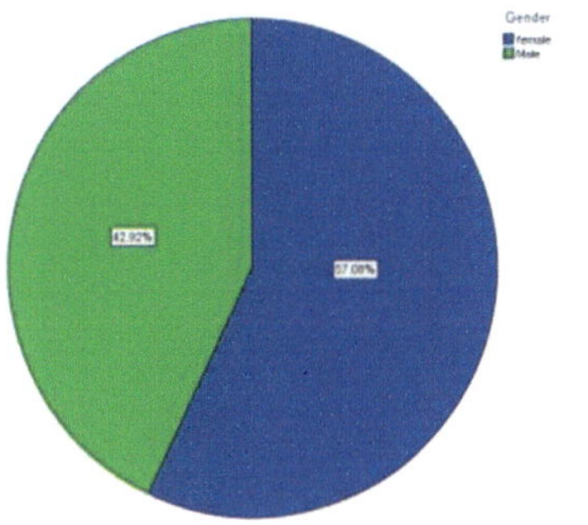

From the considered sample size of the population, the total percentage of female participants is higher than that of male participants. 42.92% of the total participants are male while the other 57.08% of participants are from the female categories. An equal number of participants for both female and male consumers from the Indian market is expected to provide the gender unbiased results from the survey. Gender neutrality in the research significantly validates the overall credibility and reliability of the survey results from the investigation.

Age

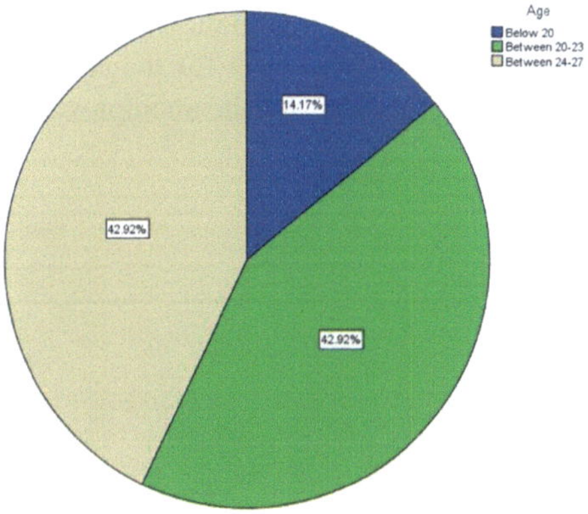

The age demographics play a critical role in determining the focus of the research, particularly the Gen Z population in India. From the above chart of age demographics, it can be observed that most of the participants are between the ages of 20–27 which is considered as an ideal age of the Gen Z population within the demographics. As the research has put the criteria of a maximum age of 27 for the Gen Z population, the investigation provides potential significance to the study. However, 14.17% of the participants are below the age of 20. It can potentially draw amateur preferences and uninformed decisions for luxury products in the market.

Occupation

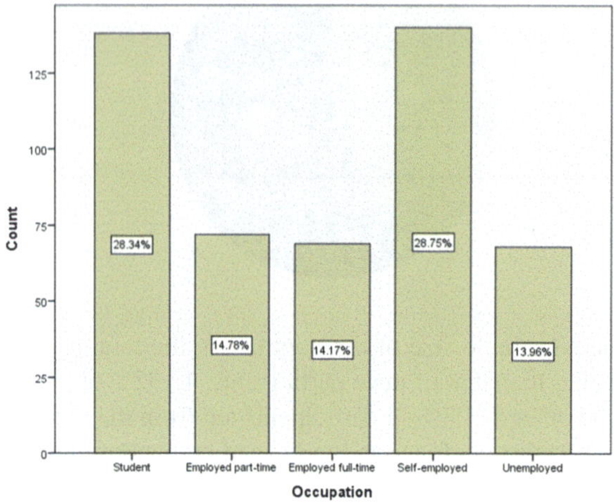

In terms of occupation, most of the participants are students (28.34%) and also potential numbers of participants are self-employed because of their parent's ongoing business or startups in the Indian market. In these two segments, potential tendencies to luxury product purchases can be seen without considering the price sensitivity of the market. The price sensitivity is relevant for the employed population either part-time (14.78%) or full-time (14.17%) and the unemployed individuals (13.96%).

Educational Degree

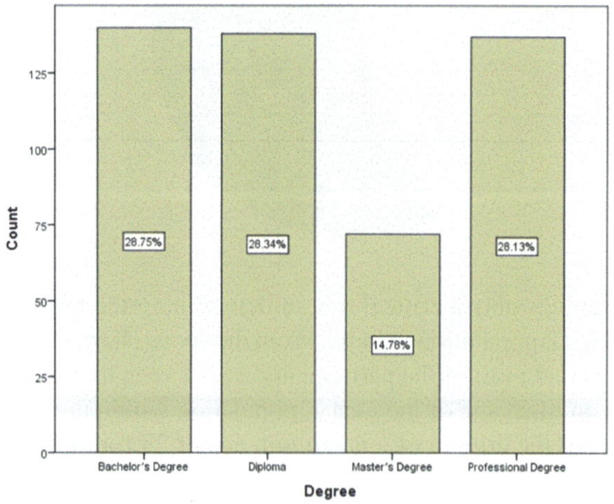

In the consumer market of India, the potential Gen Z population has come from different educational backgrounds. Most participants (28.75%) have a Bachelor's degree while 28.34% of participants have a diploma degree. The number of participants from professional degree backgrounds is almost 28.13%. However, few participants (14.78%) have pursued the Master's degree. Since most participants are educated enough in the respective field, the survey results can be potentially unbiased and reliable in the current market landscape.

6.2 Relation Between Contest of Social Media, Website Engagement, Blog Post Promotions and Gen Z's Purchase Intention

H1: There is a Potential Relationship Between the Purchasing Intention and Influencer Giveaway Platforms

Model summary

Model	R	R square	Adjusted R square	Std. error of the estimate
1	0.730[a]	0.533	0.530	0.33973

a. Predictors: (Constant), GP1.3, GP1.2, GP1.1

The above table provides the summarization of the model discussing the potential correlation values between different variables. From the table, the correlation value of the r-square is 0.533 and the adjusted value is 0.530. This value significantly depicts the stronger correlation between the variables as the values are close to the standard correlation value of 0.6. For such a significant correlation with purchase intention, potential variances of the measures can have a potential impact on purchasing decisions.

ANOVA[a]

Model		Sum of squares	df	Mean square	F	Sig.
1	Regression	63.560	3	21.187	183.569	0.000[b]
	Residual	55.746	483	0.115		
	Total	119.306	486			

a. Dependent variable: DV1.1

b. Predictors: (Constant), GP1.3, GP1.2, GP1.1

ANOVA table analysis has provided the tests for significance in explaining the correlations between these variables. By considering the three variables in the above table, the higher F-Statistic value has outlined the potential significance of the relationship while suggesting the reliability and fitness to the average intention of luxury product purchasing by the Gen Z population. Moreover, the significance value is 0 depicting the significance of the model statistically. It is known that if the significance value is less than 0.05, then the correlations defined by the variables can be considered statistically significant.

Coefficients[a]						
Model		Unstandardized coefficients		Standardized coefficients	t	Sig.
		B	Std. Error	Beta		
1	(Constant)	1.736	0.167		10.408	0.000
	GP1.1	0.191	0.044	0.174	4.327	0.000
	GP1.2	– 0.067	0.024	– 0.097	– 2.757	0.006
	GP1.3	0.503	0.033	0.649	15.188	0.000

a. Dependent variable: DV1.1

For the coefficients of this table, the multiple regression equation can be derived as the following.

*Purchase Intention = Social Media Contest * (0.174) + Fashion Brand's Website * (−0.097) + Blog Post Promotions * (0.649) + 1.736 + 0.167*

The obtained regression line has potentially justified the change in purchase intention due to the characteristics of social media giveaways, effective brand engagement through the website, and the promotion of products through quality blog content. The above equation provides significant insights about the degree of change in purchase intention and it provides that the contest for online media platforms changes with the coefficient of 0.174, website engagement at 0.097, and blog media presentation at 0.649. The significant rates of change for social media content as well as blog content are positive while the website engagement highlights the negative potential to regressions. These relations provide critical insights into both the positive and negative impacts of customer engagement through different characteristics and types of influencer marketing. The rising influence of blog content on purchase intention can be anticipated through the above regression equation.

6.3 Relation Between Various Influencer Characteristics with Gen Z's Purchase Intention

H2: There is a Significant Relationship Between Potential Characteristics of Influencers on the Giveaway Process and Purchasing Intentions Among the Gen Z Population

Model summary

Model	R	R square	Adjusted R square	Std. error of the estimate
1	0.909[a]	0.826	0.825	0.29252

a. Predictors: (Constant), IC1.3, IC1.2, IC1.1

The above model summarizes the potential influences of individual characteristics of different influencers on purchase decisions. From the above model, the R square value can be observed as 0.826 which is approximately close to the standardized correlation value of 1. Therefore, a strong correlation can be obtained through micro-influencers, admired fashion influencers, and celebrity promotions with customer's purchasing intention. For such a stronger correlation, effective relationships can be found for the purchase intention of customers with the key characteristics of influencers in the digital landscape. Due to a stronger correlation with reduced errors in the calculation, no variations could be derived from the correlations. However, further tests and regression equations could provide a potential relationship between the discussed variables.

ANOVA[a]

Model		Sum of squares	df	Mean square	F	Sig.
1	Regression	196.423	3	65.474	765.159	0.000[b]
	Residual	41.330	483	0.086		
	Total	237.754	486			

a. Dependent variable: DV1.2

b. Predictors: (Constant), IC1.3, IC1.2, IC1.1

In the above ANOVA test, the obtained value of significance is 0 defining the acceptance of the alternative hypothesis while rejecting the null hypothesis. Since the p-value of the above test reaches 0, the considered correlation through the hypothesis is statistically significant. The potential findings about the influences of social media influencers, micro-influencers, as well as celebrity influencers have strongly impacted Gen Z's buying decisions for different luxury items. With the rising trend of celebrity or influencer endorsement through brand promotions, brands have to consider the evaluation of the types of influencers as per their market segmentation and target. The expertise and follower base from the market niche for respective influencers have become crucial for the brand to potentially impact Gen Z through different Giveaway programs.

Coefficients[a]						
Model		Unstandardized coefficients		Standardized coefficients	t	Sig.
		B	Std. Error	Beta		
1	(Constant)	−4.483	0.193		−23.229	0.000
	IC1.1	−0.083	0.023	−0.083	−3.652	0.000
	IC1.2	1.253	0.030	0.888	41.423	0.000
	IC1.3	0.744	0.031	0.527	24.367	0.000

a. Dependent variable: DV1.2

From the coefficient values, the corresponding multiple regression equation can be structured as the following.

*Purchase Intention = Fashion Influencers * (−0.083) + Celebrity Endorsement * (0.888) + Micro-Influencers * (0.527) + (−4.483) + (0.193)*

The equation is structured from the obtained values of the coefficients for the respective variables in the analysis. The equation suggested that the influence of traditional fashion influencers is less impactful to the purchase intention of the customer in the Indian market drawing negative outcomes to the changes. While other two influencer types including micro-influencers and celebrity influencers through social media can effectively impact the customer's intention to purchase luxury goods. With the changes of fashion influencers' degree at 0.083 time, celebrity's influences with the degree of 0.888, and micro-influencers degree of 0.527 to the model, the purchase intention is changed proportionally.

6.4 Relations Between Different Influencer Giveaway Mechanisms and Purchase Intention

H3: There is a Potential Relation Between Customer's Purchasing Intention with Various Giveaway Mechanisms

Model summary				
Model	R	R square	Adjusted R square	Std. error of the estimate
1	0.620[a]	0.385	0.381	0.39001

a. Predictors: (Constant), GM1.3, GM1.2, GM1.1

The model has summarized the relations of purchase intention with the giveaway mechanisms through user-generated content, random ways, and tagging multiple friends. The potential correlations for these variables have been effectively obtained and summarized in the above table. As the R-squared value is 0.385, it has a weaker correlation with the purchase intention. However, potential variances might be significant in this model due to the weaker correlations.

ANOVA[a]

Model		Sum of squares	df	Mean square	F	Sig.
1	Regression	45.979	3	15.326	100.762	0.000[b]
	Residual	73.466	483	0.152		
	Total	119.446	486			

a. Dependent variable: DV1.3

b. Predictors: (Constant), GM1.3, GM1.2, GM1.1

Despite the weak correlation, the statistical significance of the relation can be justified from the above table. ANOVA test has provided the value of significance as 0 recommending for rejecting the null hypothesis and accepting the alternative hypothesis. Relations between the mechanisms of the giveaway campaign and the purchase intention of customers can be effectively defined from statistical perspectives. The ANOVA table also highlights a higher value of F-statistics suggesting the model's strategic fit with the hypothesis.

Coefficients[a]

Model		Unstandardized coefficients		Standardized coefficients	t	Sig.
		B	Std. Error	Beta		
1	(Constant)	2.419	0.259		9.335	0.000
	GM1.1	0.073	0.045	0.073	1.612	0.108
	GM1.2	− 0.284	0.043	− 0.260	− 6.616	0.000
	GM1.3	0.643	0.051	0.589	12.531	0.000

a. Dependent variable: DV1.3

Considerations of the coefficients can provide potential values for shaping the regression line for the above variables.

*Purchase Intention = UGC Based Giveaways * (0.073) + Random Giveaways * (−0.260) + Tagging Multiple Friends * (0.589) + 2.419 + 0.259*

The multiple regression line equation provides insights into the degree of change in different influential factors for giveaway mechanisms for purchasing luxury items. The giveaways focusing on UGC development for the market, multiple tagging or random giveaways have different impacts on purchase decisions. The less genuine giveaways that ask to tag multiple friends have a potential contribution to influencing purchasing decisions for Gen Z customers while the random giveaways have negative coefficients in the regression equation. However, the UGC content making in giveaways has a positive correlation to the regression.

6.5 Relation Between Different Giveaway Products with Purchase Intention

H4: There is a Potential Relationship Between the Purchasing Intention of Gen Z Customers and Giveaway Products

Model summary				
Model	R	R square	Adjusted R square	Std. error of the estimate
1	0.733[a]	0.537	0.535	0.33794
a. Predictors: (Constant), GI1.3, GI1.2				

The above model summary provides effective insights into the types of giveaway luxury items in influence the overall purchase decisions of the Gen Z population. The R-square value of 0.537 provides a significant correlation between the variables. Although the potential errors in the model significantly impact the effectiveness of the model, it outlines the potential relationship between the items with the purchase intention.

ANOVA[a]						
Model		Sum of Squares	df	Mean square	F	Sig.
1	Regression	64.032	2	32.016	280.343	0.000[b]
	Residual	55.274	484	0.114		
	Total	119.306	486			
a. Dependent variable: DV1.4						
b. Predictors: (Constant), GI1.3, GI1.2						

ANOVA test provides the significance value as 0 potentially reflecting the acceptance of the alternative hypothesis and avoidance of the null hypothesis. As the significance value is less than 0, it potentially draws the statistical significance of the dataset. Moreover, the higher F-statistics value of 280.343 suggests the overall fit and reliability of the average decision-making process of Gen Z customers in India.

Coefficients[a]

Model		Unstandardized coefficients		Standardized coefficients	t	Sig.
		B	Std. Error	Beta		
1	(Constant)	3.806	0.155		24.524	0.000
	GI1.2	0.598	0.026	0.771	23.434	0.000
	GI1.3	− 0.401	0.036	− 0.367	− 11.158	0.000

a. Dependent variable: DV1.4

The above coefficient table provides the respective coefficients of the corresponding variables.

Purchase Intention $=$ *Higher Values of Luxury Items* $* (0.771) +$ *Luxury Items from Trusted Brands* $* (-0.367) + 3.806 + 0.155$

The multiple regression equation potentially highlights the positive influence of prices for different giveaway items in shaping purchasing decisions. Justification can be obtained for potential changes in luxury brand's items and high-valued products in the giveaways with the relative changes to the purchase intention of Gen Z customers. Although the corresponding branded items are not regarded as the potential influences to purchase decisions, the higher priced products significantly value customers, particularly from India's Gen Z population. The negative coefficients for branded items and positive coefficients for high-valued products justify the above regression relations. Due to potential implications for psychographic factors of consumers, the higher valued products are preferred in the Indian market rather than the branded items in the marketplace.

7 Discussion

The significant correlation equation, ANOVA test, and model summarization can be effectively utilized to find the significant correlations between the identified variables in the research process. The summarized table for the model effectively provides r-squared value for the analysis reflecting potential strengths of the correlation between different variables in the responses. For the individual independent variables, the r-squared values highlight the significant correlations while the giveaway mechanisms build a weak correlation with the purchase intentions of Gen Z customers. ANOVA test effectively provides the measures for the statistical significance of the data reflecting the acceptance or rejection of the hypothesis. When the significance value in the ANOVA test derives below 0.05, the constructed null hypothesis has been rejected while accepting the alternative research hypothesis. For the above four independent variables, four hypotheses are created and the ANOVA test detects the acceptance of all the alternative hypotheses while rejecting the null hypothesis. Therefore, the defined hypothesis determines the relationship with the variables as the null hypotheses of the statements nullify all the relations. Identified variables including giveaway items, mechanisms, platforms, and influencer types have potential relations with customer's purchasing decisions [28].

On the other hand, the coefficient table for the corresponding variables helps in structuring the regression equations that can be considered as the significant research equation describing the behaviours of the coefficients in influencing the dependent variable. Based on the coefficients of independent variables, the equation provides the results for the dependent variable depicting the degree of changes in the current marketplace. Although the variables might have negative coefficients in the analysis, it can be determined that the respective variables do not define the purchase intention of customers. Negative coefficients of branded luxury items, random giveaways, fashion influencers, and brand website engagement have been referred to the least impact on consumer's buying behaviour for luxury products. Although the negative coefficients enable the differentiation strategies or variations in the variables, the variables with positive coefficients potentially impact the overall buying attitude toward luxury items. Moreover, the overall effectiveness of the model can be derived for explaining the potential relationship between purchase intention and influencer giveaway programs by the brands. The statistical significance of all the relationships and hypotheses is also highlighted through the analysis. The relative strengths and directions of the relationship are also obtained through utilizing the models [29].

8 Conclusion

In the evolving market of luxury brands in India, the young populations is the key contributor to its success and growth. The study has outlined the significant correlations between their mindset toward luxury purchasing and social media influencers' giveaway campaigning promoted by the brands. The implementation of Sproles's theory potentially helps in understanding the influential factors for purchasing luxury goods from different brands. Based on other secondary information reviews, potential variables including influencers' characteristics, giveaway mechanisms, giveaway items, and giveaway platforms are obtained. Since the self-concepts, self-esteem, and uniqueness of products are driving luxury purchase decisions among the targeted population, the brands are required to consider the market trends. The brands need to design effectively and focus on branding through social media platforms to leverage the unique brand identity and embrace the enhancement of self-image for its customers. Although several studies have potentially addressed the evolving behaviour of purchasing items through social media by Indian Gen Z individuals, it was limited to some niche marketplace or niche demographic segments. From the addressed variables, potential hypotheses for the study were developed to propose the framework for the study [30].

Based on the established hypotheses, the research methodology is effectively framed aligning with the objective and research problems. From the quantitative survey seeking opinions about influencer's giveaways and purchase intention, the study had planned to evaluate potential impacts on Gen Z primarily. For the survey, Gen Z individuals in India were targeted to review their perspectives towards influencer giveaways for luxury goods. The sampling table of Krejcie Morgan provided significant insights about considering minimum samples for the evaluation and investigation of the impact of giveaway campaigns on the purchasing decisions of customers. ANOVA test and regression coefficient analysis enabled finding the results of the survey reflecting potential influences of giveaways as well as influencer characteristics on luxury product purchasing in India. Statistically significant data have provided potential findings for this research [31].

In conclusion, the research has contributed to providing a significant understanding of the luxury fashion interest of Gen Z customers influencing their purchase intention. Aspiration for societal status and peer influences is driving today's Gen Z consumers to participate in different luxury item giveaways and be involved in purchasing the latest luxury fashion trends from the brand without being price sensitive about these products. With the evolving patterns of luxury purchase decisions in the Indian market, the brands are required to adopt potential strategies to partner with different digital influencers and develop effective giveaway campaigns. Influencers are required to be selected based on their potential expertise and specific niche of followers for luxury lifestyles and fashions. Moreover, the brands should incorporate best business practices to leverage influencer marketing, retain brand loyalty and improve brand engagement from Generation Z customers.

9 Future Scope and Limitation

The research study is limited because it considers only the Gen Z population rather than focusing on all the Indian consumers. The limitation of this study's approach is not only restricted to influencer giveaway programs but also bound to some luxury retail products in the market. The study has not also specified the gender-based views for specific luxury segments within the Indian context. Moreover, the rich cultural context of India for luxury items has not been evaluated in the study leaving a significant scope for further research. With the rising potential for influencer marketing, ethical concerns might emerge and further research programs could investigate the ethical implications of various giveaway mechanics and influencer selection policies. The unique potential for the significant dynamics of different social media platforms including Instagram might be leveraged in further research studies within the Indian context [32].

References

1. Thangavel, P., Pathak, P., Chandra, B.: Consumer decision-making style of gen Z: A generational cohort analysis. Glob. Bus. Rev. **23**(3), 710–728 (2022)

2. Dash, G., Kiefer, K., Paul, J.: Marketing-to-Millennials: Marketing 4.0, customer satisfaction and purchase intention. J. Bus. Res. **122**, 608–620 (2021)
3. Joshi, R., Garg, P.: Role of brand experience in shaping brand love. Int. J. Consum. Stud. **45**(2), 259–272 (2021)
4. Duffett, R.: The YouTube marketing communication effect on cognitive, affective and behavioural attitudes among Generation Z consumers. Sustainability **12**(12), 5075 (2020)
5. Agrawal, G. et al.: Abundantly rare: changing consumer trends in the luxury market. Vision, vol. 0, no. 0 (2022). 10.09722629221104201
6. Tumsah, A.: The Impact Social Media Influencers Have on Consumer Purchase Intentions: A Study Understanding the Influence Beauty Gurus Have on Younger Generations. Diss. Dublin, National College of Ireland (2021)
7. Wangshu, G., Guanhua, W.: How influencers marketing motivates consumers' buying behaviour: A focus group investigation of the impulse buying behaviour via Chinese millennials' lens (2020)
8. Macheka, T., Quaye, E.S., Ligaraba, N.: The effect of online customer reviews and celebrity endorsement on young female consumers' purchase intentions. Young Consum. (2023)
9. Rizma, F.A., Marsasi, E.G.: The effect of trustworthiness to increase brand trust and purchase intention on social media promotion based on theory of persuasion in generation Z. Jurnal Manajemen (Edisi Elektronik) **15.1**, 61–81 (2024)
10. Jorge, C.I.B.C.M.: What is the impact of influencer marketing through Instagram, by promoting fitness apparel, on female consumer behaviour?. Diss. (2023)
11. Vijaya, M.T., Yamuna, Chitra, B.: Gen Z versus Millennials, purchase intentions: a comparative study based on social media marketing strategies in India. Appl. Mark. Anal. **9.1**, 84–100 (2023)
12. Hemantha, Y.: Celebrity Endorsement of Luxury Brands in Purchasing Pattern Among Generation Z Customers. SSRN (2020)
13. Singh, N., Chakrabarti, N., Tripathi, R.: Factors determining fashion clothing interest and purchase intention: a study of Generation Z consumers in India. Fashion, Style & Popular Culture (2023)
14. Thangavel, P., Pathak, P., Chandra, B.: Millennials and Generation Z: a generational cohort analysis of Indian consumers. Benchmarking: Int. J. **28.7**, 2157–2177 (2021)
15. Agrawal, D.K.: Determining behavioural differences of Y and Z generational cohorts in online shopping. Int. J. Retail Distrib. Manag. **50**(7), 880–895 (2022)
16. Thomas, S., Bhatt, V., Patel, R.: Impact of skepticism on CRM luxury campaign participation intention of Generation Z. Int. J. Emerg. Mark. **19**(4), 964–988 (2024)
17. Sudirjo, F., et al.: The influence of generation Z consumer behavior on purchase motivation in E-Commerce shoppe. Profit: Jurnal Manajemen, Bisnis dan Akuntansi **2.2**, 110–126 (2023)
18. Lal, R., Sharma, G.: Social media influencers for online purchase behaviour: mediation by brand consciousness. J. Content Commun. Commun. **13**(7), 83–94 (2021)
19. Sharma, P., Pradhan, S., Srivastava, A.: Understanding the luxury purchase intentions of young consumers: a qualitative analysis. Asia-Pacific J. Bus. Admin. **13**(4), 452–476 (2021)
20. Munsch, A.: Millennial and generation Z digital marketing communication and advertising effectiveness: a qualitative exploration. J. Glob. Scholars Market. Sci. **31**(1), 10–29 (2021)
21. Gazzola, P., et al.: Trends in the fashion industry. The perception of sustainability and circular economy: a gender/generation quantitative approach. Sustainability **12.7**, 2809 (2020)
22. Sharma, K., Khosla, R., Kumar, Y.: Application of Morgan and Krejcie & chi-square test with operational errands approach for measuring customers' attitude & perceived risks towards online buying. Int. J. Intell. Syst. Appl. Eng. **11**(3s), 280–285 (2023)
23. Singh, R.: Dynamics of online shopping trends for fashion products: a case study of Himachal Pradesh, India. Asian J. Adv. Res. Rep. **18**(3), 29–49 (2024)
24. Tumiran, M.A.: How to deal with insufficient sample size due to non-response in surveys? Quantum J. Soc. Sci. Human. **5**(2), 70–86 (2024)
25. Garg, P., et al.: Examining the relationship between social media analytics practices and business performance in the Indian retail and IT industries: The mediation role of customer engagement. Int. J. Inf. Manag **52**, 102069 (2020)

26. JagadeeshBabu, M.K., SaurabhSrivastava, S.M.G., AditiPriya Singh, M.B.S.: Influence of social media marketing on buying behavior of millennial towards smart phones in Bangalore city. PalArch's J. Archaeol. Egypt/Egyptol. **17.9**, 4474–4485 (2020)
27. Sheoran, M., Kumar, D.: Conceptualisation of sustainable consumer behaviour: converging the theory of planned behaviour and consumption cycle. Qualitative Res. Organ. Manag.: Int. J. **17**(1), 103–135 (2022)
28. Kaur, J. et al.: Impact of materialism on purchase intention of sustainable luxury goods: an empirical study in India. Soc. Bus. Rev. **17.1**, 22–44 (2022)
29. Gokhale, A., Mishra, A., Veluchamy, R.: Factors influencing purchase decision and consumer behavior in luxury cars. Dogo Rangsang Res. J. **11.7**, 34–40 (2021)
30. Dobre, C. et al.: The common values of social media marketing and luxury brands. The millennials and generation z perspective. J. Theore. Appl. Electron. Commerce Res. **16.7**, 2532–2553 (2021)
31. Bardey, A., Mehdiratta, U., Turner, R.: Fashioning the future generation: generation Z Indian consumers' attitudes towards Western and Indian Fashion. Fashion Marketing in Emerging Economies Volume II: South American, Asian and African Perspectives, pp. 19–48. Springer International Publishing, Cham (2022)
32. Garg, S., Gautam, S.: Factors affecting smartphone purchase decisions of generation-Y and generations-Z: The Indian Scenario. Abhigyan **39**(4), 22–33 (2022)

Innovative Fusion: Attention-Augmented Support Vector Machines for Superior Text Classification for Social Marketing

Raghavendra M. Devadas, Vani Hiremani, J. Praveen Gujjar, N. Shobha Rani, and K. R. Bhavya

Abstract Text classification is a constantly growing area, which requires new approaches to achieve better, faster, and smoother classifier performance. This paper introduces a novel architecture combining classical Support Vector Machines (SVM) with attention mechanisms, that dramatically improves text classification performance. The fusing of SVMs with attention mechanisms offers a few possibilities for text categorization. This integration capitalizes the individual strengths of both paradigms defeating limitations to classify a wider range of documents in an enhanced quality. This increases the adaptability of SVMs and provides attention enhancement to assign different importance levels to the words or features in the input text. This variation allows highlighting only the most important features, not paying as much attention to the rest, and provides an exhilarating opportunity to be able to distinguish fine-grained nuances and compound relations between data streams. Our evaluation goes beyond traditional accuracy metrics to be based on precision, recall and F1-score hence providing a comprehensive assessment of the model performance. The hyper-parameter tuning through GridSearchCV is ensuring that our attention-augmented SVM is perfectly well configured. Rigorous testing and validation then portray the impact on training and testing sets by incorporating attention mechanisms into SVMs

R. M. Devadas
Department of Information Technology, Manipal Institute of Technology Bengaluru, Manipal Academy of Higher Education, Manipal, India

V. Hiremani (✉)
Symbiosis Institute of Technology, Pune Campus, Symbiosis International (Deemed University), Pune, India
e-mail: vani.hiremani@gmail.com

J. P. Gujjar
Faculty of Management Studies, JAIN (Deemed-to-be University) Bengaluru, Bengaluru, India

N. S. Rani
Department of Artificial Intelligence and Data Science, GITAM School of Technology, GITAM University, Bengaluru, India

K. R. Bhavya
Department of Computer Science and Engineering, GITAM School of Technology, GITAM University, Bengaluru, India

S. Dutta et al. (eds.), *Advances in Data Analytics for Influencer Marketing: An Interdisciplinary Approach*, Information Systems Engineering and Management 9,
https://doi.org/10.1007/978-3-031-65727-6_17

for text classification. In application, the proposed methodology would be applicable in the domain of Social Marketing wherein a subtle decoded understanding of textual content would make a major difference regarding targeted campaigns and also for enhancing audience participation. On the other side, there are constraints about potential difficulties in handling massive-size datasets and also the computational cost implicated through attention mechanisms. Results show precision of 81.59% from which we can infer that SVM model is suitably attuned with attention broadly tested and validated. The model has proved to reduce false positives with a rate of precision of up to 60%. Looking at the 40% recall rate examination is an evaluation of its ability to identify true positives. The overall result of precision in addition to recall is the F1-score, which stands at about 48% performance in general. These results extend the range of text classification approaches, presenting a different point of view on how attention mechanisms can be embedded with classical machine learning models. The results of the research exceed existing text classification methods as well as open the way for future findings in attention mechanisms beyond neural networks. Technological advancement has all indications to revolutionize tools, which in turn gets to be a turning point for experts dealing with natural language processing and machine learning professionals asserting extensive applications over independent domains.

Keyword Text classification · SVM · Attention mechanism · Machine learning

1 Introduction

Text classification relates to an important issue in the natural language processing (NLP) framework whose activities include copying and building collections from electronic documents. So, it's a form of machine learning where a model studies a labeled dataset, aiming to identify connections and patterns between the characteristics of the text and its associated category. The skill is practically applied to spam filtering, sentiment analysis, topic categorization, and a lot more. The text classification problem may be broadly divided into two phases: pre-processing of the text and classification using machine learning models. Common strategies consist of Bag-of-Words (BoW): a method that portrays a document as a disordered collection of words devoid of contextual grammar and word sequencing. Term Frequency-Inverse Document Frequency: Assigns weight to the importance of words by their frequency in a single document compared to how recurrently they show in all documents. Once the text is represented, then features are therefore extracted. This requires transforming the text between numerical features suitable for machine learning models to understand. The BoW and TF-IDF representations stand as feature vectors. Generally, various machine learning algorithms can be used in classifying text. Common models include:

- Naive Bayes: Based on Bayes' theorem, assumes independence between features.

- Support Vector Machines (SVM): Constructs a hyperplane to separate classes with a maximum margin.
- Logistic Regression: Utilizes the logistic function to represent the probability of being part of a specific category.
- Deep Learning Models (e.g., LSTM, CNN): Neural network architectures accomplished by learning intricate patterns in sequential data.

The selected model undergoes training on a dataset featuring distinct labels, associating each document with its relevant category. During training, the model is refined by adjusting its parameters to curtail the difference involving its predicted categories and the actual ones. Text classification models are evaluated using various metrics, including:

- Accuracy is established by calculating the ratio of correctly predicted instances to the total number of instances.
- Precision, on the other hand, is the proportion of accurately predicted positive observations assessed to the total foretold positives.
- Recall, in this context, represents the proportion of correctly predicted positive observations in relation to the actual positives.
- Lastly, the F1-score is assessed as the harmonic mean of precision with recall.

1.1 Challenges in Text Classification

- Tackling the Challenge of Imbalanced Data: Imbalanced data, where one class is represented more heavily than others, can result in biased models. To overcome this issue, various techniques such as oversampling, undersampling, or incorporating balanced class weights have proven effective.
- Overcoming the Obstacle of Unknown Vocabulary: Text classification models can struggle with vocabulary that was not included in the training data. Employing methods like embedding layers or sub-word tokenization can aid in handling out-of-vocabulary words.
- Comprehending Context: Certain tasks require models to grasp context, sarcasm, or subtle linguistic nuances. Deep learning models with attention mechanisms or contextual embeddings strive to tackle these complexities.

1.2 Applications of Text Classification

- Spam detection involves categorizing emails as either spam or non-spam depending on their content.
- Sentiment Analysis: Determining the sentiment (positive, negative, neutral) expressed in a piece of text.

- Topic Categorization: Assigning predefined categories to news articles, blog posts, or social media content based on their topics.
- Intent Classification: Understanding the intent behind user queries, is crucial for chatbots and virtual assistants.
- Authorship attribution entails identifying the author of a text by analyzing their distinctive writing style.

Text classification remains a effective tool for structuring documents by their meaning and significance. This process entails matching words to a hierarchy to gain insights from texts and create descriptive categories. Popular machine learning models, such as Naive Bayes, Logistic Regression, and Support Vector Machine, are often utilized for text classification, along with advanced natural language feature extraction techniques. The ultimate aim is to recognize when a teacher should intervene in a conversation that has strayed from its intended topic [1]. Sentiment analysis, a type of text classification, empowers us to gain insight into people's emotions and viewpoints. It leverages natural language processing to interpret texts in various fields, such as product reviews and social media analysis. Machine learning methods such as Random Forest, K-means clustering, and voting classifiers are utilized for sentiment analysis [2, 3]. Python's powerful text classification techniques are employed to label unstructured text and to train supervised learning models for real-time scenarios [4]. Additionally, text summarization techniques enhance the accuracy of text classification results [5]. The novel fusion paradigm that this study introduces is the combination of classical support vector machines (SVM) with attention mechanisms for advancing text classification tailor-made for Social Marketing. In the domain of Social Marketing, audience involvement becomes predominant, and understanding the subtleties of textual content as a backbone to create precise campaigns toward targeted marketing. The dynamism associated with the influential factors forming Social Marketing has been discussed in the paper especially quoting about the need for sophisticated classification models. Models for handling social media data have to be better in discerning the intricate relationships that form the textual content. Our novel approach aims at increasing text classification application through incorporating an attention mechanism into SVMs hence increasing their ability to adapt to a considerably dynamic social media discourse environment. This report contributes with a fresh perspective towards the fusion of attention mechanisms to established machine learning models, presenting transformational implications toward Social Marketing strategies. Text classification is used in social media marketing to analyze and understand user sentiments and emotions [6]. It involves training classifiers on benchmark datasets and assessing their performance in terms of accuracy [7]. The objective is to comprehend emotions and feelings towards online campaigns and threads [8].

Machine learning falls within the domain of artificial intelligence whose major preoccupation lies in the creation of models by algorithms that were initially trained on data sets thereby allowing machines to do what at first only humans could do—ranging from cataloging photos through analyzing data up to modeling predictions of price fluctuations. Machine learning is data-driven technology whereby it lies on the basis that patterns and relationships are found in large amounts of data, not by explicit instructions.

Supervised learning is the most familiar class of machine learning where an algorithm learns from label data, i.e. data that has a known output or target variable. Suppose, for example, an algorithm can be placed under a supervised one that classifies emails as spam or not spam given the characteristics from the email and the human-generated label. Supervised learning may be broken further down into classification and regression. Classification is the duty of appointing a discrete label to an input instance—for example, spam or not spam. Regression, instead, calls up predicting a continuous value for example the price of a house or the temperature in a room.

Unsupervised learning is a kind of machine learning in which an algorithm discovers data that is not labeled, i.e., not classified or categorized. An unsupervised learning algorithm, for instance, trying to learn how to cluster customers based on their purchase behavior might discover that there are clusters of customers with similar behavior, but not necessarily having the ones in one group with similar demographics or preferences as those in another group. These are clustering, dimensionality reduction, anomaly detection, and association rule learning. Clustering diverges similar inputs in a set based on some measure of similarity or distance. Dimensionality reduction refers to the process of reconstructing and mapping the dataset to a new one with reduced features or dimensions while preserving the most important information within the data. These outputs are further detected as outliers or abnormal inputs by these systems and usually, it is referred to as anomaly detection such as fraudulent transactions or faulty sensors. Association rule mining is the task of locating an association or a relationship among various items of data.

Reinforcement learning is a kind of machine learning, in which an algorithm learns from its actions and feedback instead of data. For example, a video game may be taught to play a game in a reinforcement-learned way if it has to try online and then get rewards or punishment based on the behavior. Reinforcement learning is interpolated in many areas like control, optimization, game playing, and robotics. What differentiates reinforcement learning from the supervised as well as unsupervised types is the fact that it has no rigid data set or output that is defined in advance. Instead, it possesses an agent, an environment, as well as policy. The agent is that who performs actions and learns the feedback. The environment refers to the system or scenario within which the agent operates and exerts influence through its interactions. It is a strategy or rule being governed by which the agent should take action in any state so that it becomes the best possible.

Machine Learning is an area of high evolution in expanding scientific discipline finding application in many areas of life as well as offering several interesting challenges. Modern preferences and topics of machine learning are deep learning, natural language processing, computer vision, generative models, explainable AI, and ethical AI. Deep learning is a subdivision of machine learning that relies on artificial neural networks, which, in turn, are inspired by the representation of biological neurons and their operations through interconnected layers of nodes. Natural language processing

(NLP) is a dedicated area in computer science besides artificial intelligence (AI) dedicated to the comprehension and generation of human language. It encompasses tasks like speech recognition, machine translation, sentiment analysis, and the development of chatbots. Computer vision is a sector of machine learning focused on analyzing and understanding visual data, including tasks such as recognizing faces, detecting objects, segmenting scenes, and generating captions for images. Generative models refer to machine learning models that can generate some new data or content such as an image, text, music, or video but based on the learned distribution of the existent one. It is one of the branches of machine learning which aims to bring transparency in the model along with interpretability of the outcomes yielded by the machine learning models, i.e., why a certain prediction has been made or how, internally, a certain model works. Ethical AI is a concern stemming from machine learning that delves into the ethical, social, and legal challenges of the machine learning systems viz-a-viz their impact on human society such as fairness, privacy, accountability, and trust.

Machine learning is an exciting and powerful field that could embolden many domains and industries to witness some radical transformation across the board. But machine learning also carries a lot of challenges and risks including data quality, capability to scale, security exposure, robustness requirement, and bias liability. Therefore, machine learning calls for cautious design, estimation, and deployment with collegial collaboration between researchers, practitioners, and stakeholders to guarantee it is used responsibly hence benefiting all involved.

Support Vector Machines (SVM) are highly impactful supervised machine learning sets of rules distinguished for their efficiency in both classification furthermore regression tasks. Pioneered by Vladimir Vapnik and his group in the 1990s, SVM research has gained considerable popularity, and it has proven successful in a variety of applications such as image classification, text categorization, and bioinformatics. SVMs effectively separate data points belonging to different classes by locating a favorable hyper-plane in some multi-dimensional realm. It is this simple yet conceptually powerful idea of a hyper-plane that forms the foundation upon which SVMs stand, as it allows one to define a decision boundary maximizing the separation between classes. SVMs will give the best results when appropriately working with data that may be perfectly separated by a straight line into four parts in two dimensions, or by a plane into six parts in three dimensions or hyperplane in higher numbers of dimensions. Our empirical is to identify the separating hyperplane that expands the margin, which is the distance involving this hyperplane and the closest data point from each class. These points are indicated to as support vectors, then they lie on the decision boundary strongly influencing the placement of the hyperplane, thus dictating the margin. SVMs can be used to address non-linear data relationships with a special type of function known as a kernel. These transform the input data into multidimensional space in such a way that with a hyperplane, classes can be separated with more precision. The most commonly utilized kernel functions include linear, polynomial, radial basis function, and sigmoid. One striking feature of the SVMs is that they emphasize maximizing the margin between classes. A wider margin ordinarily results in better generalization performance, as it reduces misclassifications of carelessness in new data. The value of the regularization parameter C signifies this trade-off between a larger margin and a greater number of misclassifications. With a smaller value of C, the model will have a larger margin but it will allow a greater number of training errors which can lead to overfitting. A higher value of C will make the margin tighter but penalize misclassifications even more, which will lead to a more complex decision boundary. SVMs are very efficient at finding their way through difficult, high-dimensional spaces and as such they are very powerful in facing tasks with many amounts of features. This combined focus on maximization of margin along with regularization reduces the problem of overfitting, particularly in such spaces. Furthermore, different kernel functions allow handling not only linearly separable but also non-linearly separable problems. However, the choice of those parameters such as kernel type and C value should be judiciously made for it can affect performance to a large extent. This tuning process of these parameters might be time-consuming, especially in cases involving large datasets or with a lot of features. Most recently improvements in optimization algorithms fixed this to some extent. SVMs are generally chosen to classify text because of their effectiveness. It can be used in cases of both regression as well as classification, but still, this classifier has been very successful in the case of the latter. Many pieces of evidence have used SVM for text classification, which includes those done while analyzing e-commerce product reviews from customers [9], emotion analysis [10], and grouping informal academic-related messages derived from pages on Facebook [11, 12]. SVM was able to achieve classification accuracy varied from 81.6 to 94.29% across different types of text in this corpus collection. The researchers have experimented with the different SVM kernels e.g. Linear, Sigmoid, RBF, besides Polynomial, with the Sigmoid kernel outperforming the others to achieve the highest result of accuracy in organizing informal academic-related messages [13]. Nonetheless, with the complexity and variety of textual data in all these applications growing drastically, new methodologies are required to enhance all these models. This research will present the way for exploring a new paradigm in text classification by mediating between attention mechanisms and Support Vector Machines, presenting an "Innovative Fusion" for enhanced superior results. The focus is centered on the popular 20 Newsgroups data, which is a flexible document collection on a broad domain.

1.3 Motivation

This research is driven by the desire to maximize the power of both attention mechanisms and SVMs. SVMs have very strong representation abilities in high dimensional spaces and are easily able to deal with sparse data as well. Hence, with the usage of a few SVMs, it will be seen that text classification will also be done properly. This is expected to make the SVMs more flexible since they will be able, through attention mechanisms to recognize and give way to important features in the text.

It may help them incorporate many complex relationships between dimensions of textual information.

1.4 Research Gaps

- While attention mechanisms have shown success in neural networks, much can still be gleaned about how they may intersect with more traditional machine learning models like SVMs. Future work should shed light on the possible synergy that can be developed between attention mechanisms and SVMs and how they can complement each other towards yielding better performance in text classification.
- Most research in the last two decades has been on classical generic text classification, leaving a wide gap in attention-augmented SVM performance for domain-specific challenges. The study of the application of these techniques and also observing how adaptable they are in particular domains or industries is work that holds promise for the future.
- This domain should look for the best designs of attention mechanisms for SVMs in this domain as the performance of SVMs with text classification could largely depend on them for the selection and mathematical representation of attention weights.

1.5 Research Objectives

The overarching objective of this research is to pioneer an "Innovative Fusion" by combining attention-augmented mechanisms with SVMs for text classification. Specifically, the goals include:

- Novel Integration: To introduce a novel integration of attention mechanisms with traditional SVMs, exploring the synergy between these two distinct paradigms in machine learning.
- Enhanced Text Representation: To enhance the representation of textual data by leveraging attention mechanisms, enabling the model to focus on crucial elements within the input.
- Improved Classification Performance: To achieve superior text classification results, measured through system of measurement such as accuracy, precision, recall, coupled with F1-score, on the challenging 20 Newsgroups dataset.

2 Literature Review

Attention mechanisms are commonly used in text classification to assign weights to different words based on their importance in expressing the sentence's meaning. Several papers propose different approaches to attention mechanisms in text classification. An innovative BiLSTM-Attention model for sentiment classification on Chinese mixed text has been proposed by [14]. This model utilizes the attention mechanism to extract key information from the text. Additionally, a BERT-based interactive attention mechanism has been proposed for sentiment classification, taking into account the interdependence among words in a sentence [15]. For automatic ICD coding, the paper introduces a pseudo-label-wise attention mechanism that consolidates similar ICD codes and calculates a single attention mode for them, effectively decreasing redundancy and enhancing accuracy. In [16] authors explore the incorporation of attention mechanisms into short Arabic text classification, particularly when combined with features derived from Wikipedia. A study made by [17] introduces a groundbreaking series of Task-Scaling (TaSc) techniques that enhance the credibility of attention-based explanations for text classification. A study [18] delves into the application of attention mechanisms in different recurrent neural network architectures for natural language processing tasks, although it does not specifically concentrate on text classification. Moreover, the work by [19] puts forth a novel knowledge-supported attention mechanism called KW-ATTN for text classification, which integrates advanced theories from external acquaintance bases keen on neural network models. According to [20], a groundbreaking study explores a new approach to enhancing attention in text classification, surpassing the results of other existing methods. Meanwhile, [21] delves into the development of a bi-level attention neural network model (TBAM) that utilizes both vocabulary and sentence-level attention to construct document representations for text classification. However, in contrast to the focus on text classification, the study by [22] delves into the effects of attention mechanisms on speech emotion recognition.

3 Theoretical Background

At the core of Support Vector Machine (SVM) lies the fundamental idea of identifying the most favorable hyperplane that can efficiently separate data points belonging to different classes by extending the margin linking them. The subsequent section provides an overview of the theoretical foundation of SVM, accompanied by mathematical formulations:

3.1 Hyperplane Equation

Consider a set of training data points (x_1, y_1),(x_2, y_2),...,(x_n, y_n) where x_i is the attribute vector and y_i stands the class label ($y_i \in \{-1,1\} \in \{-1,1\}$), the hyperplane equation corresponds to:

$$f(x) = w \cdot x + b \tag{1}$$

where,

w is the weight vector.

X is the Input Attribute Vector.

B is the bias term.

3.2 Decision Function

The decision function is employed to classify a data point corresponding towards the sign of f(x). If f(x) is greater than 0, the data point is categorized as one class, while if f(x) is less than 0, it is categorized as the other class.

$$\text{Prediction} = \text{sign}(f(x)) \tag{2}$$

3.3 Margin and Support Vectors

The margin signifies the spatial gap between the hyperplane and the nearest data points since every class. SVM endeavors to expand this margin. Support vectors being the specific data points positioned either on the margin or on the side opposite to their correct classification according to the hyperplane.

$$\text{Margin} = \frac{2}{\|w\|} \tag{3}$$

3.4 SVM Optimization Problem

The aim of SVM is to locate w besides b that minimize:

$$\frac{1}{2}\|w\|^2 + C\sum_{i=1}^{n} = \max(0, 1 - y_i(w \cdot x_i + b)) \tag{4}$$

where,

$\frac{1}{2}\|w\|^2$ is the regularization term that penalizes the magnitude of the weight vector.

The parameter C functions as the regularization parameter that governs the balance between minimizing training error and maximizing the margin.

The summation term enforces that each data point is arranged on the accurate side of the margin.

Attention Mechanism Basics.

3.5 *Attention Score Calculation*

Given a set of input vectors h_1, h_2,...,h_n (where n is the number of input vectors), the attention mechanism calculates attention scores (e_i) for each input vector. These scores represent the importance of each input concerning a context vector c:

$$e_i = \text{score}(\text{query}(s_t), \text{key}(h_i)) \tag{5}$$

The score function can take various forms, such as a dot product, a scaled dot product, or a learned function.

3.6 *Attention Weights*

The attention scores are transformed into attention weights (α_i) using a softmax function to ensure that the weights sum to 1:

$$a_i = \frac{\exp(e_i)}{\sum_{j=1}^{n}\exp(e_j)} \tag{6}$$

3.7 *Context Vector Calculation*

The context vector c is the weighted sum of the input vectors based on their attention weights:

$$c = \sum_{j=1}^{n} a_i \cdot h_i \tag{7}$$

4 Proposed Methodology

The goal of this study is to explain how attention mechanisms can be integrated with Support Vector Machines (SVMs) for improved text classification.

4.1 Attention-Augmented SVM for Text Classification

Let us recall the linear SVM decision function:

The function $f(x) = w \cdot x + b$ represents the decision function in Support Vector Machines (SVM).

To introduce attention mechanisms, we modify the SVM decision function by incorporating attention weights. Let α_i represent the attention weight associated with the ith feature in the input vector:

$$f(x) = \sum\nolimits_{i=1}^{n} \alpha_i (w_i \cdot x_i) + b \tag{8}$$

Here, w_i denotes the attention-weighted version of the original weight vector w.

4.2 Attention Score Calculation

The attention score e_i can be determined through dot-product attention score.

$$e_i = u \cdot \tanh(W \cdot (h_i + b)) \tag{9}$$

Here, u, W, and b are learnable parameters, and h_i represents the input feature.

4.3 Innovative Fusion and SVM Training

The attention-augmented SVM is trained by optimizing a combined loss function that incorporates both the traditional SVM hinge loss and an additional attention regularization term:

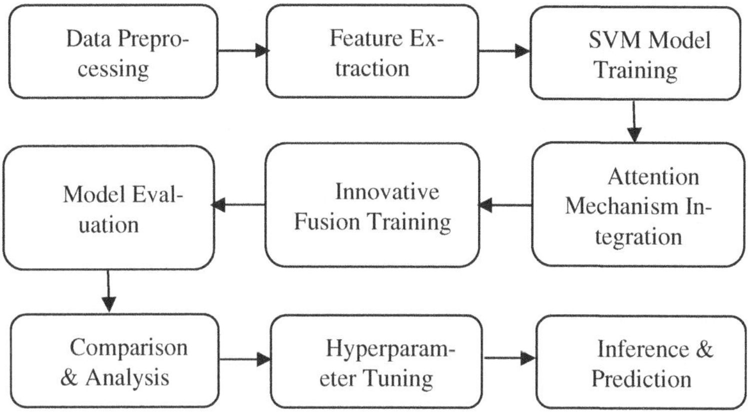

Fig. 1 Proposed methodology

$$\text{Total Loss} = \text{SVM Hinge Loss} + \lambda \cdot \text{Attention Regularization Term}.$$

Here, λ regulates the strength of the attention regularization.

Figure 1 represents the block diagram of the proposed study.

Steps:

(1) Data Preprocessing: Preprocess the text data by using the provided code to remove headers, footers, and quotes from the newsgroups dataset. Tokenize, clean, and convert the text into a suitable format.
(2) Feature Extraction: Use the CountVectorizer to convert the preprocessed text data into Bag-of-Words representations. This creates feature vectors for training the SVM.
(3) SVM Model Training: Train a traditional SVM model using the Bag-of-Words features. This serves as the baseline SVM for comparison.
(4) Attention Mechanism Integration: Integrate an attention mechanism into the SVM model using the provided SVMWithAttentionKernel class. This class modifies the SVM decision function to include attention weights.
(5) Innovative Fusion Training: Train the attention-augmented SVM model using the combined loss function. This combines the traditional SVM hinge loss and an attention regularization term.
(6) Model Evaluation: Evaluate the performance of both the traditional SVM and the attention-augmented SVM on a held-out validation set, using accuracy as a metric. The accuracy_score from the scikit-learn library is used for this purpose.
(7) Comparison and Analysis: Compare the accuracy of the traditional SVM and the attention-augmented SVM. Analyze the impact of the attention mechanism on classification accuracy and model interpretability.

Table 1 Dataset description

Aspect	Description
Dataset name	20 newsgroups dataset
Source	scikit-learn (Python library)
Purpose	Text classification and clustering benchmark dataset
Categories	20 newsgroups covering diverse topics, e.g., alt.atheism, comp.graphics, rec.sport.baseball, sci.med
Documents per category	Varies; typically, multiple documents per newsgroup
Format	Plain text documents
Data size	Approximately 20,000 documents across all newsgroups
Challenges	Diverse topics, potential overlap between categories, requires models to generalize well
Relevance	Widely used for testing text classification algorithms
Code-specific usage	The code uses the 'fetch_20newsgroups' function and focuses on categories like alt.atheism, comp.graphics, rec.sport.baseball, and sci.med. Headers, footers, and quotes are removed during preprocessing

(8) Hyperparameter Tuning: Fine-tune hyperparameters, including the strength of the attention regularization term (λ), using GridSearchCV.
(9) Inference and Prediction: Use the trained attention-augmented SVM for making predictions on new, unseen text data. Evaluate its real-world applicability and performance.

Table 1 illustrates the description of the dataset applied intended for the study.

The algorithm below outlines the steps involved in training and evaluating the attention-augmented Support Vector Machine (SVM) for text classification.

Algorithm 1. Innovative fusion SVM with attention for text classification

Input:
- Training text data (X_train)
- Training labels (y_train)
- Testing text data (X_test)
- Testing labels (y_test)

1. Preprocessing:
   a. Clean and preprocess the raw text data (X_train, X_test).
   b. Tokenize and convert the text into a suitable format for feature extraction.

2. Feature Extraction:
   a. Use the CountVectorizer to convert text data into Bag-of-Words representations.
   b. Create feature vectors for both training and testing data.

3. Traditional SVM Model Training:
   a. Train a traditional SVM model on the Bag-of-Words features using X_train and y_train.
   b. Use linear kernel or other suitable kernel based on the problem.

4. Attention Mechanism Integration:
   a. Define the SVMWithAttentionKernel class, extending the SVM model with attention.
   b. Implement the attention kernel function within the class, modifying the decision function.

5. Innovative Fusion Training:
   a. Train the attention-augmented SVM model using the combined loss function.
   b. Combine the traditional SVM hinge loss and an attention regularization term.

6. Model Evaluation:
   a. Assess the performance of both the traditional SVM and the attention-augmented SVM on the validation set.
   b. Use system of measurement like accuracy, precision, recall, besides F1-score for evaluation.

7. Hyperparameter Tuning:
   a. Fine-tune hyperparameters, including the strength of the attention regularization term, using techniques like GridSearchCV.

8. Inference and Prediction:
   a. Use the trained attention-augmented SVM for making predictions on the testing data (X_test).
   b. Obtain classification results.

9. Results and Conclusion:
   a. Analyze and compare the results of the traditional SVM and the attention-augmented SVM.
   b. Summarize findings and draw conclusions about the impact of attention mechanisms on text classification performance.

End.

Output:
- Trained attention-augmented SVM model
- Classification results

5 Results and Discussion

Below are the four metrics which was used to assess the performance of the proposed study.

5.1 Accuracy

Definition: The fraction of correctly predicted cases to the entire cases.

Formula: Accuracy = Correct Predictions/Total Predictions.

Interpretation: Accuracy gauges overall correctness, with higher values indicating better performance.

An accuracy score of 0.81 was achieved for this study.

5.2 Precision

Definition: The model's accuracy in identifying true positive instances among those anticipated as positive.

Formula: Precision = True Positives/True Positives + False Positives.

Precision reflects the model's ability to minimize false positives.

A precision of 0.6 was found for this study.

5.3 Recall (Sensitivity)

Definition: The model's capacity to distinguish genuine positive occurrences between all existing positive occurrences.

Formula: Recall = True Positives/True Positives + False Negatives.

Interpretation: Recall emphasizes the model's capacity to minimize false negatives.

0.4 recall score was found in this study.

5.4 F1-Score

Definition: The balanced harmonic average of precision also recall.

Formula: F1-score = 2 × Precision × Recall/Precision + Recall.

Interpretation: The F1-score offers a balanced evaluation metric, especially valuable in scenarios where there's an imbalance between positive and negative instances.

F1-score of 0.48 was found during the performance evaluation.

Interpretation Highlights:

Trade-off: Precision and recall often involve a trade-off; enhancing one may reduce the other.

The F1-score is valuable for imbalanced class distributions as it provides a single metric that examines into account both false positives moreover false negatives.

Figure 2 Illustrates the performance values of the above four parameters

From the above graph, it is evident that the accuracy is about 0.8, which means that 80% of the occurrences stayed correctly projected by the model and the precision is about 0.6, which means that 60% of the cases that remained foreseen as positive by the model were positive. The recall is about 0.4, which means that 40% of the positive instances were predicted as positive by the model. An F1-score of approximately 0.48 indicates that the model achieves a moderate balance between precision and recall. Confusion matrix is essential to evaluate the execution of the classification algorithm. It establishes the correct versus incorrect predictions on a class basis, thus allowing for determining accuracy, precision, recall, likewise F1-score. Additionally, the matrix helps in pointing out the most frequently confused classes and diagnosing the biases

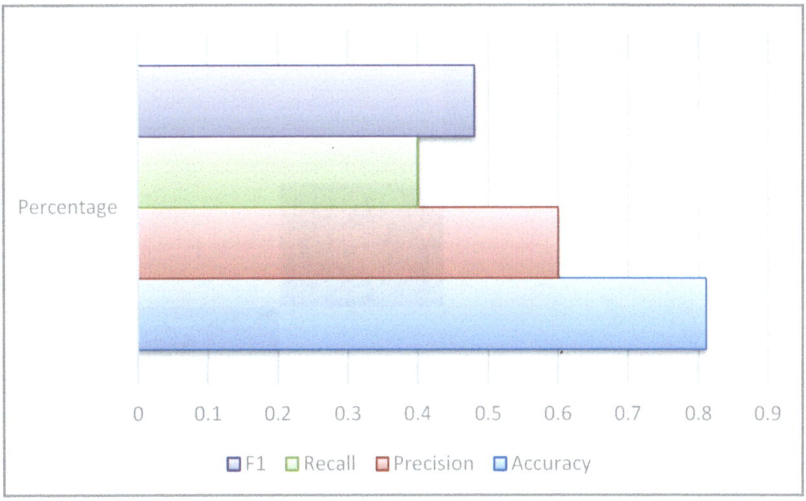

Fig. 2 Performance metrics

present in an imbalanced dataset. Besides, through visual heatmap representation, it is possible to interpret the behavior of the model because it provides valuable insight into potential improvements—Table 2. Represents a confusion matrix with numerical values indicating the number of predictions for each category Fig. 3. illustrates the confusion matrix of this study.

The cells along the diagonal display the number of accurate forecasts within each classification, whereas the cells on the diagonal demonstrate the number of inaccurate forecasts. The greater the numerical value, the more intense the coloration of the cell.

The Receiver Operating Characteristic (ROC) curve is a graphic version that point out the performance of a binary classification model across various thresholds. Within the realm of multi-class classification, the ROC curve can be generated for individual classes in comparison to the rest of the classes (one-vs-rest) or for pairs of classes (one-vs-one). When dealing with a multi-class classification scenario involving more than two classes (in this instance, there are four classes), distinct ROC curves can be constructed for each class to assess the model's effectiveness about that specific

Table 2 Number of predictions

Actual\Predicted	alt.atheism	comp.graphics	rec.sport.baseball	sci.med
alt.atheism	84	2	21	6
comp.graphics	5	83	13	4
rec.sport.baseball	4	2	113	5
sci.med	9	3	14	83

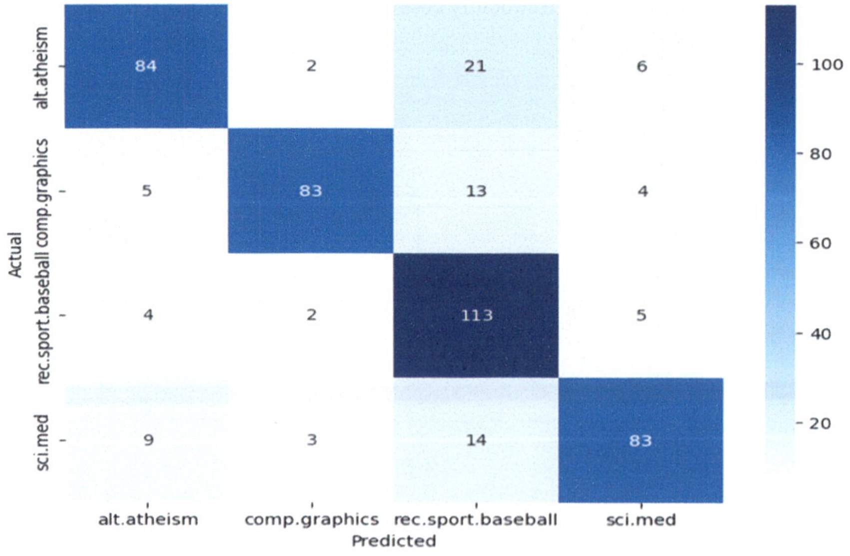

Fig. 3 Confusion matrix

class and the others. The Area Under the Curve (AUC) is calculated for each ROC curve. AUC offers a single scalar value that encapsulates the performance of the model across different thresholds.

A higher AUC generally indicates better overall classification performance. The Receiver Operating Characteristic (ROC) curve serves as a visual depiction of the balance between the true positive rate, also known as sensitivity, and the false positive rate, which is calculated as 1 minus specificity, across various threshold levels. A curve positioned nearer to the upper-left corner of the graph signifies the enhanced ability to discern between different classes. Figure 4. Represents the ROC curve of this study.

Figure 4 is a graph of the Receiver Operating Characteristic (ROC) Curve for four different classes. The numerical values that can be estimated from the image are the Area Under Curve (AUC) values for each class, which measure how well the classifier can distinguish between the positive and negative cases. The AUC values are:

Class 0: AUC = 0.17.
Class 1: AUC = 0.33.
Class 2: AUC = 0.63.
Class 3: AUC = 0.86.

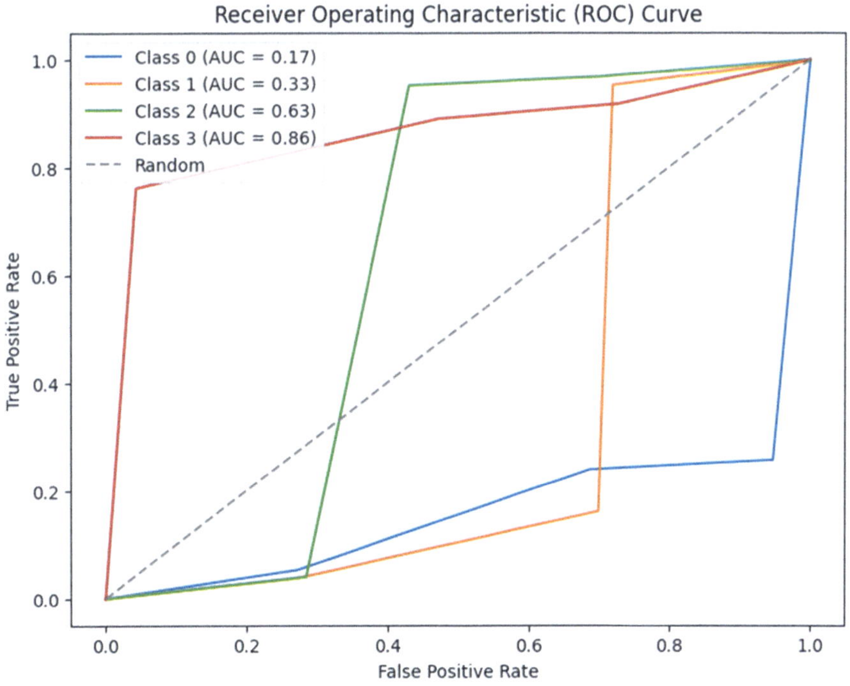

Fig. 4 ROC curve

The higher the AUC, the better the classifier performance. Class 3 has the highest AUC and Class 0 has the lowest AUC. The random prediction line has an AUC of 0.5, which means it cannot discriminate between the classes at all.

6 Conclusion

This research work presents a novel fusion approach that employs classical SVM with the attention mechanism in a bid to enhance the text classification over the 20 Newsgroups dataset. The attention mechanism integrated with SVM improves on adaptability of an SVM by giving importance to the different words in different proportions hence improving the quality of classification. The results depict an accuracy of about 80%, signaling a successful attention-boosting of the SVM model. Precision is 60%, raising hope for a check on false positives, while recall is 40%, showing an accurate detection of true positives. On the score of the value of the F1-score that considers both precision and recall, it shows an overall performance of about 48%. The research broadens the horizons of methods for text classification, which may provide a new perspective on the mechanisms for exam concentration and common machine learning models. The present fusion not only advances everything that exists but also provides prospects for the search for Terra Nova in the field of attention mechanisms outside neural networks. This research aligns with the broader landscape of machine learning advancements, as it has the potential to revolutionize natural language processing tools. The application of social marketing principles could further enhance its relevance in real-world contexts, offering valuable insights for sentiment analysis, brand management, and content classification in the social media domain. In conclusion, the attention-augmented SVM model depicts a novel way of text classification that adds to the evolution and definition of machine learning techniques with the ability to promise future developments for applications in social marketing and beyond.

References

1. Vivod, J.: Using Machine Learning and Natural Language Processing Techniques to Analyze and Support Moderation of Student Book Discussions (2020). arXiv:abs/2011.11712
2. Singh, J., Sharma, G.: Sentiment analysis study of human thoughts using machine learning techniques. Int. Conf. Disrup. Technol. (ICDT) **2023**, 776–785 (2023)
3. Hassan, F., Qureshi, N.A., Khan, M.Z., Khan, M.A., Soomro, A.S., Imroz, A., Marri, H.B.: Performance evolution for sentiment classification using machine learning algorithm. J. Appl. Res. Technol. Eng. (2023)
4. Shi, Q., Zhu, Y., Fang, C., Yang, D.: TLWSR: weakly supervised real-world scene text image super-resolution using text label. IET Image Process. **17**, 2780–2790 (2023)
5. Kaddour, A., Zellal, N., Sayad, L.: Improving text classification using text summarization. In: 2022 2nd International Conference on New Technologies of Information and Communication (NTIC), pp. 1–8 (2022)

6. Kumar, P., Wahid, A.: Social media analysis for sentiment classification using gradient boosting machines. Algorithms Intell. Syst. **923–934**,(2021). https://doi.org/10.1007/978-981-16-3246-4_70
7. Kachamas, P., Chandrachai, A., Sinthupinyo, S.: The influence of text classification on Facebook with AISAS method. Int. J. Bus. Inf. Syst. **35**(3), 401 (2020). https://doi.org/10.1504/ijbis.2020.111419
8. Reyes, M.: Consumer behavior and marketing. BoD—Books on Demand (2020)
9. Kausar, M.A., Fageeri, S.O., Soosaimanickam, A.: Sentiment classification based on machine learning approaches in amazon product reviews. Eng. Technol. Appl. Sci. Res. (2023)
10. Parmar, P., Dubey, S., K, S.M., Kadwal, M.: Evaluation of feelings of smart phone product review using SVM technique of sentiment analysis. Int. J. Res. Appl. Sci. Eng. Technol. **11**(4), 829–840 (2023). https://doi.org/10.22214/ijraset.2023.50082
11. Ebora, J.G., Espanol, J.C., Padilla, D.A.: Text classification of Facebook messages using Multi-class support vector machine. In: 2022 13th International Conference on Computing Communication and Networking Technologies (ICCCNT) (2022). https://doi.org/10.1109/icccnt54827.2022.9984554
12. Utami, N.A., Maharani, W., Atastina, I.: Personality classification of Facebook users according to Big Five personality using SVM (Support vector machine) method. Procedia Comput. Sci. **179**, 177–184 (2021). https://doi.org/10.1016/j.procs.2020.12.023
13. Yamasari, Y., Qoiriah, A., Rochmawati, N., Suartana, I., Putra, O.V., Nurhidayat, A.I.: Exploring the kernel on SVM to enhance the classification performance of students' academic performance. In: 2022 Fifth International Conference on Vocational Education and Electrical Engineering (ICVEE) (2022). https://doi.org/10.1109/icvee57061.2022.9930405
14. Xiaoyan, L., Raga, R.C.: BiLSTM model with attention mechanism for sentiment classification on Chinese mixed text comments. IEEE Access **11**, 26199–26210 (2023). https://doi.org/10.1109/access.2023.3255990
15. Verma, S., Kumar, A., Sharan, A.: IAN-BERT: combining post-trained BERT with interactive attention network for aspect-based sentiment analysis. SN Comput. Sci. **4**(6) (2023). https://doi.org/10.1007/s42979-023-02229-7
16. Berrimi, M., Oussalah, M., Moussaoui, A., Saidi, M.: Attention mechanism architecture for Arabic sentiment analysis. ACM Trans. Asian Low-Res. Lang. Inf. Proc. **22**(4), 1–26 (2023). https://doi.org/10.1145/3578265
17. Cheng, Z., Zhou, J., Wu, W., Chen, Q., & He, L. (2023). Tell model where to attend: Improving Interpretability of aspect-based sentiment classification via small explanation annotations. In: ICASSP 2023–2023 IEEE international conference on acoustics, speech and signal processing (ICASSP). https://doi.org/10.1109/icassp49357.2023.10096952
18. Kim, D.: Text classification based on neural network fusion. Tehnički glasnik **17**(3), 359–366 (2023). https://doi.org/10.31803/tg-20221228154330
19. Li, Y., Wang, H., Li, X., Xiu, P.: Research on sensitive text classification based on knowledge base and hybrid network. In: 2022 IEEE 8th International Conference on Computer and Communications (ICCC) (2022). https://doi.org/10.1109/iccc56324.2022.10065790
20. Feng, H., Lin, Z., Ma, Q.: Perturbation-based self-supervised attention for attention bias in text classification. IEEE/ACM Trans. Audio Speech Lang. Proc. **31**, 3139–3151 (2023). https://doi.org/10.1109/taslp.2023.3302230
21. Li, X., Li, Z., Luo, X., Xie, H., Lee, X., Zhao, Y., Wang, F.L., Li, Q.: Recurrent attention networks for long-text modeling. Findings of the Association for Computational Linguistics: ACL 2023 (2023). https://doi.org/10.18653/v1/2023.findings-acl.188
22. Liang, Y., Li, H., Guo, B., Yu, Z., Zheng, X., Samtani, S., Zeng, D.D.: Fusion of heterogeneous attention mechanisms in multi-view convolutional neural network for text classification. Inf. Sci. **548**, 295–312 (2021). https://doi.org/10.1016/j.ins.2020.10.021

Fuzzy Artificial Intelligence as a Technique to Find Relative Desirability for Fake Review Detection

A. Firos and Seema Khanum

Abstract This chapter delves into the innovative application of Fuzzy Artificial Intelligence (FAI) as a methodological approach for detecting fake reviews, a growing concern in the digital age where online consumer feedback significantly influences purchasing decisions. The core of this study lies in employing FAI to assess the relative desirability of reviews, distinguishing between genuine and fabricated content. By integrating fuzzy logic with artificial intelligence, this technique enables a nuanced analysis of review authenticity, leveraging the inherent vagueness and ambiguity of human language to identify potential falsifications. The methodology builds upon the principle that genuine reviews exhibit certain identifiable patterns and characteristics that can be quantitatively analyzed and compared against the expected norms of authentic feedback. FAI's strength lies in its ability to process and evaluate linguistic variables and their degrees of truth, allowing for a more flexible and adaptable detection system compared to traditional binary approaches. This chapter outlines the development of a FAI-based model that assesses reviews across multiple dimensions—such as sentiment, specificity, and consistency with known product attributes—to calculate a desirability score indicating the likelihood of authenticity. The experimental results, derived from a dataset comprising both real and artificially generated reviews, demonstrate the efficacy of the FAI approach in identifying fake reviews with a high degree of accuracy. The FAI model not only outperforms conventional binary classification models but also provides insights into the characteristics that most strongly indicate review authenticity, offering valuable guidelines for further refinement of fake review detection systems. This chapter discusses the implications of FAI in the broader context of online trust and consumer

A. Firos
Department of Computer Science and Engineering, Rajiv Gandhi University, Rono Hills, Doimukh 791112, India
e-mail: firos.a@rgu.ac.in

S. Khanum (✉)
Indian Computer Emergency Response Team (ICERT), MeitY, Electronics Niketan, 6, CGO Complex, Lodhi Road, New Delhi 110003, India
e-mail: seem.kh@gmail.com

S. Dutta et al. (eds.), *Advances in Data Analytics for Influencer Marketing: An Interdisciplinary Approach*, Information Systems Engineering and Management 9,
https://doi.org/10.1007/978-3-031-65727-6_18

behavior, highlighting its potential to enhance the reliability of online review ecosystems. By providing a robust tool for distinguishing genuine feedback from deceptive content, FAI contributes to the integrity of online marketplaces and supports informed decision-making by consumers. This chapter presents a compelling argument for the adoption of FAI in combating the challenge of fake reviews. It offers a significant advancement in the field of digital trust, showing how the fusion of fuzzy logic and artificial intelligence can create a more secure and trustworthy online environment for consumers and businesses alike.

Keywords Artificial neural networks · Product selection · Fuzzy system · Parameterized fuzzy measures · Decision making system

1 Introduction

1.1 Generative Models for Assessing Relative Desirability for Fake Review Detection

Generative models have emerged as a powerful tool in the realm of artificial intelligence, particularly in the detection of fake reviews [1]. These models, which can generate new data instances that resemble the training data, are being increasingly applied to assess the relative desirability of online reviews, distinguishing between genuine and fabricated content. This capability is crucial in maintaining the integrity of online platforms where user-generated reviews significantly influence consumer behavior.

The application of generative models in fake review detection centers around their ability to learn the complex patterns and nuances of genuine reviews. By training on a dataset composed of authentic reviews, these models can grasp the subtleties of human language, including sentiment, syntax, and style, which are often challenging for fake reviewers to replicate accurately. This understanding allows generative models to generate synthetic reviews that serve as a benchmark for assessing the authenticity of new reviews [2].

One of the key strengths of generative models in this context is their ability to deal with the ambiguity and variability inherent in human-generated content. Traditional binary classification models often struggle with the fuzzy boundaries between genuine and fake reviews, as the latter become increasingly sophisticated. Generative models, however, can navigate this complexity by evaluating the likelihood of a review being real based on its similarity to the learned patterns of authentic reviews.

Another significant advantage is the continuous learning capability of generative models. As they are exposed to more data over time, including newly identified fake reviews, they refine their understanding of what constitutes a genuine review. This dynamic learning process ensures that the models remain effective even as fake review tactics evolve, maintaining a high level of accuracy in the detection process.

The implementation of generative models also brings challenges, such as the need for substantial and diverse training data to avoid biases and ensure the model's generalizability [3]. Additionally, the computational complexity of training generative models requires significant resources, which might be a barrier for some applications. Despite these challenges, the potential benefits in terms of improving the reliability of online review ecosystems are considerable.

In conclusion, generative models represent a promising approach to enhancing the detection of fake reviews by assessing the relative desirability of user-generated content. Their ability to learn and mimic the complexity of genuine reviews, adapt to new patterns of deception, and navigate the ambiguity of human language positions them as a valuable tool in safeguarding the authenticity of online review platforms. As these models continue to evolve, they will play a critical role in fostering trust and transparency in the digital marketplace [4].

1.2 Deep Neural Network for Giving Suggestions

Deep neural networks (DNNs) are revolutionizing the way we approach the detection of fake reviews, leveraging their ability to learn complex patterns and nuances in data. These networks, which are inspired by the structure and function of the human brain, have shown remarkable success in various tasks, including natural language processing, making them particularly suitable for analyzing textual data such as customer reviews. By utilizing DNNs for fake review detection, we can significantly enhance the accuracy and reliability of identifying fraudulent content on online platforms [5].

The architecture of a DNN for fake review detection typically involves multiple layers of neurons, each capable of learning different aspects of the data. The initial layers might learn basic patterns such as word frequencies and sentence structures, while deeper layers can recognize more complex features like sentiment inconsistency, subtle linguistic cues, and context that might indicate a review is fake [6]. This hierarchical learning process enables DNNs to capture the essence of genuine reviews and distinguish them from fabricated ones with high precision.

Training a DNN for this purpose involves feeding it a large dataset of labeled reviews, where each review is marked as either genuine or fake. The network adjusts its weights through backpropagation, minimizing the difference between its predictions and the actual labels. This process requires substantial computational resources, especially for large datasets, but the investment pays off in the network's ability to accurately identify fake reviews once it is adequately trained.

One of the key strengths of using DNNs for fake review detection is their ability to learn from unstructured text data without the need for extensive manual feature engineering [7]. Traditional machine learning models rely heavily on the quality of features extracted from the text, which can be a labor-intensive process requiring domain expertise. DNNs, however, can automatically learn to identify relevant features directly from the data, making them more adaptable and easier to scale across different domains and languages.

Despite their advantages, DNNs also pose challenges, such as the need for large labeled datasets for training and the risk of overfitting, where the model learns the training data too well and performs poorly on unseen data. Furthermore, the "black box" nature of DNNs can make it difficult to interpret why the model classifies a review as fake, which can be a drawback for applications where explainability is important [8].

In conclusion, deep neural networks represent a powerful tool for detecting fake reviews, offering improved accuracy over traditional methods through their ability to learn complex patterns in data. As the technology continues to evolve, it is expected that DNNs will play an increasingly central role in maintaining the integrity of online review systems, helping to protect consumers and businesses from the harmful effects of fraudulent content.

1.3 Fuzzy AI System for Fake Review Detection

As artificial intelligence (AI) technology develops rapidly, fuzzy AI systems are attracting attention in the field of fake review detection. Fuzzy AI systems are based on fuzzy logic and have the ability to handle ambiguity and uncertainty. This is very useful for analyzing and understanding the complex and diverse nature of online reviews. The development of fuzzy AI systems for detecting fake reviews can play an important role in improving the trustworthiness and transparency of online platforms.

The fuzzy AI system comprehensively analyzes various factors such as the linguistic characteristics of reviews, emotions, and user behavior patterns. This system quantifies the uncertainty of each element through fuzzy logic to determine whether a review is real or fake. In this process, Fuzzy AI provides a relative assessment of the review's trustworthiness, which is greatly helpful in filtering out fake reviews [9].

The key to developing a fuzzy AI system for detecting fake reviews is building an efficient fuzzy logic model [10]. This model sets various rules and criteria to evaluate the authenticity of reviews. These rules are created taking into account several aspects, including the content of the review, the author's profile, and review writing patterns. Based on these rules, the fuzzy AI system calculates a trust score for each review, which is used to assess the authenticity of the review.

The fuzzy AI system also deepens its understanding of reviews through continuous learning and self-improvement mechanisms. This is because the online review environment is constantly changing, and the technology for writing fake reviews continues to evolve. The fuzzy AI system continuously improves detection capabilities by analyzing new review data and adjusting its decision rules.

However, effective implementation of fuzzy AI systems involves several challenges. Examples include obtaining accurate and diverse training data, designing and optimizing complex fuzzy logic rules, and balancing the processing speed and accuracy of the system. Nevertheless, fuzzy AI systems appear to be a very promising approach for detecting fake reviews [11].

1.4 Fuzzy Based Technique to Find Relative Desirability for Fake Review Detection

The digital marketplace is increasingly relevant on user reviews for guiding consumer decisions, making the detection of fake reviews a critical concern for maintaining trust and integrity. The development of a fuzzy-based technique for assessing the relative desirability of reviews represents a significant advancement in the ongoing battle against fake review manipulation. This technique leverages the principles of fuzzy logic to handle the inherent ambiguity and subjectivity in textual reviews, providing a nuanced approach to identifying fraudulent content [12].

Fuzzy logic, with its ability to process imprecise or ambiguous data, is particularly well-suited for analyzing the complex and often subtle linguistic features that characterize fake reviews. Traditional binary classification models can struggle with the nuanced differences between genuine and fake reviews, as they require clear-cut definitions that do not always exist in natural language [13]. However, a fuzzy-based system assesses reviews based on degrees of truth or membership to certain linguistic categories, allowing for a more flexible and accurate analysis.

One of the core components of this technique involves establishing a set of linguistic variables that can be indicative of a review's authenticity. These variables might include sentiment inconsistency, overuse of superlatives, and unnatural frequency of brand mentions. Each review is then scored based on its degree of membership to these predefined categories, with the fuzzy logic system aggregating these scores to determine an overall desirability score. This score reflects the likelihood that a review is genuine, providing a more refined tool for fake review detection.

The application of fuzzy logic in this context also allows for the incorporation of user behavior and review context, further enhancing the detection process [14]. By analyzing patterns such as the timing of reviews and the historical credibility of the reviewer, the system can adjust the desirability score to reflect these additional dimensions of trustworthiness. This multi-faceted approach is crucial for adapting to the sophisticated tactics employed by entities generating fake reviews.

Implementing this fuzzy-based technique poses certain challenges, including the need for extensive training data to accurately define the linguistic variables and membership functions [15]. Additionally, the dynamic nature of language and consumer behavior requires continuous updates to the system to maintain its effectiveness. Despite these challenges, the potential benefits of a more reliable and nuanced fake review detection system are immense for both businesses and consumers.

The fuzzy-based technique for assessing the relative desirability of reviews represents a promising direction for enhancing the detection of fake reviews. By embracing the complexity and ambiguity of natural language, this approach offers a more sophisticated and adaptable solution to a problem that undermines the integrity of online marketplaces. As this technology evolves, it holds the potential to significantly improve the trustworthiness of online review ecosystems, benefiting both consumers and businesses [16].

The following outlines the key components and steps of a fuzzy-based technique for this purpose:

- Data Collection: Gather a dataset of reviews from various sources, including e-commerce platforms or review websites, containing both genuine and fake reviews.
- Feature Extraction: Extract relevant features from the review data, such as text sentiment, word frequency, reviewer behavior patterns, and linguistic cues.
- Fuzzy Logic Model Design: Develop a fuzzy logic model that can effectively capture the uncertainty and imprecision inherent in human language and behavior. This model should include linguistic variables, fuzzy rules, membership functions, and fuzzy inference mechanisms.
- Rule Base Creation: Create a rule base that maps input linguistic variables (e.g., review sentiment, word frequency) to output variables (e.g., likelihood of a review being fake). These rules should be derived from domain knowledge or expert input and encoded into the fuzzy logic system.
- Membership Function Definition: Define membership functions for each linguistic variable to quantify the degree of membership of input values to fuzzy sets. These functions represent the fuzzy relationships between input and output variables.
- Fuzzy Inference: Apply fuzzy inference techniques, such as Mamdani or Sugeno methods, to process the fuzzy rules and input data and generate fuzzy output values.
- Defuzzification: Convert the fuzzy output values into crisp values using defuzzification methods, such as centroid or weighted average, to make a binary decision (i.e., genuine or fake review).
- Model Evaluation: Evaluate the performance of the fuzzy-based fake review detection model using metrics such as accuracy, precision, recall, and F1-score. This step involves testing the model on a separate validation dataset to assess its effectiveness in identifying fake reviews accurately.
- Optimization and Refinement: Fine-tune the fuzzy logic model parameters, including membership functions and rule base, based on the evaluation results to improve detection accuracy and minimize false positives/negatives.
- Deployment and Integration: Deploy the optimized fuzzy-based fake review detection model into a practical application or system, integrating it with existing review platforms or e-commerce websites to automatically identify and flag suspicious reviews in real-time.

By incorporating these key components and following these steps, a fuzzy-based technique becomes a powerful and adaptable tool for the nuanced assessment of relative desirability in fake reviews detection.

2 The Background

2.1 The Fake Review Detection System

The Fake Review Detection System represents a critical technological advancement aimed at preserving the integrity and reliability of online review platforms. In an age where consumer decisions are heavily influenced by online reviews, the proliferation of fake or manipulated reviews poses a significant threat. These systems leverage sophisticated algorithms and data analysis techniques to differentiate genuine reviews from those that are fabricated or misleading, ensuring that users receive accurate information [17].

At its core, the Fake Review Detection System utilizes natural language processing (NLP) and machine learning algorithms to analyze the textual content of reviews [18]. By examining linguistic patterns, frequency of specific terms, sentiment analysis, and writing style, the system can identify anomalies that suggest a review might not be authentic. For instance, an unusually high frequency of positive adjectives or a lack of specific details about the user experience could raise red flags. In addition to linguistic analysis, behavioral analysis plays a crucial role in detecting fake reviews. The system examines patterns such as the timing of posted reviews, the frequency of reviews by a single user, and the relationship between reviewers and businesses. Unnatural patterns, such as a sudden influx of positive reviews following a period of negative feedback, can indicate coordinated efforts to manipulate a business's online reputation.

The effectiveness of a Fake Review Detection System relies heavily on its training data. The system must be trained on a large and diverse dataset of genuine and fake reviews to accurately learn the characteristics of each. This process involves supervised learning, where the algorithm iteratively adjusts its parameters to minimize the difference between its predictions and the actual labels of the training data. However, the challenge of adapting to evolving tactics used by those creating fake reviews remains. As detection methods become more sophisticated, so do the strategies employed to generate convincing fake reviews [19]. This necessitates continuous updates and refinements to the detection algorithms, incorporating the latest linguistic and behavioral patterns associated with deceptive practices.

The development and deployment of advanced Fake Review Detection Systems are essential for maintaining consumer trust in online platforms. By ensuring that reviews accurately reflect genuine customer experiences, these systems help guide consumers to make informed decisions while protecting businesses from unfair competition and reputation damage. The ongoing battle against fake reviews underscores the importance of innovation and vigilance in the digital age [20].

2.2 Advantages of Fuzzy ANN in Fake Review Detection Systems

The integration of Fuzzy Logic with Artificial Neural Networks (ANN), creating Fuzzy Artificial Neural Networks (Fuzzy ANN), presents a significant advancement in the field of fake review detection [21]. This hybrid approach combines the strengths of both fuzzy logic's ability to handle ambiguity and uncertainty with ANN's powerful data processing capabilities, offering several advantages in identifying and filtering out deceptive reviews in online platforms. The Fuzzy ANN approach enhances the system's ability to deal with the inherent vagueness and subjectivity of natural language found in reviews. Traditional binary classification systems often struggle with the nuances of human language, whereas fuzzy logic can model these uncertainties more naturally. When fused with an ANN, the system can learn and adapt to these nuances, improving its accuracy in distinguishing between genuine and fake reviews [22].

The adaptability of Fuzzy ANN systems is a significant advantage. The learning capability of neural networks, coupled with the flexible rule-based nature of fuzzy logic, allows the system to continuously evolve and adapt to new patterns and tactics used by those posting fake reviews [23]. This adaptability ensures that the detection system remains effective over time, despite the ever-changing landscape of online reviews. Fuzzy ANNs can process a vast amount of unstructured data efficiently. Online reviews are diverse and unstructured, posing a challenge for conventional data processing techniques. The combination of fuzzy logic and ANN is adept at handling this diversity, enabling the system to analyze large datasets of reviews with varying levels of detail, sentiment, and authenticity [24].

The interpretability of Fuzzy ANN systems offers an advantage. While neural networks are often criticized for being "black boxes," the integration of fuzzy rules can provide insights into the decision-making process [25]. This transparency is crucial for stakeholders who wish to understand the basis on which reviews are flagged as fake, fostering trust in the detection system. Fuzzy ANN systems can achieve higher accuracy with less computational complexity compared to other machine learning models. The fuzzy logic component reduces the need for extensive pre-processing of data by managing ambiguity directly within the model. This efficiency is particularly beneficial in real-time applications where quick and accurate detection of fake reviews is essential.

The versatility of Fuzzy ANN systems makes them suitable for various domains beyond fake review detection, such as fraud detection, sentiment analysis, and customer behavior prediction [26]. This versatility underscores the broad applicability and potential of Fuzzy ANNs in tackling complex problems across different industries. Fuzzy Artificial Neural Networks offer a compelling solution for detecting fake reviews, capitalizing on their ability to handle linguistic ambiguity, adapt to new patterns, efficiently process large datasets, provide interpretability, and maintain high accuracy with less complexity. As online platforms continue to grow in importance, such advanced systems will play a crucial role in ensuring the integrity and trustworthiness of user-generated content.

2.3 *Artificial Neural Networks for Automation*

Artificial Neural Networks (ANNs) have emerged as a powerful tool in automating the detection of fake reviews, a challenge that has grown with the proliferation of online marketplaces and review platforms. ANNs mimic the neural structure of the human brain, enabling them to learn and make intelligent decisions based on the data fed into them. This capability makes them particularly suited to the complex and nuanced task of identifying fake reviews, which often require the analysis of subtle linguistic cues and patterns that might elude simpler, rule-based algorithms [27].

The strength of ANNs lies in their ability to process and learn from vast amounts of unstructured text data, such as customer reviews. By training on datasets comprising both genuine and fake reviews, ANNs can learn to identify the distinguishing features of each. This training involves adjusting the weights of the connections between the neurons in the network, optimizing the network's ability to classify reviews accurately. The more data the network is exposed to, the better it becomes at discerning the nuanced differences between genuine and deceptive content [28].

One of the key advantages of using ANNs for fake review detection is their flexibility. They can be trained to recognize a wide variety of deceptive tactics, from the subtle manipulation of sentiment to the strategic placement of keywords [29]. Moreover, ANNs can continuously learn and adapt to new patterns of deception, making them highly effective against evolving strategies employed by those looking to game the system.

However, the effectiveness of an ANN in detecting fake reviews is heavily dependent on the quality and diversity of the training data. A well-curated dataset that accurately reflects the complexity of human language and the myriad ways in which it can be manipulated is crucial for training a robust model. This necessitates ongoing efforts to collect and label data, a task that can be both time-consuming and challenging.

Despite these challenges, ANNs offer a scalable and efficient solution for the automation of fake review detection. Their ability to process large volumes of data quickly means they can provide real-time monitoring of reviews across multiple platforms, alerting businesses to potential fake reviews before they can do significant damage. This is particularly valuable in today's fast-paced online marketplace, where reputation can be significantly impacted by the perceived authenticity of user-generated content.

In conclusion, while challenges remain in training and fine-tuning ANNs for the specific task of fake review detection, their potential benefits are undeniable. As these systems continue to evolve and improve, they will play an increasingly vital role in safeguarding the integrity of online review platforms, ensuring that consumers can make informed decisions based on trustworthy and authentic reviews [30].

2.4 Fuzzy Logic and ANN Based Technique to Find Relative Desirability

Integrating Fuzzy Logic with Artificial Neural Networks (ANN) to establish a technique for evaluating the relative desirability in fake review detection offers a powerful tool in the fight against fraudulent online content. This hybrid approach leverages the strengths of both fuzzy systems, known for their ability to handle uncertainty and ambiguity, and ANNs, renowned for their learning capabilities and adaptability. The outcome is a sophisticated system capable of nuanced analysis and interpretation of complex patterns in data, particularly useful in distinguishing genuine reviews from fake ones.

Fuzzy logic introduces an element of human-like reasoning into the system, allowing it to process imprecise or ambiguous information commonly found in natural language. For example, when analyzing reviews, the system can handle subjective expressions and gradations in sentiment, which are often challenging for traditional binary logic systems. This capability makes the technique particularly adept at interpreting the subtleties of human language, a critical factor in effective fake review detection [2].

On the other hand, ANNs contribute their powerful pattern recognition capabilities, learning from examples to improve their performance over time. By training on a dataset of known genuine and fake reviews, the neural network learns to identify the distinguishing features of each. This learning process enables the system to adapt to new strategies used by individuals or entities generating fake reviews, maintaining its effectiveness even as deceptive tactics evolve.

The integration of fuzzy logic with ANNs creates a dynamic system that can evaluate the relative desirability of reviews with remarkable accuracy. By assigning degrees of credibility rather than making binary true/false judgments, the system provides a more nuanced assessment of each review's authenticity. This approach is particularly valuable in cases where the veracity of a review is not immediately clear, allowing for a more refined detection process.

Moreover, this hybrid technique offers scalability and efficiency, essential features for modern online platforms hosting millions of user-generated reviews. It can quickly process large volumes of data, identifying potential fake reviews in real-time and thus protecting consumers and businesses from the detrimental effects of fraudulent content.

In conclusion, the fusion of fuzzy logic and ANN into a technique for assessing the relative desirability in fake review detection represents a significant advancement in the field. This approach not only enhances the accuracy and reliability of fake review detection systems but also offers a flexible and adaptive solution capable of responding to the ever-changing landscape of online review manipulation.

3 Proposed Model

The proposed model to enhance fake review detection leverages the integration of Preference-Leveled Evaluation Functions (PLEFs) with an Artificial Neural Network (ANN). This model aims to not only identify fake reviews with high accuracy but also to understand the underlying patterns and preferences that characterize such reviews, using a novel approach that combines the strengths of both quantitative analysis and neural learning.

1. System Overview

   The core of the proposed model is a multi-layer Artificial Neural Network, designed to learn from and adapt to the complex patterns found in review data. The ANN is tasked with processing text data, extracting features, and classifying reviews as either genuine or fake. To augment the ANN's capabilities, Preference-Leveled Evaluation Functions are integrated into the system. PLEFs are mathematical functions that quantify the level of preference or sentiment expressed in a review, based on linguistic cues and contextual understanding.
2. Data Preprocessing and Feature Extraction

   Before feeding data into the ANN, reviews undergo preprocessing to clean and normalize text. This includes removing irrelevant characters, correcting typos, and stemming. Next, feature extraction occurs, where both traditional NLP techniques and PLEFs are used to quantify various aspects of the reviews, such as sentiment intensity, subjectivity levels, and specific word patterns associated with deceptive practices.
3. Integration of PLEFs

   PLEFs are integrated into the feature extraction phase, providing a nuanced understanding of the preferences and sentiments expressed in reviews. Each review is evaluated based on a series of PLEFs, which assess the authenticity of sentiment expression. This integration allows the model to capture not just the presence of certain words or phrases, but the genuine or deceptive intention behind them.

4. ANN Architecture and Training

   The ANN architecture is designed to be deep enough to capture complex patterns but streamlined to ensure efficient processing. It includes input layers corresponding to the extracted features, hidden layers for pattern recognition and learning, and an output layer for classification. The network is trained on a labeled dataset of genuine and fake reviews, using backpropagation to adjust weights and minimize classification errors.
5. Model Evaluation and Refinement

   After training, the model is evaluated using a separate test dataset to assess its accuracy, precision, recall, and F1 score. Continuous refinement is conducted by adjusting the ANN structure, redefining PLEFs based on emerging patterns in fake reviews, and enriching the training dataset to cover a wider array of deceptive tactics.
6. Implementation and Real-World Application

   The final step involves implementing the model in a real-world environment, such as an e-commerce platform, where it can monitor and flag potentially fake reviews in real-time. Continuous monitoring and periodic retraining ensure the model remains effective against evolving deceptive practices.

By integrating PLEFs with an ANN, this proposed model offers a sophisticated approach to fake review detection, capable of understanding not just the textual content of reviews, but the subtler nuances of human sentiment and preference expression, making it a powerful tool in maintaining the integrity of online review ecosystems.

This designed model integrates Preference-Leveled Evaluation Functions with Artificial Intelligence, specifically Fuzzy Logic, to evaluate the relative desirability of fake reviews. By merging structured preferences, fuzzy reasoning, and continuous learning, the system becomes both robust and adaptable, aiming to enhance decision support and optimize the accuracy of fake review detection. A block diagram of the proposed system, which utilizes Preference-Leveled Evaluation Functions for assessing the desirability of fake reviews, is depicted in Fig. 1, with the corresponding steps outlined in Algorithm 1. The implementation employs example features of a fake review dataset, as shown in Table 1. This synergy between preference-based evaluations and machine learning capabilities improves the model's ability to discern nuanced value states and inform interventions that contribute to a more effective fake review detection process.

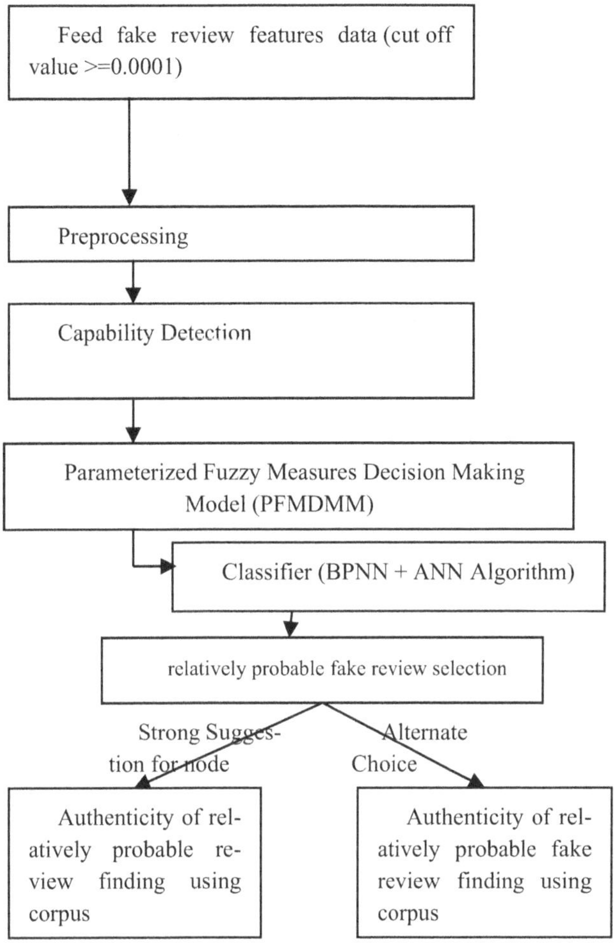

Fig. 1 Block diagram of proposed PFMDMM based on preference leveled evaluation functions for finding relative desirability of fake review

Table 1 Examples features of fake review dataset [31]

Time	Use x[0]	Gen x[1]	Preference−1 x[2]	Preference−2 x[3]	Preference−3 x[4]	Preference−4 x[5]	Preference−5 x[6]	Preference−6 x[7]	Preference−7 x[8]
1.45E + 09	0.932833	0.003483	0.932833	3.33E-05	0.0207	0.061917	0.442633	0.12415	0.006983
1.45E + 09	0.934333	0.003467	0.934333	0	0.020717	0.063817	0.444067	0.124	0.006983
1.45E + 09	0.931817	0.003467	0.931817	1.67E-05	0.0207	0.062317	0.446067	0.123533	0.006983
1.45E + 09	1.02205	0.003483	1.02205	1.67E-05	0.1069	0.068517	0.446583	0.123133	0.006983
1.45E + 09	1.1394	0.003467	1.1394	0.000133	0.236933	0.063983	0.446533	0.12285	0.00685
1.45E + 09	1.391867	0.003433	1.391867	0.000283	0.50325	0.063667	0.447033	0.1223	0.006717
1.45E + 09	1.366217	0.00345	1.366217	0.000283	0.4994	0.063717	0.443267	0.12205	0.006733
–	–	–	–	–	–	–	–	–	–

Algorithm 1: PFMDMM Based on Preference Leveled Evaluation Functions to assess the relative desirability of fake review.

Input:
Input x features of fake review Dataset.
Output:
Categorizes the fake review for node and Alternate Choice.
Start
1. Input x features of data of fake review dataset
2. PFMDMM to recognize accurate review for the node; arranged in correct format to feed into the ANN for classification.
3. *Training stage*: weights of the Feed Forward Neural Network were given by some arbitrary value as per Table 1 and is then tuned for optimal during the iterative learning procedure with the help of back propagation algorithm.
4. *Testing stage*: the neural network is tested against a variety of test samples, to ensure whether the acquired system correctly categorizes the review to best probable value and other value parts.
5. Categorizes the value to best suggestion range and alternate opinion.

Stop

4 Conclusion

This work introduces a groundbreaking BPNN and ANN model that leverages the PFMDMM data classification system for determining the optimal desirability of fake reviews for a node. The results from the trials illustrated that the parameterized fuzzy measures decision-making model for superior fake review detection showcases promising performances in identifying the most accurate fake reviews. This model is constructed upon preference-leveled assessment functions.

To the best of our knowledge, this study is the pioneering effort to utilize the preference-leveled evaluation functions-based parameterized fuzzy measures decision-making model for data categorization in the context of fake review detection.

The main outcomes of this research, specifically, include:

- The proposed method is capable of delivering decision-making on clustering within limited time frames.
- This study introduces a novel application of the Parameterized Fuzzy Measures Decision Making Model Based on Preference-Leveled Evaluation Functions for the classification of fake reviews using a BPNN architecture.
- Regarding the evaluation, the provided fake review Dataset has been considered to test the proposed methodology. This dataset comprises robust fake review measurements that were collected to construct a parametric model for fake review classification.

- Specifically, by employing the decision-making capabilities of the Parameterized Fuzzy Measures Decision Making Model Based on Preference-Leveled Evaluation Functions, this study developed a novel automated system for detecting fake reviews.

References

1. Hu, K.H., Chen, F.H., Hsu, M.F., Tzeng, G.H.: Identifying key factors for adopting artificial intelligence-enabled auditing techniques by joint utilization of fuzzy-rough set theory and MRDM technique. Technol. Econ. Dev. Econ. **27**(2), 459–492 (2021)
2. Hu, Q., Gois, F.N.B., Costa, R., Zhang, L., Yin, L., Magaia, N., de Albuquerque, V.H.C.: Explainable artificial intelligence-based edge fuzzy images for COVID-19 detection and identification. Appl. Soft Comput. **123**, 108966 (2022)
3. Bouramdane, A.A.: Cyberattacks in smart grids: challenges and solving the multi-criteria decision-making for cybersecurity options, including ones that incorporate artificial intelligence, using an analytical hierarchy process. J. Cybersecur. Priv. **3**(4), 662–705 (2023)
4. Nwakanma, C.I., Ahakonye, L.A.C., Njoku, J.N., Odirichukwu, J.C., Okolie, S.A., Uzondu, C., Kim, D.S.: Explainable artificial intelligence (xai) for intrusion detection and mitigation in intelligent connected vehicles: a review. Appl. Sci. **13**(3), 1252 (2023)
5. Abid, A., Khan, M.T., Iqbal, J.: A review on fault detection and diagnosis techniques: basics and beyond. Artif. Intell. Rev. **54**(5), 3639–3664 (2021)
6. Aylan, O., Alkabaa, A.S., Alqabbaa, H.S., Pamukçu, E., Leiva, V.: Early prediction in classification of cardiovascular diseases with machine learning, neuro-fuzzy and statistical methods. Biology **12**(1), 117 (2023)
7. Band, S.S., Yarahmadi, A., Hsu, C.C., Biyari, M., Sookhak, M., Ameri, R., ... Liang, H.W.: Application of explainable artificial intelligence in medical health: a systematic review of interpretability methods. Inf. Med. Unlocked 101286 (2023)
8. Zhao, S., Blaabjerg, F., Wang, H.: An overview of artificial intelligence applications for power electronics. IEEE Trans. Power Electron. **36**(4), 4633–4658 (2020)
9. Bordoloi, M., Biswas, S.K.: Sentiment analysis: A survey on design framework, applications and future scopes. Artif. Intell. Rev. **56**(11), 12505–12560 (2023)
10. Basurto-Hurtado, J.A., Cruz-Albarran, I.A., Toledano-Ayala, M., Ibarra-Manzano, M.A., Morales-Hernandez, L.A., Perez-Ramirez, C.A.: Diagnostic strategies for breast cancer detection: from image generation to classification strategies using artificial intelligence algorithms. Cancers **14**(14), 3442 (2022)
11. Zadmirzaei, M., Hasanzadeh, F., Susaeta, A., Gutiérrez, E.: A novel integrated fuzzy DEA–artificial intelligence approach for assessing environmental efficiency and predicting CO2 emissions. Soft. Comput. **28**(1), 565–591 (2024)
12. Temel, F.A., Yolcu, O.C., Turan, N.G.: Artificial intelligence and machine learning approaches in composting process: a review. Biores. Technol. **370**, 128539 (2023)
13. Al Radi, M., AlMallahi, M.N., Al-Sumaiti, A.S., Semeraro, C., Abdelkareem, M.A., Olabi, A.G.: Progress in artificial intelligence-based visual servoing of autonomous unmanned aerial vehicles (UAVs). Int. J. Thermofluids **21**, 100590 (2024)
14. Osamy, W., Khedr, A.M., Salim, A., AlAli, A.I., El-Sawy, A.A.: Recent studies utilizing artificial intelligence techniques for solving data collection, aggregation and dissemination challenges in wireless sensor networks: a review. Electronics **11**(3), 313 (2022)
15. Lee, M., Kwon, W., Back, K.J.: Artificial intelligence for hospitality big data analytics: developing a prediction model of restaurant review helpfulness for customer decision-making. Int. J. Contemp. Hosp. Manag. **33**(6), 2117–2136 (2021)

16. Teslyuk, V., Kazarian, A., Kryvinska, N., Tsmots, I.: Optimal artificial neural network type selection method for usage in smart house systems. Sensors **21**(1), 47 (2020)
17. Osamy, W., Khedr, A.M., Salim, A., Al Ali, A.I., El-Sawy, A.A.: Coverage, deployment and localization challenges in wireless sensor networks based on artificial intelligence techniques: a review. IEEE Access **10**, 30232–30257 (2022)
18. Boahen, J.K., Elsagheer Mohamed, S.A., Khalil, A.S., Hassan, M.A.: Application of artificial intelligence techniques in modeling attenuation behavior of ionization radiation: a review. Radiat. Detect. Technol. Methods **7**(1), 56–83 (2023)
19. Suhag, A., Daniel, A.: Study of statistical techniques and artificial intelligence methods in distributed denial of service (DDOS) assault and defense. J. Cyber Secur. Technol. **7**(1), 21–51 (2023)
20. Kar, T., Kanungo, P., Mohanty, S.N., Groppe, S., Groppe, J.: Video shot-boundary detection: issues, challenges and solutions. Artif. Intell. Rev. **57**(4), 104 (2024)
21. Janga, J.K., Reddy, K.R., Kvns, R.: Integrating artificial intelligence, machine learning, and deep learning approaches into remediation of contaminated sites: a review. Chemosphere 140476 (2023)
22. Alsalem, M.A., Alamoodi, A.H., Albahri, O.S., Dawood, K.A., Mohammed, R.T., Alnoor, A., Al-Obaidi, J.R.: Multi-criteria decision-making for coronavirus disease 2019 applications: a theoretical analysis review. Artif. Intell. Rev. **55**(6), 4979–5062 (2022)
23. Kaur, R., Gabrijelčič, D., & Klobučar, T.: Artificial intelligence for cybersecurity: literature review and future research directions. Inf. Fus. 101804 (2023)
24. Behara, R.K., Saha, A.K.: Artificial intelligence methodologies in smart grid-integrated doubly fed induction generator design optimization and reliability assessment: a review. Energies **15**(19), 7164 (2022)
25. Mahbooba, B., Timilsina, M., Sahal, R., Serrano, M.: Explainable artificial intelligence (XAI) to enhance trust management in intrusion detection systems using decision tree model. Complexity **2021**, 1–11 (2021)
26. Mustapha, M.T., Ozsahin, D.U., Ozsahin, I., Uzun, B.: Breast cancer screening based on supervised learning and multi-criteria decision-making. Diagnostics **12**(6), 1326 (2022)
27. Wood, D.A.: Real-time monitoring and optimization of drilling performance using artificial intelligence techniques: a review. Sustainable Natural Gas Drilling, pp. 169–210 (2024)
28. Manhas, J., Gupta, R.K., Roy, P.P.: A review on automated cancer detection in medical images using machine learning and deep learning based computational techniques: challenges and opportunities. Arch. Comput. Methods Eng. **29**(5), 2893–2933 (2022)
29. Gull, S., Akbar, S.: Artificial intelligence in brain tumor detection through MRI scans: advancements and challenges. Artif. Intell. Int. Things 241–276 (2021)
30. Soni, S., Seal, A., Mohanty, S.K., Sakurai, K.: Electroencephalography signals-based sparse networks integration using a fuzzy ensemble technique for depression detection. Biomed. Signal Process. Control **85**, 104873 (2023)
31. Salminen, J.: Fake Reviews Dataset (2020). https://osf.io/tyue9/

A Study on the Influence of Artificial Intelligence on Brain Plasticity with Reference to Consumer Behaviour and Workplace Dynamics

K. Umashankar and H. G. Charitra

Abstract The study did an extensive review in understanding the relationship between Artificial Intelligence and Brain Plasticity aligned with consumer behaviour and workplace dynamics. The study conducted an intensive review of various dimensions of neuromarketing and workplace dynamics, and the role of AI in influencing these two variables. Further, the workplace dynamics include factors like employee engagement, preventive stress management, and brand loyalty. The dimension of consumer behaviour comprises neuromarketing and brand loyalty. Besides, the study has demographic variables like age, gender, and culture as moderators, and this will help understand the role of these intervening variables between AI and Brain Plasticity (Neuromarketing and Workplace Dynamics). The study has formulated five assumptions and these assumptions have been validated through the existing literature. The study concludes with an inference that the goals of influencing consumers' buying intentions and nurturing a strong sense of citizenship behaviours among the employees. The study's findings imply that adopting AI to influence consumer behaviour will enhance the intensity of neuromarketing strategies and strengthen brand loyalty among customers. In continuation, the agents of Artificial Intelligence have become potentially consequential in devising innovative employee engagement strategies, and preventive stress management initiatives.

Keywords Artificial intelligence · Neuromarketing · Brain plasticity · Brand loyalty · Preventive stress management

---

K. Umashankar (✉)
Department of Behavioural Sciences, Manipal Academy of Higher Education, Bengaluru, India
e-mail: shankartagore@gmail.com

H. G. Charitra
School of Arts, Humanities, and Social Sciences, REVA University, Bengaluru, India

S. Dutta et al. (eds.), *Advances in Data Analytics for Influencer Marketing: An Interdisciplinary Approach*, Information Systems Engineering and Management 9,
https://doi.org/10.1007/978-3-031-65727-6_19

1 Introduction

Artificial Intelligence has been powering the disruptive technologies in the contemporary world and it has opened diversified evolutionary paths for global businesses. Besides, the tentacles of Artificial Intelligence have been assisting human intelligence to explore novel solutions to existing problems. Before the advent of Artificial Intelligence, there were many unexplored dimensions of knowledge that human intelligence was unable to penetrate. As Artificial Intelligence started evolving to new heights, human intelligence started exploring diversified areas like business technology, medicine, education, human psychology, and many other areas of importance. In connection to the *Glocal* (Global–local) business landscape, Artificial Intelligence propelled the disruptive changes in the consumer market and compelled organizations to rethink their business strategies and redesign their business models to pace the competition [1, 2]. Machine Learning has been playing a crucial role in replacing humans in certain areas of employment that involve monotonous work patterns and they are adding incremental value to the service of the human world. The pseudo-human functions are performed through applications of machine learning with high efficacy. However, there have been mixed responses to the consequences of applying machine learning trends as it is leading to human replacement and a rise in unemployment crises. There has been substantial evidence that supports that humanoids powered by machine learning have substantially increased human employment in various domains [3, 4].

According to the existing literature review, Artificial Intelligence has been evolving at a rapid pace spreading itself in various forms like machine learning, deep learning, search algorithms and many more and each dimension of Artificial Intelligence has its unique value addition to the human undertakings. Besides, the aspiration of business landscapes across the globe has indirectly propelled the evolutionary pace of AI in all its forms. Simultaneously, computers that were meant only for the storage of data became a harbinger of *Big Data* that served as a massive repository for Artificial Intelligence and derived meaningful conclusions or judgements about the contemporary problems of future predictions [5]. The inception of the concept of *Logic Theorist* [6] played a pivotal role in the expansion of Open AI. It became the first computer programming venture to prove that computers have the power of logical reasoning and can do a better mathematical or numerical job than human intelligence. Thinking beyond the mathematical calculation, the *Logic theorist* ventured into human expressions, and it was programmed on two dimensions of memory that are Storage and Working memory. Besides, the pivotal programmers proved that computers are not only used in mathematical calculations but also to perform basic functions of human interface [7]. Besides, the efficacy of computers in dealing with human emotions and feelings was further studied and it was assumed that these machines can mimic appropriate human emotions or feelings derived from the storage memory. The study portrayed these machines as impeccable and influencing storytellers about the human journey in various paths of life. It was implied that the intelligence of these machines goes beyond human estimation and creates exceptional landscapes that can marvel at the entire human civilization. Indeed, it is asserted that AI can percolate deeper into the mindscapes and reproduce relevant solutions for mind-matter problems. The mind-matter problems include the attention required in the fields of health, psychology, and behavioural/cognitive-related interventions [8].

As the studies continue to explore new insights from the evolutionary path of Artificial intelligence, the studies in cognitive neuroscience and psychiatry initiated many scientific investigations on the functioning of the brain-mind concept. These scientific investigations led by psychiatrists and cognitive psychologists started unravelling the unexplored mindscapes of the human brain. These insights contradict existing assumptions that the human brain ceases to evolve after a certain chronological age. Besides, the studies implied that the brain not only reflects the genetic imprints but also the impressions acquired through various experiences throughout life episodes [9]. According to developmental psychology, the human embryo in the mother's womb is genetically aligned with the maternal environment and the genetic structures are influenced by the mother's personal or social experiences. However, the studies were confined to the influence of genetics on the human brain, and they implied that brain development ceases during adolescence. Besides, the studies during the twentieth century inferred that the genetic expressions of the brain can be altered, moderated, and manipulated through external social or personal factors The recent medical prognosis through various experiments inferred that the synaptic changes occur irrespective of age and these synaptic changes are inflicted by various factors like personal, social, and other external experiences. However, the density of synapses may vary from one age group to another [10]. One of the studies implied that the brain is a dynamic organ and the neural circuitry or neural connection keeps growing as the individuals are exposed to new experiences in their external environment. Besides, these synapses (Repositories of long-term memory) have less to do with chronological age. According to the study, the child at birth may have 7500 connections in each neuron and it keeps expanding as the child evolves through various walks of life the studies do not confirm the actual age regarding the cessation of expansion in neural networks [11].

The cited studies in brain plasticity (neuroplasticity) imply that the human brain is highly receptive to external stimuli and these external stimuli have the potential to generate new neural networks in the brain. The neuro-cognitive flexibility has leveraged multifarious development in neuromarketing powered by Artificial Intelligence. It is inferred that the insights from brain plasticity have boosted many business portfolios, and these portfolios are emerging in a new direction to redefine their approach or interaction with the consumer market. New knowledge landscapes like neuroeconomics, consumer neuroscience, and neuromarketing started emerging. Accordingly, neuroimaging techniques were employed to study consumers' buying patterns and brand positioning in the global or local business market. The study implied that Artificial Intelligence augmented the deployment of these brain imaging techniques as it could interpret the big data procured from the market. However, the study also presented the limitations of using these strategies as most of the organizations could not afford these practices [12]. One of the experimental studies on the influence of advertising on the emotional brain inferred that the consumers' emotional connection with the brand is influenced by the techniques of neuromarketing. The study employed brain imaging techniques assisted by Artificial Intelligence, and this was a pivotal study in determining the efficacy of the three applied knowledge dimensions Brain plasticity, Neuromarketing, and Artificial Learning [12]. Exploration of the dynamic mechanism of brain propelled the businesses across the world to align the insights with their brand-building initiatives and influence consumers' brand loyalty and their buying behaviours. Advancement in Artificial Intelligence was a blessing in disguise that helped neuroscientists, psychologists, and marketing professionals.

2 Operational Definition of Terms

Artificial Intelligence: It is defined as the ability of computer-assisted agents like robots that accomplish the tasks of humans with precision and these computers are programmed by human intelligence to perform certain tasks that involve reasoning, pseudo-emotions, innovation, or/and problem-solving [13].

Machine Learning: Machine Learning is a statistically empowered agent that can collect big data and derive meanings from the available data, besides, these agents play a crucial role in interpreting data and making possible predictions based on the behaviour of data and it is widely used in the areas of psychology, marketing, philosophy, etc.... [14].

Brain Plasticity: The term brain plasticity is defined as a process wherein new neural networks are formed in the nervous system as the individual gets stimulated by the new environment, and new experiences, and accordingly new neural pathways are created in the brain. Besides, brain plasticity is also called neuroplasticity [15].

Neuromarketing: Neuromarketing is a combination of marketing and cognitive neuroscience that studies the link between marketing strategies and consumer buying behaviour, and how these marketing strategies can influence the unconscious inclination towards the brand. Besides, neuromarketing studies various cognitive triggers that can influence consumers' brand loyalty and buying behaviours [16].

Brand Loyalty: Brand loyalty may be defined as a psychological or emotional connection of a consumer with an organization or its products that lasts for a long term, and the consumers' intention of buying the same brand products is very high. Besides, the consumers' brand loyalty is also associated with various intangible factors that are not focused on in this study [17].

Employee Engagement: Employee Engagement comprises the behavioural, attitudinal, and emotional connection of employees with the job roles or organization that positively contribute to the business goals, and highly engaged employees tend to display robust levels of energy and passion while fulfilling their job charters while embracing organizational core values. Besides, it is assumed that highly engaged employees translate the vision and mission of the organisation through their workplace behaviours [18].

Preventive Stress Management: Contrary to stress management, the concept of preventive stress management may be defined as an enabling mechanism that keeps employees informed about futuristic stressful episodes and equips employees with appropriate strategies to deal with occupational stress. In this study, preventive stress management has been perused in connection with the role of Artificial Intelligence and its role in mitigating stress [19].

3 Objective of the Study

The objective of the study is to understand the influence of Artificial Intelligence on Brain Plasticity and its impact on consumer behaviour and workplace dynamics of an organization. Besides, the study intends to study organisational strategies to optimize business performance and nourish the internal organisational health index by adopting Artificial Intelligence.

4 Need for the Study

Artificial Intelligence has been one of the major disruptive powers in the global business ecosystem and big fortune companies have been resorting to consulting the marketing firms to position their brands in the consumer markets. Besides, there has been a monumental shift from conventional marketing to contemporary marketing trends due to high volumes of brand inflow into the consumer market. Hence it is very crucial to adopt new ways of positioning their brand deeper into the psyche of consumers and instilling brand loyalty in them. To navigate through the tempests of global competition, it is indispensable for any organization to embrace Artificial Intelligence and improvise on neuromarketing strategies. In addition, organizations have been keen to nurture the psychological/emotional well-being of employees as the employees' well-being is directly aligned with the organizational effectiveness. Hence, organizations need to embrace AI to improve the emotional or psychological well-being of employees. Indeed, organizations may explore new ways of using AI to create virtual simulations that may reengineer employees' thinking patterns and it may help implement the practices of neurofeedback.

5 Scope of the Study

The conceptual study focuses on various assumptions validated by the existing literature. The study is based on two major assumptions. The first assumption examines the cause-effect relationship between AI and Brain Plasticity wherein AI is treated as the independent variable and brain plasticity as the dependent variable. Further, the study comprises the demographic variable as a moderator that influences the direction of the relationship. Besides, the conceptual study focuses on establishing the relevance of Artificial Intelligence in influencing the consumer behaviour and employee dynamics of an organization. The study may provide valuable insights into the recent trends in Open AI and the factors that can boost organizational effectiveness.

6 Artificial Intelligence and Brain Plasticity

As the landscapes of technology and cognitive neuroscience started emerging from their limitations, the global business platforms felt the need to combine these two forces for a better understanding of the VUCA world (Volatile, Uncertain, Complex, and Ambiguous). The business arena has been encountering tremendous challenges since the 1990s and the established business models have become obsolete to deal with unpredictable and constant changes. Besides, the digitization of many services has added fuel to the disruptive nature of business rituals across the globe [20]. Therefore, business corporations have been proactive in rethinking their existing business models and reengineering their business models to align with the existing consumers' buying patterns. In this regard, the organizations have been diving deeper into the market trends. The employment of Artificial Intelligence and cognitive neuroscience to run micro/macro analysis of data and make meaningful business predictions has become a ritualistic and ongoing activity. One of the major trends in the business world is to strengthen brand engagement using the strategies of artificial intelligence and brain plasticity. According to neurocognitive scientists, the neuro adaptation to external stimuli is triggered by the constant repetition of an episode or incident and gradually the repeated stimuli become a part of long-term memory and can no longer surprising element be lost. Therefore, neuromarketers with the use of Artificial Intelligence intend to desensitize the consumers' neuro adaptation that happened over frequent time intervals and they try to position their brand through innovative or novel cognitive ads. Cognitive scientists imply that cognitive ads enhance consumers' cognitive or emotional engagement with the brand [21]. Besides, Artificial Intelligence and Brain Plasticity have become inseparable domains of knowledge and can contribute to the internal and external ecosystem of an organization. It is assumed that AI has become a potential and crucial tool that can materialize the prognosis of Brain Plasticity and interpret the big data that can add meaning to business solutions or business goals across the globe. An *Artificial Neural Network* (ANN) has been one of the recent inventions of Artificial Intelligence wherein AI scientists are recreating artificial neural networks that can mimic the basic functions of the brain. Although the human neural network is the most complex communication system that transmits information in the form of electric messages, AI intends to create basic neural networks that can perform day-to-day functions and the same has been infused into humanoids. The humanoids powered by AI have been infused with artificial neural networks to perform basic human functions and considerable research has been initiated to overcome the limitations of AI while confronting the complex mechanisms of the human brain. ASIMOV Robotics Pvt Ltd (2022) displayed a humongous breakthrough when it sent a human-simulated robot to submit loan documents to the Federal Bank in Kerala [22].

7 Demographic Variables Mediating Between AI and Brain Plasticity

One of the studies intended to assess the emotional or psychophysiological responses to various brands presented on the public media. The study embraced the technological tools to measure the responses and these tools were Electroencephalogram (EEG), Electronic Design Automation (EDA), and eye tracking device. Indeed, AI was an integral part that helped in deriving interpretable information from the exhaustive data [23]. During the laboratory experiments with a defined number of samples, the study inferred that the neuroimaging data (EEG) and the eye tracking mechanism (ET) did not reveal any substantial evidence that can influence consumers' buying decisions. The study inferred that the psychophysiological response measured through EDA influences the buying behaviours of consumers. As a conclusion, the study implied that these psychological, emotional, or psychosomatic responses may be diverse depending on the age factor [24–26]. One of the studies adopted an occasional sampling method to assess the gender differences during the process of brand receptivity in some of the defined occasions. In this study, the cerebral responses were recorded while the select population were exposed to specific brands. It is assumed that the neurological responses differ based on gender and even the intention to purchase gets influenced [27]. As the study concluded with its research outcome, crucial inferences were derived. It is said that the neurological nature and the process of brain plasticity may vary from gender to gender. It is assumed that consumer perception and purchase intention differ between men and women, and their cerebral triggers also differ from one another. According to the study, women are assumed to be more sensitive and receptive to the strategies of neuromarketing and their cerebral activation during the cognitive ads is higher than men [28].

8 Research Framework

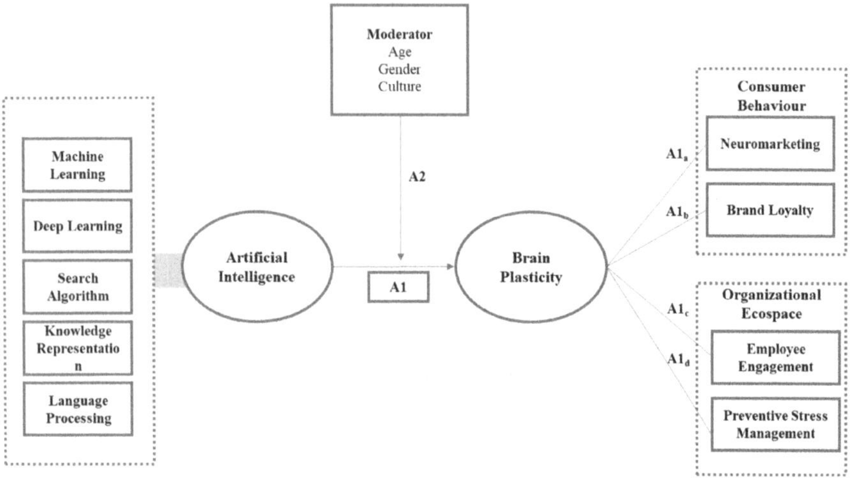

9 Assumptions of the Study

The study has adopted a conceptual model wherein it addresses various assumptions based on the existing literature. These assumptions have been derived by the study's extensive review of literature and the assumptions are formed by observing the data on the interface between Artificial Intelligence and Brain Plasticity. The following assumptions further guide the conceptual study and will derive meaningful outcomes.

$A1_a$—Artificial Intelligence will influence brain plasticity during the initiatives of Neuromarketing.

$A1_b$—Artificial Intelligence will strengthen brand loyalty of consumers/customers by reengineering brain plasticity.

$A1_c$—Artificial Intelligence will augment the employee engagement initiatives at the organisational level.

$A1_d$—Artificial Intelligence will add incremental value to the employees' preventive stress coping strategies.

$A1_5$—Age, gender, and culture moderates the relationship between artificial intelligence and brain plasticity.

10 Conceptual Interpretation of Assumptions

$A1_a$—Artificial Intelligence will influence brain plasticity during the initiatives of Neuromarketing

The contemporary business world is going through a phenomenal phase wherein the consumer market changes rapidly and constantly. The transformation in marketing strategies powered by digitization and machine learning has led to a new renaissance in the areas of marketing and sales. The advent of machine learning has added value to the domain of neuromarketing [29]. One of the major contributions of Artificial Intelligence is the introduction of Artificial Neural Networks (ANN) and this brought phenomenal changes in problem-solving and imitated the functions of the human brain. Unlike algorithms, the ANN was designed to perform the functions of the human brain and understand the information processing in the human brain. This revolutionised neuromarketing strategies and evolved new trends in cognitive ads. These cognitive ads were focused on captivating the attention through various visuals, and this pattern was assessed through an eye-tracking system [30].

A study on the influence of neuromarketing on the human brain inferred that humans' purchase intentions can be assessed by observing brain waves, skin response, and eye tracking. The study used an fMRI (functional magnetic resonance imaging) device to study brain activity while exposed to various brands. Based on the derived data through these experiments, the marketing consultants have been designing advertisements that can manipulate brain waves and infuse their brand in the customers' memory [31]. These neural predictors assessed through the fMRI were widely adopted by the portals of neuromarketing. The adoption of AI to assess neural behaviour, neural adaptations, and the interpretation of data enhanced the efficacy of neuromarketing. Due to enormous developments in the field of OpenAI, it has become a ritual for Fortune companies to adopt these techniques to monopolise the business market. Further, the assessment of neural behaviours and the prediction of purchase intention were assessed based on the reports generated by the neural AI devices. The study implied that the AI-stimulated audio/visual could influence the customers to form various impressions of the same brand as the same brand was presented through various representations, accordingly, this enhanced the purchase intention of the customers [32].

Due to advancements in the latest technologies AI and Virtual Reality, there is a disruptive transformation in the strategies of neuromarketing. Indeed, these AI tools help in understanding user or customer experience with the brand, and they enable the agents to capture the consumers' emotions during their interface with different brands. Accordingly, AI tools like *Machine Learning, Knowledge Representation, and Language Processing* are oriented towards influencing the consumer's neural network and this also forms a new set of neural paths with manipulated or newly formed schemas. AI combined with Virtual Reality can create an enriched customer experience and help the intending customers to make the right or informed choices about the product or brand [33]. Human emotions have been the centre of contemporary neuromarketing ventures and the elements of OpenAI have been employed to influence the emotional interface with the customers and the strategies to emotionally position a brand in the customers' long-term memory. Besides, it is carefully observed how the frequent stimulation of certain emotions can influence the neural networks in the brain. Indeed, AI-influenced neuromarketing strategies have a significant contribution to the formation of new neural networks and the efficacy of these neuromarketing strategies has been updated and improvised to manifest the human neural system in the machines. These efforts have been intensifying the influence of AI-led neuromarketing on consumers' buying decisions [34–36].

$A1_b$—Artificial Intelligence will strengthen brand loyalty of consumers/customers by reengineering brain plasticity

The contemporary practices in neuroplasticity are directed towards strengthening brand loyalty of the exhibiting customers/consumers through various cognitive ads driven by Artificial Intelligence. Some of the portals like Google ads and Meta Ads empowered by AI create a personalized connection between the brand and the consumers and this increases the longevity of customers' relationship with specific brands. One of the studies on Advertising and the role of AI segregated cognitive advertisement into four stages consumer insight discovery, ad creation, media planning and buying, and ad impact evaluation. The study implied each stage of the cognitive ad played a significant role in understanding the consumers' cerebral response to the brands and cerebral conditioning for creating long-term impressions in the consumers' long-term memory. The study suggests that using Temporal Automation Algorithms can decipher the outcome of user interaction with the brand. It is noted that emotional conditioning towards brand preferences does play a crucial role in neuromarketing [37]. One of the studies implies that OpenAI augments the efficacy of *Programmatic Creative* by improving various stages of media production and producing emotionally charged stories on digital platforms. It is observed that programmatic creative helps in identifying premade digital assets like audio files, pictures, and other video clips, and these assets are combined to give a personalized experience to the media consumers [38]. A systematic study into the effectiveness of AI-driven cognitive ads on brain plasticity revealed that the conversion of digital content into emotive advertisements must include various elements like a model evaluating the influence of web content, the sentiment analysis of words or phrases used in the advertisements, and the generation of advertising copy based on the given templates. These processes solely depend on Artificial Intelligence and the efficacy of these processes and implementation are measured by Steady-state topography which assesses and measures the brain activity while consumers meet various brands in the market. These *personalised advertising copies* have been highly successful in tailoring emotive advertisements to create brand awareness and positioning the specific brand into the long-term memory of consumers [39].

Artificial Intelligence has been one of the major interventions that are boosting organizational performance. Besides, the interface between AI and Humans has led to a new paradigm in the areas of cognitive neuroscience and this propelled studies on the influence of neuromarketing on brain plasticity. Accordingly, the augmented versions in the ecosystem of AI were embraced to stimulate and record the brain activity of employees during their interaction with a specific brand. The studies imply that AI played a crucial role in influencing the neural networks causing modifications in the emotional spaces of employees. Employees [40]. The ecosystem of the consumer market is a nexus of emotions and feelings that are intrinsically aligned with the brand; the source for these emotions or feelings is neural spaces. Similarly, the neural spaces of consumers may be influenced, and the target emotion or trait may be inseminated through appropriate strategies adopting AI. A study adopted *NRM* technology (Neuro Magnetic Resonance) and it inferred that it is essential to study the prevailing gap between consumers' subconscious state and their behavioural response to the advertisements. However, the study proposes the adoption of a new mechanism to study the nuances of information processing in the brain stimulated by various elements of advertisements like colour, odour, sound, and the presentability of the product or services. The *NRM* technology was used to record the neural responses in the human brain while interacting with the advertised brand. Further, it may be inferred that these cognitive ads can create new neural networks aligned to the brand and gradually the brand becomes a source of customers' experience [41].

$A1_c$—Artificial Intelligence will augment the employee engagement initiatives at the organisational level.

Due to changing the organizational landscapes, it has been a herculean task for HR professionals to keep the employees engaged consistently and strengthen their psychological contract with the organization. However, the advent of Artificial Intelligence heralded a new phase in the domain of human resource management and enhanced the quality of employee engagement enterprise. Gamification, one of the AI tools, has been widely used among Fortune companies for employees' behavioural or cognitive modification. One of the studies integrated *BIS* (Business Intelligence System) with the Gamification modules to study the relationship between variables like customer experience and the business performance of employees. The gamification modules derived included affordances like visualization, consistency/persistence, and editability. The study concluded that gamification with multiple gaming genres influenced the business intelligence of employees. Further, it is asserted that gamification can augment the efficacy of employee engagement leading to a better occupational environment [42]. It is essential to reflect on the importance of employee well-being and neutralize the factors that cause deviations in the employees' behavioural transactions or inflict cognitive agitations. It is very crucial to devise strategies that can minimize workplace incidents (Emotional/psychological) and augment a progressive work culture. To achieve the purpose of boosting employee engagement, a study focused on the intervention of algorithm management that can accumulate employees' behavioural data and construct insightful stories that can offer tangible solutions to the problems connected with employee engagement. Although it is implied that AI can moderate aggravated behavioural transactions, the management must be mindful of certain challenges like trust factors, perceived challenges, and fairness in handling workplace incidents. Besides, AI management systems are employed for behavioural conditioning through various approaches like directing, monitoring, rewarding, or punishing. The concept of these behavioural articulations or modifications is suggested in McGregory's Theory X and Y [43]. The AI-mediation in the process of employee engagement has been one of the significant breakthroughs in the domain of Artificial Intelligence and Behavioural modification. One of the longitudinal studies observed the influence of chatbots embedded with augmented employee voice implying that the employees were better engaged with the help of chatbots, and it was successful in creating an enriched employee experience. In addition, the chatbots were employed to create positive cognitive patterns that added to workplace efficacy [44].

It is assumed that neuroplasticity or brain plasticity cannot be influenced through stereotyped employee engagement training programs and these engagement programs have not shown any significant improvement. Besides, it is inferred that psychological safety and the matter of trust played a crucial role in stimulating neural networks in the brain and these neural activities in the brain were captured using multiple medical devices. One of the experimental studies inferred that the training programs may not enhance employee engagement or augment workplace performance as the gap of indirectness mediates between the two. However, the explicit assessment and analysis of ergonomic behaviours and the factors influencing these behaviours will help design better training strategies. Besides, the flexibility of cognitive response to the training must be mediated by various psychological factors [45].

One of the studies assumes that AI-stimulated empathy training and compassion training have their intricacies. The people who have been trained on empathy associated with others' pain revealed that the *Anterior Insula* is associated with empathy and the overactivation of this region may result in distress among the people. However, the study implied that compassion training through AI-simulated modules may do better compared to empathy training. Compassion training may include good-will factors like affection, love, and award, and these factors stimulate the brain regions like the insula, ventral striatum, and medial orbitofrontal cortex. This helps in coping with distress among the people and positive relationships are built in social or personal spaces. It can be inferred that integrating AI-simulated compassion training with employee engagement may result in a positive outcome [46].

$A1_d$—Artificial Intelligence will add incremental value to the employees' preventive stress coping strategies

Preventive stress coping strategies play a crucial role in the organizational ecosystem wherein occupational stress is directly linked with employee well-being and work performance. Therefore, it is inevitable to explore the strategies that can prevent episodes of occupational burnout. The emergence of Artificial Intelligence and its contribution to the business performance of organizations and enhancing the efficacy of human factors are commendable. The natural language processing agents of AI have been widely used as chatbots that offer stress-coping strategies to people and they can also assess mental conditions and suggest preventive measures to overcome stress. One of the experimental studies in a medical facility implied that people could interact with the chatbots regarding their stressful episodes, and the chatbots in return offer viable solutions to the shared concern. It is emphasised that the usage of chatbots in preventive stress coping strategies to reduce cognitive overload on human stress coping resources [47]. Digital workspaces have been one of the major transforming agents in the organizational context and these digital workspaces have helped employees in adopting stress coping strategies. Despite the employees being exposed to technostress, the studies infer that there must be some well-defined AI policies in the organisations that can create awareness of AI intervention in their workspaces, and the agents of AI can be effectively adopted in preventing occupational burnout among the employees [48]. Employees belong to various diversified emotional or psychological backgrounds and some of them may have undiagnosed psychosomatic conditions that may hamper workplace productivity. Besides, it is essential to pursue preventive strategies in coping with stress that can reduce the episodes of burnout or other psychosomatic incidents in future. In this regard, Artificial Intelligence helps in identifying the neural responses to external stimuli and this can predict future episodes of occupational burnout and offers workable preventive strategies. Employing a Machine Learning Model in assessing psychosomatic conditions and promoting employee psychological well-being has been one of the major contributions of Artificial Intelligence [49]. Neuropsychological issues may be aggravated if the employees are exposed to stress for a prolonged period. If the occurrences of anxiety or depression are not treated or avoided, the employees' psychological well-being and work productivity may be compromised. Burnout symptoms like psychological/mental exhaustion, cynical mindset or behaviour, and self-doubt must be identified and treated before they become a chronic medical condition. Therefore, the employment of AI agents may be crucial to neutralize the factors that contribute to occupational burnout. In addition to the traditional protocols or strategies in stress management therapies, agents like Natural Language Processing and Machine Learning Models can employed to achieve the purpose of implementing preventive stress coping strategies [50, 51].

$A1_5$ –Age, gender, and culture moderates the relationship between artificial intelligence and brain plasticity

Gender, age, and culture may have a moderating role in Artificial Intelligence and Brain plasticity. These factors have been biologically or culturally influenced and they have a pattern of influence on the human brain [52]. However, there is a contrasting assertion from other studies that implies that the cited factors may not have significant moderation between the two variables. The studies imply that the differences may be due to the sex-dependent endocrine system influenced by synapses. It is implied that these physiological differences do prevail in different sexes and the socially or culturally assigned gender like men and women do have some implications on the intensity of brain plasticity. The agents like external stimuli may also diversify in terms of influencing brain plasticity [53].

11 Conclusion

The study did an extensive study in understanding the relationship between Artificial Intelligence and Brain Plasticity aligned with organizations' external and internal affairs. Besides, the study intended to explore the influence of AI on brain plasticity based on five assumptions the first assumption is justified with extensive literature that AI can influence brain plasticity through the techniques of neuromarketing, and the cognitive advertisements were influential in effecting brain plasticity. Besides, contemporary marketing agents have been using AI to promote their brands by manipulating consumers' buying behaviours or purchase intentions. The second assumption on establishing brand loyalty through AI was justified, and it is inferred that brand loyalty relates to human cognition wherein the cognitive ads comprising some defined set of emotions can easily connect the brand with the consumers. The emotional connection between the brand and consumer assures the consistent or enhanced business performance. Assumption 3 focused on understanding the role of AI in driving employee engagement initiatives and how these initiatives can influence brain plasticity. The study identified multifarious employee engagement initiatives driven by Natural Language Processing (Chatbots) and Machine learning Models. Assumption 4 contemplated on adopting AI to predict stressful events in the future and initiate appropriate preventive coping strategies. The study acknowledges the role of AI in taking effective coping strategies and enhancing employee well-being in the occupational environment. However, the fifth assumption was study could not be established as there are limited studies that studied the moderation role of intrinsic variables between AI and Brain plasticity. Indeed, contemporary organizations have been promoting inclusive and diversified culture, and the employees belonging to gender, sexual, or cultural orientations have been working under one roof. Therefore, it is necessary to understand the roles of these demographic variables and their possible influence on brain plasticity.

12 Recommendations of the Study

Artificial Intelligence cannot be analysed with the mindset of Frankenstein syndrome and the awareness of brain plasticity plays a crucial role in driving business and employee engagement. According to the established assumptions in the study, Artificial Intelligence may be adopted during the brand positioning and initiate new strategies in capturing the consumer market. Besides, the business intelligence of an organization must be diagnosed periodically using contemporary AI agents, and necessary corrective or augmented actions may be initiated.

Human Resource Capital plays a major role in the materialization of organizational goals, and it is inevitable for any organization to keep their employees highly engaged in their job roles. However, the traditional or stereotyped learning and development ventures may not give sustainable outcomes as they cannot personal the training programs to the diversified employees. Hence, the authorized departments of the organizations must think like game changers and embrace multiple or relevant agents of AI to influence the workplace behaviours of employees. In addition, it is essential to acclimatize oneself to the concept of brain plasticity that is aligned with the employees' behavioural dynamics.

Declaration: The authors have no conflict of interest to declare that are relevant to this article.

References

1. Ghosh, S., Singh, A.: The scope of Artificial Intelligence in mankind: a detailed review. J. Phys.: Conf. Ser. (2019)
2. Raj, R., Kos, A.: Artificial intelligence: evolution, developments, applications, and future scope. Przeglad Elektrotechniczny 1–13 (2023)
3. Anantrasirichai, N., Bull, D.: Artificial intelligence in the creative industries: a review. Artif. Intell. Rev. **55**, 589–656 (2022)
4. Raj, R., Kos, A.: Artificial intelligence: evolution, developments, applications, and future scope. Przeglad Elektrotechniczny 1–23 (2023)
5. Anyoha, R.: Can Machines Think? The History of Artificial Intelligence 28 Aug. 2017
6. Gugerty, L.: Newell and Simon's logic theorist: historical background and impact on cognitive modeling. In: Proceedings of the Human Factors and Ergonomics Society Annual Meeting, vol. 50, no. 9, pp. 880–884 (2006)
7. Crevier, D.: AI: The Tumultuous History of the Search for Artificial Intelligence. BasicBooks, NewYork (1993)
8. Pamela, McCorduck: Machines Who Think—A Personal Inquiry into the History and Prospects of Artificial Intelligence. CRC Press, London (2004)
9. Kolb, B., Gibb, R.: Brain plasticity and behaviour in the developing brain. J. Can Acad. Child. Adolesc. Psychiatry **20.4**, 265–276 (2011)
10. Kolb, B., Gibb, R., Gorny, G.: Experience-dependent changes in dendritic arbor and spine density in neocortex vary with age and sex. Neurobiol. Learn. Mem. **79**, 1–10 (2003)
11. Drigas, A.S., Karyotaki, M., Skianis, C.: An integrated approach to neuro-development, neuroplasticity and cognitive improvement. Int. J. Recent. Contrib. Eng., Sci. IT **6.3**, 4–18 (2018)
12. Nick, L., Broderick Amanda, J., Laura, C.: What is 'neuromarketing'? A discussion and agenda for future research discussion and agenda for future research. Int. J. Psychophysiol. **63.2**, 199–204 (2007)
13. Copeland, B.J.: Artificial intelligence. Britannica—Science and Tech 20 Nov. 2023
14. Jafar, A., Nayyar, A., Kuma, A.: Machine learning from theory to algorithms: an overview. J. Phys.: Conf. Ser. (2018)
15. Anonymus.: American Psychological Association. 19 04 2018. https://dictionary.apa.org/neural-plasticity?s2=P1136618603_1683331208458181917
16. Morin, C.: Neuromarketing: the new science of consumer behavior. Soc (Springer Link) **48**, 131–135 (2011)
17. Kato, T.: Brand loyalty explained by concept recall: recognizing the significance of the brand concept compared to features. J. Mark. Anal. **9**, 185–198 (2021)
18. Macey, H.W., Schneider, B.: The meaning of employee engagement. Ind. Organ. Psychol. **1**, 3–30 (2008)
19. Anonymus.: APA Dictionary of Psychology. 04 19 2018. https://dictionary.apa.org/preventive-stress-management

20. Johansen, B., Euchner, J.: Navigating the VUCA World. Res. Technol. Manag. **56**(1), 10–15 (2013)
21. Yahia, M., Hassan, A.: Using Neuromarketing and AI to collect and analyse consumer's emotion: literature review and perspectives. Int. J. Bus. Econ. Strat. **12**, 34–38 (2019)
22. Anonymus.: This company brings robot to a bank branch for getting loan sanctioned. CNBCTV18.com 8 09 2022
23. Gorn, G., Pham Tuan, M., Sin Yatming, L.: EEG frontal asymmetry predicts product purchase differently for national brands and private labels. J. Consum. Psychol. **11.1**, 43–55 (2001)
24. Bosshard, S.S., et al.: Established liked versus disliked brands: brain activity, implicit associations and explicit responses. Cogent Psychol. **3.1** (2016)
25. Davidson, R.J., et al.: Withdrawal and cerebral asymmetry: emotional expression and brain physiology. J. Pers. Soc. Psychol. **58**(2), 330–341 (1990)
26. Garczarek-Bąk, U., et al.: A comparative analysis of neuromarketing methods for brand purchasing predictions among young adults. J. Brand Manag. **28**, 171–185 (2021)
27. Birknerová, Z., et al.: 'Gender differences in perception of advertising in the context of expectations of advertising. J. Appl. Econ. Sci. **12**(6), 1751–1758 (2017)
28. Stefko, R., et al.: Consumer purchasing behaviour and neuromarketing in the context of gender differences. J. Mark. Res. Case Stud. **1.11** (2021).
29. Syam, N., Sharma, A.: Waiting for a sales renaissance in the fourth industrial revolution: machine learning and artificial intelligence in sales research and practice. Ind. Mark. Manage. **69**, 135–146 (2018)
30. Pieters, R., Wedel, M.: Attention capture and transfer in advertising: brand, pictorial, and text-size effects. Am. Mark. Assoc. **68.2** (2004)
31. Knutson, B., et al.: Neural predictors of purchases. Neuron **53**(1), 147–156 (2007)
32. Glimcher, P.W., Rustichini, A.: Neuroeconomics: the consilience of brain and decision. Science **306**, 447–452 (2004)
33. Adeola, O., et al.: The future of marketing: artificial intelligence, virtual reality, and neuromarketing. In: Adeola, O., Hinson, R.E., Sakkthivel, A.M. (eds.), Marketing Communications and Brand Development in Emerging Economies Volume I. Palgrave Studies of Marketing in Emerging Economies, pp. 73–86. Palgrave Macmillan, London (2022)
34. Kumar, V., et al.: Understanding the role of artificial intelligence in personalized engagement marketing. Calif. Manage. Rev. **61**, 135–155 (2019)
35. Pantano, E., Scarpi, D.: I, Robot, You, Consumer: measuring artificial intelligence types and their effect on consumers emotions in service. J. Serv. Res. **25.4** (2022)
36. Rainer, M.: Human-Aided artificial intelligence: or, how to run large computations in human brains? Towards a media sociology of machine learning. New Media Soc. **22**(10), 1868–1884 (2020)
37. Hairong, L.: Special section introduction: artificial intelligence and advertising. J. Advert. **48**(4), 333–337 (2019)
38. Chen, G., Tianfu, W., Xie, P.: Understanding programmatic creative: the role of AI. J. Advert. **48**(4), 347–355 (2019)
39. Deng, S., et al.: Smart generation system of personalized advertising copy and its application to advertising practice and research. J. Advert. **48**(4), 356–365 (2019)
40. Prentice, C., Lopes, S.D., Wang, X.: Emotional intelligence or artificial intelligence–an employee perspective. J. Hosp. Mark. Manag. 1–27 (2019)
41. Dragolea, L.-L., Cotirlea, D.: Neuromarketing—between influence and manipulation. Pol. J. Manag. Stud. **3**, 79–89 (2011)
42. Fischer, T.C.: Gamification and affordances: how do new affordances lead to gamification in a business intelligence system? In: Stolfa, J., Stolfa, S., O'Connor, R., Messnarz, R. (eds.), Systems, Software and Services Process Improvement. Systems, Software and Services Process Improvement Springer, Cham (2017)
43. Hughes, C., et al.: Artificial intelligence, employee engagement, fairness, and job outcomes. In: Hughes, C., et al.: Managing Technology and Middle- and Low-skilled Employees (The Changing Context of Managing People), pp. 61–68. Emerald Publishing Limited, Leeds (2019)

44. Dutta, D., Mishra, S., Tyagi, D.: Augmented employee voice and employee engagement using artificial intelligence-enabled chatbots: a field study. Int. J. Hum. Res. Manag. **34.12**, 2451–2480 (2022)
45. Kokubun, K., et al.: Brain condition may mediate the association between training and work engagement. Sci. Rep. **10**, 6848 (2020)
46. Olga, M. Klimecki, et al. "Differential pattern of functional brain plasticity after compassion and empathy training." *Social Cognitive and Affective Neuroscience* 9.6 (2013): 873–879.
47. Mauriello, M.L., et al.: A suite of mobile conversational agents for daily stress management (Popbots): mixed methods exploratory study. JMIR Publ. **5.9** (2021)
48. Nishtha, M., Shalini, N.T., Arpan, K., Shivam, G.: Impact of artificial intelligence on employees working in industry 4.0 led organizations. Int. J. Manpow. (2021)
49. Morande, S.: Enhancing psychosomatic health using artificial intelligence-based treatment protocol: a data science-driven approach. Int. J. Inf. Manag. Data Insights **2.2** (2022)
50. Alarco, G.M.: A meta-analysis of burnout with job demands, resources, and attitudes. J. Vocat. Behav. **79**, 549–562 (2011)
51. Madigan, J.D., Kim, E.L., Glandorf, L.H.: Interventions to reduce burnout in students: A systematic review and meta-analysis. Eur. J. Psychol. Educ. (2023)
52. Breda, T., et al.: Gender stereotypes can explain the gender-equality paradox. PNAS (2020)
53. Hyer, M.M., Philips Lind, L., Neigh Gretchen, N.: Sex differences in synaptic plasticity: hormones and beyond. Front. Mol. Neurosci. **11** (2018)

Influential Asanas: An In-Depth Study Exploring the Dynamics of Yoga Influencers

Shilpa Sarvani Ravi, Shikha Bhagat, and Rashmi Rai

Abstract Tracking the number of clicks, likes, and shares is more than just a pastime in this social media-fueled world, Thus Influencer Marketing has become the key to every brand's digital reach. Influencer marketing is not just the buzzword but the most happening ecosystem in 2024. Beyond merely creating content, influencers in India have developed into powerful economic catalysts in boosting the country's economy. The Country is anticipated to have a value of 28 billion Indian rupees by 2026, having reached 12 billion in 2022. The most popular category covered by influencers is lifestyle (14.32%) which is projected to rise at a CAGR of 25% during the next five years. Considering many individuals are interested in adopting healthier lifestyles, fitness influencers are becoming a powerful force on most social media platforms. This chapter is an extensive study exploring the realm of yoga influencers by examining the mutually beneficial relationship between influencers and companies on major social media platforms like Instagram and YouTube in the health and fitness segment influencers. This chapter also explores the motivations of influencer marketing on the expansion of yoga-related goods and services, as well as the total engagement and reach attained through influencer partnerships through qualitative methodologies using Nvivo and ATLAS.ti software. Ultimately, this chapter contributes to a broader conversation on the evolving nature of marketing strategies in the digital era and its implications for brands, marketers, yoga influencers, consumers, the academic or research community, marketing agencies and government bodies.

Keywords Influencer marketing · Yoga influencers · Instagram · Lifestyle · Fitness and wellness · YouTube · Brand recognition · Social media platforms · Trend analysis

S. S. Ravi (✉) · S. Bhagat · R. Rai
School of Business and Management, Christ University, Bangalore 560076, India
e-mail: shilpa.sarvani@christuniversity.in

S. Dutta et al. (eds.), *Advances in Data Analytics for Influencer Marketing: An Interdisciplinary Approach*, Information Systems Engineering and Management 9,
https://doi.org/10.1007/978-3-031-65727-6_20

1 Introduction

The Bhagavad Gita, or **"Song of the Lord",** is an ancient Sanskrit text mentioned in Chap. 3, verse 8 states that, "*Niyatam kuru karmatwam, karmajayo hrukarmanaha*" which translates as "desisting from action, you cannot even maintain your body" that phrase emphasis for the maintenance of physical health through activity and exercise. In general, fitness refers to the physical aspects of health, such as strength, flexibility, endurance, and cardiovascular health. Exercises that improve cardiovascular health, balance, and coordination include yoga, aerobics, strength training, and flexibility exercises [1]. Yoga is widely practiced as a kind of exercise and for its physical health advantages. This spiritual book provides enduring guidance and teachings about our nature and life's purpose (Fig. 1).

Yoga is the path of unselfish deeds without expecting the consequences or end results [2]. It also teaches us how to maintain harmony between our mind and spirit through theoretical content supported by instances [3]. The Gita comprises the primary components of yoga, including bhakti (devotion), gnana (knowledge), karma (obligation), and raja which is a practical manual for elevating awareness (meditation).

As per Lord Krishna, doing the asanas and pranayams of yoga is not the only aspect of it. It also refers to a person's active, conscious, and voluntary participation in their life as well as their simultaneous pursuit of the ideal, healthy existence on a mental, physical, and spiritual level. Yoga also teaches us to appreciate the beauty all around us and to react to it with love and astonishment [4]. It is about serving others, being unselfish, and carrying out one's responsibilities happily and peacefully. A yogi has a link to the entire universe because, just as the world is a part of them, so are they a part of the world. For those who either currently practice yoga for a living or would love to, being a social media influencer seems like the ultimate desire. Just as likely to come true as realizing one morning that one can perform the two-footed

Fig. 1 Fitness or Yoga. *Source* Generated by the author using generative AI

staff pose in an upward-facing position [5]. The top yoga influencers in India who are leading this revolutionary movement are crucial as yoga develops further. Influencers are social media users who are compensated by brands to attempt and persuade their followers to purchase goods or engage in certain experiences. It should come as no surprise that the most prominent influencers on the internet are individuals who reportedly make "six-figure sums" from endorsing products to massive audiences consisting of millions of followers.

These influencers span the gamut of yoga, fitness, and wellbeing, from business owners to Hollywood stars' trainers to yoga enthusiasts. Social media has indeed emerged as a useful tool for yoga instructors to connect with the community and share their knowledge and enthusiasm. Because it allows practitioners to showcase their skills in both still photos and time-lapse dynamic videos, yoga is particularly well-suited for visual social media platforms. As the social status of yoga instructors rises, astute marketers are forming partnerships with up-and-coming yogi stars who can help them reach a wider range of consumers. The need for fitness professionals who utilize social media to provide advice, knowledge, and inspiration to mobile and digital audiences will increase in tandem with the need for health and fitness apps [6].

Influencers who have as little as 1,000 followers are eager to market items on social media. One of the things that make them amiable is their lack of celebrity. On Instagram, their recommendations for shampoo, lotion, or furniture brands come across as sincere as those of a friend. They are easy to deal with, which is one of the reasons brands appreciating dealing with them. The study also emphasized that firms give free products or a tiny profit in exchange for their words [5].

1.1 The Rise of Social Media

With its emergence in 1997, social media is a relatively new phenomenon [7]. There have been 4.6 billion social media users active worldwide, according to Statista [8]. The ease of communication on social media has led people to believe that it has shrunk the world and fostered greater interconnectedness, thereby promoting interdependence [9]. This has strengthened ties between individuals worldwide and has made it possible for them to interact and exchange beliefs [10, 11]. Social media is acknowledged as a mediator that improves minority self-representation, culture, and identity on the internet [12]. Thus, it has been demonstrated that its influence on society is growing [13].

However, social media quickly developed into one of the most significant technologies of our day and continues to undergo technological advancements while also taking user needs into account [14]. The quantity of individuals utilizing social media is on the rise [15, 16]. This also has opened the door for persuasion [17], for instance by encouraging impulsive behavior [18, 19]. The advent of the COVID-19 pandemic too imposed limits on physical activity which was accompanied by a predictable rise in demand for online exercise material [20]. It was also possible that this demand had an impact on the characteristics of the video content being created.

Marketers may take advantage of the emotional bond and relationship that seven influencers have with their audience by utilizing them as middlemen in communications with consumers. Because of their perceived legitimacy, influencers have a halo effect that makes followers feel compelled to follow them around. Influencers are being observed, discussed, and engaged [21].

2 Unveiling the Yogis

2.1 YouTube for Yoga Influencers

Yoga is an age-old practice that has gained a lot of popularity in contemporary culture, particularly with the advent of social media. This research review explores the ways in which yoga influencers use social media to promote well-being, shape opinions, and spread yoga-related content. Yoga influencers can now effectively reach global audience because of contemporary media platforms like YouTube, Instagram, and TikTok [22]. On these platforms, these influencers—who frequently have sizable fan bases—share advice, yoga positions, and lifestyle content [23].

According to research, people's opinions about yoga and their motivation to practice it are greatly influenced by yoga influencers [24]. Influencers supply aspirational photos and videos of yoga that appeal to customers looking to better their well-being and self-improvement by carefully selecting aesthetically appealing content [25]. Academics worry that yoga could be commodified and turned into a social media commodity [26]. Reference [27] Influencers frequently endorse branded goods, experiences, and educational programs, which may make it difficult to distinguish between sincere practice and business objectives. Additionally, YouTube yoga influencers have come under fire for promoting a limited view of yoga practitioners and maintaining unattainable physical standards [28].

Viewers who do not identify as conventional yogis may be put off by some films' emphasis on performance and aesthetics [29]. In conclusion, because they offer customized, easily accessible content to a vast global audience, YouTube influencers have a significant influence on the way yoga is taught and practiced. Despite the fact that their work has increased accessibility to yoga for all, there are still problems with accuracy, authenticity, and representation that require more research.

2.2 Instagram for Yoga Influencers

Based on research conducted by [30] it has been found that over 500 million accounts engage with Instagram stories every day. Among the various types of video content, "how-to" tutorials are highly favored, surpassing both news segments and interviews in popularity. It has been proposed that the COVID-19-related limitations also aided in the quick movement in the industry toward the online distribution and promotion of yoga in terms of the availability of video content expressly linked to yoga on digital platforms [31]. Post-COVID online health and fitness companies have also seen a sudden and extraordinary rise in the consumption of yoga content in 2020 [32]. Through their Instagram profiles, Instagram influencers have gained notoriety and established themselves as authorities on one or more relevant subjects [33]. In Some of the most talked-about subjects are fashion, fitness, food and beauty, and female influencers handle these subjects the most [34]. Digital information functions as a form of "public pedagogy" [35], serving as a domain where institutional and symbolic power can be extended through the monitoring and shaping of social perceptions surrounding gendered and abled bodies, their capabilities, and health-related matters [36, 37]. Platforms like Instagram facilitate the rapid dissemination of content within user networks, fostering a dynamic, multidirectional exchange where users actively engage in sharing information about healthy bodies and health-seeking behaviors.

Additionally, Instagram is the social network with the quickest rate of growth in the world and is now seen by young consumers as an empowering platform for social networking [38]. Businesses can connect with customers through Instagram by using visual content rather than only textual presentations. If done well, this approach will result in customer engagement [39]. Men and women who participate in social media fitness influencers, YouTube training sessions and Instagram stories who view these fitness videos as essential to their way of life have greater intent when it comes to using them for their workouts. Influencers who are perceived as credible, knowledgeable, attractive, and possessing significant motivational influence might effectively encourage individuals to participate in physical activities. (e.g., working out at home with social media fitness influencers), which could enhance users' physical well-being and health through digital media.

3 Unpacking the Dynamics of Influencer-Marketing

Influencer marketing is an effective marketing strategy that helps the influencers to establish the brand and increase traffic to their blog, social media page, storefront, and yoga company website, as well as signups and purchases. Influencer marketing is a major deal; it's using social media to their advantage to use influencers to help expand a business or brand. Influencer marketing might obtain a well-known figure in the industry or niche to recommend or endorse their company's goods (Fig. 2). Even for brand-new yoga instructors, it is a growingly helpful company growth technique for startups or small enterprises. For instance, if someone pictures with a favorite celebrity or yoga expert mentioning the forthcoming online yoga course, yoga blog, or retreat. That might be a significant deal that helps the business's bottom line [40]. It's also among the greatest strategies to expand the brand or business on social media for yoga.

By collaborating with the appropriate influencers, they can:

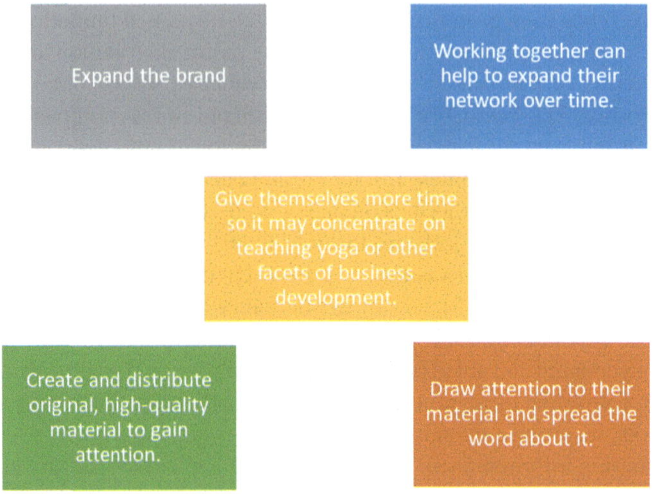

Fig. 2 Different aspects of Yoga influencer marketing. *Source* Compiled by the author

3.1 Fitness Influencer-Money Making Methods

As a fitness influencer, there are numerous avenues for earning money. Influencers have the ability to produce their goods or services, advertise, or engage in affiliate marketing. They can also market their goods and services on social media sites like Facebook, Instagram, and YouTube (Fig. 3). They can assist people in maintaining an active and healthy lifestyle while earning a good living by utilizing the following money strategies [41].

4 Research Methodology

The research methodology used in this study is an extensive approach within the context of influencer marketing in the yoga industry. This book chapter's research utilizes a comprehensive methodology that consists of three unique steps to investigate the subtleties of influencer marketing, specifically in relation to yoga influencers.

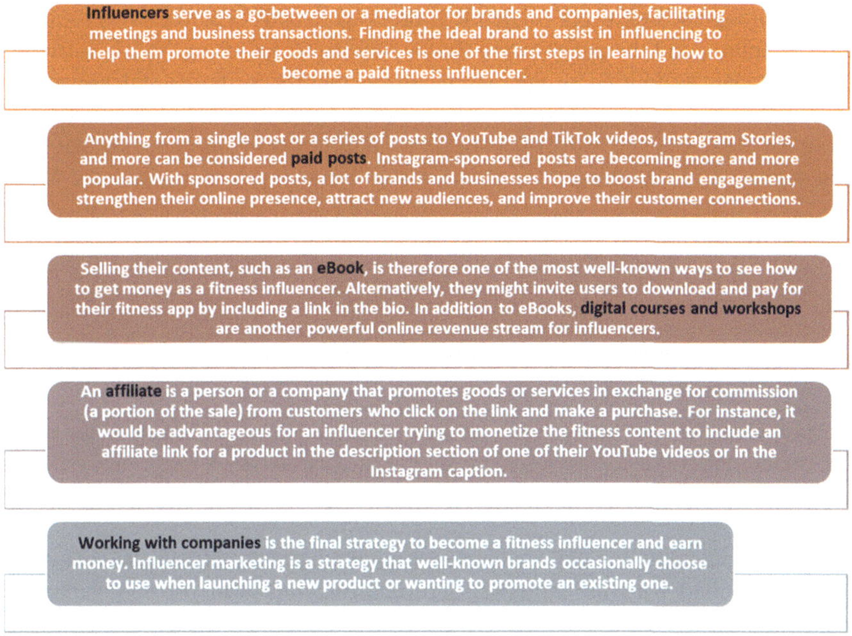

Fig. 3 Money-making methods. *Source* Compiled by the author

4.1 Stage One: Influencer Marketing Secondary Data Analysis

In the first part of the study, secondary data is analyzed with an emphasis on influencer marketing in the yoga and fitness industries. The goal of this phase is to gather studies, industry analysis, and current research on influencer marketing strategies unique to the yoga and fitness sectors. In the context of yoga and fitness, this stage seeks to provide a comprehensive understanding of the current trends, tactics, and obstacles in the field of influencer marketing.

An extensive literature search and an examination of pertinent trade periodicals revealed significant themes and insights into the mechanics of influencer marketing in the yoga and fitness industries. At this stage of the research process, significant background information was analyzed to identify the trends of Yoga Influencers.

4.2 Stage Two: Scopus Scholarly Study Evaluation

Using the insights gathered from the secondary data analysis, the second phase of the research focuses on assessing scholarly research papers concerning influencer marketing in yoga and fitness. Ninety-five pertinent research publications are found after a thorough search using academic sources as Scopus. The data was acquired using Ncapture, and the papers were thoroughly examined using NVivo software. Thematic analysis is used to identify recurring themes, patterns, and theoretical frameworks relevant to influencer marketing in the yoga and fitness domains. The emotional tone associated with many of the influencer marketing elements that have been discussed in the literature is also assessed through sentiment analysis. Word cloud analysis is used to visually portray and assess the frequency and significance of key terms and concepts within the corpus of scholarly research.

4.3 Stage Three: User Generated Content Analysis

The last phase of the study looks at user-generated material about yoga influencer marketing. Eighteen user-generated articles—including webpage analyses and blogs—are chosen for examination to represent a range of viewpoints within the yoga community.

The user-generated content is subjected to qualitative analysis using the ATLAS.ti software. To find patterns, attitudes, and recurring topics in the user-generated content, co-occurrence analysis, sentiment analysis, and word-cloud techniques are used.

The study attempts to provide a thorough knowledge of influencer marketing dynamics within the yoga sector by integrating findings from all three stages and adding user viewpoints and scholarly research insights. A thorough examination of the patterns, mindsets, and factors influencing the phenomenon of yoga influencer marketing is made possible by this multifaceted approach.

5 Data Analysis and Discussions

Yoga has grown in popularity, and fitness influencers play a critical role in shaping attitudes and behaviors surrounding it. This study examines 95 articles that are indexed by Scopus and taken from the www.scopus.com homepage to identify similar themes and viewpoints in discussions around yoga-focused fitness influencers. Deliberate and strategic consideration went into selecting NVivo software to analyze the sentiment and theme elements of the Scopus-indexed articles about fitness influencers who lean towards yoga. One well-known qualitative data analysis tool for researchers doing in-depth, nuanced studies of unstructured data is NVivo, which was created by QSR International [42]. NVivo was selected for this investigation due to its potent capabilities for handling qualitative data, carrying out in-depth topic and sentiment evaluations, and creating visual representations [43].

5.1 *Thematic Analysis for Scopus Research Data*

The thematic analysis of 95 Scopus-indexed publications as shown in Fig. 4 on fitness influencers with a leaning towards yoga has shown a rich tapestry of recurring themes. The frequency distribution of the main topics provides valuable insight into the emphasis and focus of this corpus. 'Influencers' is a highly prominent topic with a frequency of 245, emphasizing the critical role that influencers play in the discourse. This suggests that the consequences, actions, and impact of fitness influencers on the yoga community are the subject of a sizable portion of the publications. "Social," the theme with 229 references, highlights the relationship between yoga and social dynamics. This theme probably explores the ways in which social media is used by influencers in the yoga and fitness industries, the social implications of their activities, and the ways in which social media shapes public opinion towards these fields. A central topic in the debates is the concept of "Fitness," which has 195 references. This is consistent with the articles' main focus and suggests a thorough investigation of the fitness benefits of yoga and the influencers' role in advancing physical health. With 89 mentions, "health" stands out as a distinct theme, indicating that a sizable amount of the discussion explores the advantages and effects of yoga on one's health. This might be talking about how yoga and fitness influencers affect both mental and physical wellness. 'Yoga' as a theme appears with 70 references, even though it is the subject matter. This suggests that although yoga serves as the overall context, the articles focus on particular facets, techniques, and viewpoints associated with yoga rather than offering a comprehensive summary.

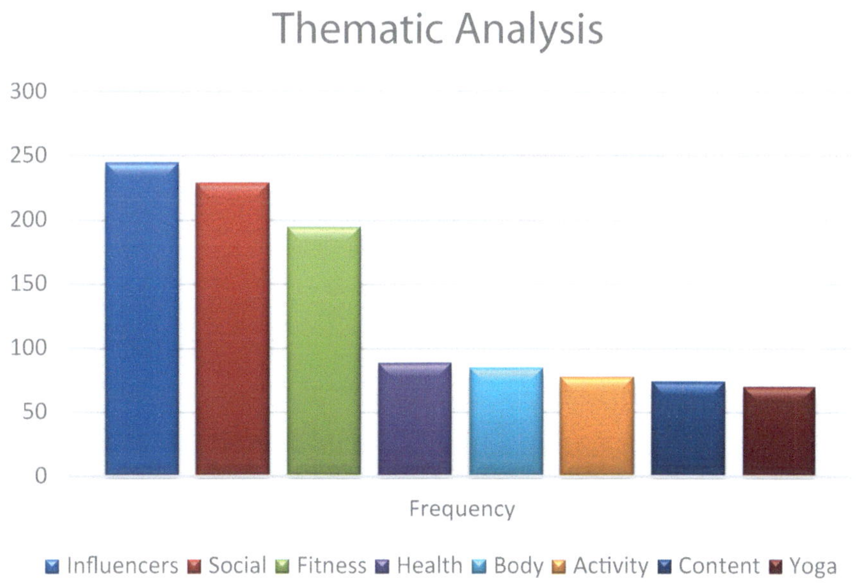

Fig. 4 Thematic analysis for scopus research data. *Source* Nvivo output

To summarize, the findings of the thematic analysis indicate a comprehensive investigation across the articles, primarily focusing on the impact of fitness influencers on the social and fitness aspects of yoga. The aforementioned topics' prevalence highlights the intricate interplay between yoga, fitness, and social dynamics within the framework of wellness led by influencers.

5.2 *Sentimental Analysis*

This study used data from a thorough sentiment analysis of 95 Scopus-indexed articles to examine how customers view yoga influencers on social media (Fig. 5). Positive, neutral, negative, and mixed attitudes are categorized by the study, which throws light on the emotional tenor of conversations around yoga influencers on social media.

Positive Reinforcement: A profusion of optimism points to an opportunity for marketers to leverage the positive and inspirational attributes associated with yoga gurus. Material creators may experience a rise in engagement and brand loyalty if they offer content that supports these favorable points of view.

Resolving Issues: Influencers and brands can enhance consumer satisfaction, transparency, and authenticity by recognizing and addressing unfavorable sentiments. Over time, a more favorable customer perception can be developed via open communication and prompt problem-solving.

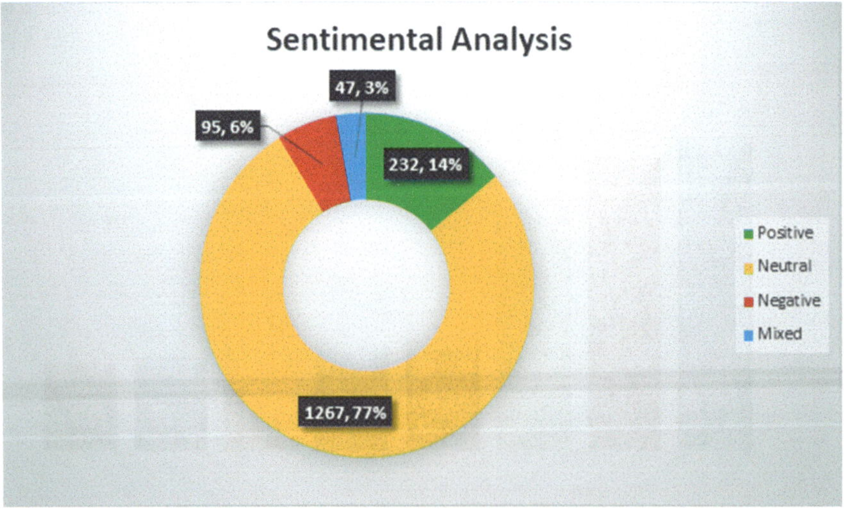

Fig. 5 Sentiment analysis scopus-indexed articles. *Source* Nvivo output

Fig. 6 Word cloud from scopus articles analysis. *Source* Nvivo output

Diverse Messaging: People feel a variety of emotions, often in opposition to one another, thus it's critical to customize marketing messaging for each target audience. Communication tactics can become more focused and efficient when they take into account the different perspectives that the customer may hold.

In conclusion, marketers, influencers, and academics can all benefit from the sensitive examination of the views that yoga influencers have on social media. The thorough insight gained from the inquiry may motivate tactics meant to maximize consumer engagement, establish credibility, and create a supportive and genuine online community around yoga and fitness celebrities. Visual aids are crucial in data analysis since they help translate complex information into insights that are easy to comprehend. One such useful visualization technique is the word cloud, an eye-catching word cloud that shows a snapshot of the terms that occur most frequently in a dataset. A word cloud enlightens us into a visual world where important themes stand out as we conclude our investigation after the sentiment and thematic analysis of 95 Scopus-indexed publications on fitness influencers leaning towards yoga (Fig. 6).

5.3 Data Analysis on Webpage Articles Using ATLAS.ti

Within the great digital void, where knowledge is freely exchanged, webpages serve as a storehouse of ideas, viewpoints, and stimulating material. Upon commencing the data analysis journey, this chapter explores the complex articles and blog posts on eighteen webpage articles to identify patterns, themes, and significant narratives that can be extracted from the abundance of available data on "Yoga Influencers".

This chapter's data analysis endeavor aims to comprehend the pulse of the digital presence of "Yoga Influencers" rather than merely decoding numbers. This chapter analysis is to decipher the stories entwined within the paragraphs, identify the sentiments, and distill the core of the information to impart through content analysis.

5.4 Introduction to ATLAS.ti Software

The appetite for advanced technologies that can decipher and make sense of the enormous and complex web of information has increased due to the exponential growth of digital content on social media platforms [44]. ATLAS.ti is a strong and flexible software solution in the field of sentiment analysis and word cloud building for social media analysis. ATLAS.ti acts as an effective software tool in interpreting sentiments and creating word clouds from the ever-changing social media world.

ATLAS.ti assists to turn data into wisdom and use analysis to bring the stories contained inside various web pages to life with strong data analysis. This analysis interprets the language used in various articles and webpage materials, exposes the sentiment stories, and opens a new dimension in the trends of yoga influencers.

5.5 Sentimental Analysis Using Sankey Chart

A Sankey chart is an effective graphical tool for illustrating how values or quantities move across several categories. The chart becomes a dynamic tool for demonstrating the distribution and transitions between neutral, positive, and negative attitudes in the context of sentiment analysis [45]. The Below Fig. 7 is an effective visual tool for displaying the sentiment flow in a dataset is a Sankey chart. This identifies the following values in the context of evaluating attitudes toward yoga influencers on ATLAS.ti: Neutral (1208), Positive (239), and Negative (106).

Fig. 7 Sentimental analysis using sankey chart. *Source* ATLAS.ti output

Sentiment-Neutral (1208): The Sankey chart starts with a sizable set of 1208 examples classified as "Neutral" feelings. This abundance of objectivity indicates that a substantial percentage of the content about yoga influencers has an impartial or well-balanced emotional tone. It could signify the existence of factual data or descriptions devoid of overtly favorable or negative language.

Sentiment-Positive (239): The graphic skillfully shows a shift from the neutral sentiment pool to the positive sentiment pool. The fact that 239 of the examples fall into the "Positive" sentiment category indicates a significant change in the emotional tone. This implies that there is a significant amount of positive sentiment—such as respect, support, or appreciation—in the information about yoga influencers.

Sentiment-Negative (106): Simultaneously, the Sankey chart displays an additional stream that signifies the shift from neutral to negative emotions. A total of 106 incidents have been categorized as "Negative," indicating the presence of unfavorable or critical perspectives towards yoga influencers in the digital content. Criticisms, anxieties, or challenges pertaining to the influences are some examples of these unpleasant emotions. Researchers utilizing ATLAS.ti for sentiment analysis can learn some new facts from this comprehensive explanation. Because neutral attitudes predominate, there appears to be a plethora of objective content; this makes ATLAS.ti a useful tool for deciphering and analyzing such data. Researchers can find patterns of support and gratitude for yoga influencers by utilizing the plethora of positive thoughts. Unfavorable feelings indicate potential areas for worry or critique, which helps academics understand the difficulties or contentious issues related to yoga impacts. In conclusion, the Sankey chart used in the ATLAS.ti analysis of the dataset offers a visual representation of the sentiment shifts. This comprehensive understanding enabled through ATLAS.ti interpret that there is a neutral and emotional terrain of yoga impacts.

5.6 *Word Cloud Analysis of Yoga Influencers*

Analyzing how often words appear visually in the analyzed data is necessary to interpret a word cloud produced by ATLAS.ti output [46]. A word cloud (Fig. 8) offers a quick, visual representation of the terms that appear most frequently in the articles. The word cloud for the eighteen articles analyzed is represented in the below picture with central themes identified as *Yoga, Fitness, Market, Influencer, Health, Ware, Products and Brands*. Smaller phrases, on the other hand, like "*Community*" "*Follow*" *or* "*Virtual*" could point to themes where people have offered helpful critiques or recommended changes.

Fig. 8 Word cloud analysis of Yoga influencers. *Source* ATLAS.ti output

6 Conclusion

This multi-phase study investigated the complex dynamics of influencer marketing in the yoga and fitness industries using academic and user-generated content analysis. A number of significant themes that demonstrated the widespread impact of social media influencers on fitness and health-related material were revealed by thematic analysis of Scopus study data. Important topics that emerged from the thematic analysis included "Influencers," "Social," "Fitness," and "Health," highlighting the importance of these elements in influencing the conversation around influencer marketing in the yoga and fitness industries.

Sentiment analysis of the academic papers revealed that the majority of the sentiment was neutral 77% with 14% of the articles being positive, 6% being negative, and 3% being mixed this makes the case that influencer marketing ought to be fairly discussed in scholarly research, with a focus on objective evaluation and analysis. Because it showed the frequency of terms like "Social," "Health," "Influencers," and "Media," which match the primary concepts and keywords discussed in the scholarly discussion, the word cloud analysis was notable. After user-generated content was analyzed using the ATLAS.ti tool, insightful information about common attitudes and themes among consumers and yoga enthusiasts was obtained. In the sentiment analysis, a significant proportion of neutral sentiments (1208) were discovered in addition to positive (239) and negative (106) thoughts.

Word clouds built from user-generated content revealed themes such as "Market," "Yoga," "Health," and "Wellness," suggesting a strong relationship between influencer marketing and users' general health and well-being. The research's conclusions have significant implications for anyone working in the yoga and fitness industries. Companies and brands can enhance the genuineness and impact of their marketing campaigns by carefully choosing influencers to strategically collaborate with who align with their beliefs and target demographic. Creating captivating content on subjects like Social Connection, Health Promotion, and Holistic Well-being can boost consumer engagement and promote brand loyalty. It takes ethical considerations about transparency and responsible content creation, as well as continual monitoring and assessment of influencer marketing initiatives, to maintain consumer confidence and credibility. Stakeholders will be more prepared to employ influencer marketing to foster deeper connections with their target audience and support sustainable, lucrative business growth if they adhere to these realistic recommendations.

References

1. Kalra, B., et al.: Coping with illness: insight from the Bhagavad Gita. Indian J. Endocrinol. Metabolism **22**(4), 560–564 (2018). https://doi.org/10.4103/ijem.IJEM_228_17
2. Yoga Basics Homepage. https://www.yogabasics.com/learn/bhagavad-gita/. Last Accessed 15 Mar 2024
3. Medium Homepage. https://medium.com/@spkulsari/the-importance-of-yoga-in-bhagavad-gita-8688b2f0abad. Last Accessed 24 Mar 2024
4. Yogacentral Homepage. https://yogacentral.in/2017/11/10/importance-yoga-bhagavad-gita/. Last Accessed 28 Mar 2024
5. OmYoga Magazine Homepage. https://www.ommagazine.com/becoming-a-social-media-yoga-influencer/. Last Accessed 29 Mar 2024
6. Social media Today Homepage. https://www.socialmediatoday.com/social-networks/ericdahan/2015-10-14/rise-yoga-influencers-instagram. Last Accessed 29 Mar 2024
7. Heidemann, J., Klier, M., Probst, F.: Online social networks: a survey of a global phenomenon. Comput. Netw. **56**(18), 3866–3878 (2012). https://doi.org/10.1016/j.comnet.2012.08.009
8. Statista Homepage. https://www.statista.com/topics/1164/social-networks/. Last Accessed 29 Mar 2024
9. Zaw, H.T.: The impact of social media on cultural adaptation process: study on Chinese government scholarship students. Adv. J.Ism Commun. **6**(03), 75 (2018). https://doi.org/10.4236/ajc.2018.63007
10. Kapoor, K.K., et al.: Advances in social media research: past, present and future. Inf. Syst. Front. **20**, 531–558 (2018). https://doi.org/10.1007/s10796-017-9810-y
11. Kizgin, H., et al.: The impact of social media on consumer acculturation: current challenges, opportunities, and an agenda for research and practice. Int. J. Inf. Manag. **51**, 102026 (2020). https://doi.org/10.1016/j.ijinfomgt.2019.10.011
12. Jafari, A., Visconti, L.M.: New directions in researching ethnicity in marketing and consumer behaviour: a well-being agenda. Mark. Theory **15**(2), 265–270 (2015). https://doi.org/10.1177/1470593114552582
13. Scholar Commons Homepage. https://scholarcommons.scu.edu/engl_176/2/. Last Accessed 29 Mar 2024
14. Appel, G., et al.: The future of social media in marketing. J. Acad. Mark. Sci. **48**(1), 79–95 (2020). https://doi.org/10.1007/s11747-019-00695-1
15. Alalwan, A.A., et al.: Social media in marketing: a review and analysis of the existing literature. Telemat. Inform. **34**(7), 1177–1190 (2017). https://doi.org/10.1002/cb.1782
16. Monika, W., Nasution, A.H., Nasution, S.: The role of social media on intercultural communication competencies, 483–491 (2020). http://repository.uir.ac.id/id/eprint/21771

17. McGraa, K.L.K.: The effects of persuasive motivational text messaging on adherence to diet and exercise programs across different personality traits. Fielding Graduate University, Diss (2010)
18. Khan, U., Dhar, R.: Licensing effect in consumer choice. J. Mark. Res. **43**(2), 259–266 (2006). https://doi.org/10.1509/jmkr.43.2.259
19. Wilcox, K., Stephen, A.T.: Are close friends the enemy? online social networks, self-esteem, and self-control. J. Consum. Res. **40**(1), 90–103 (2013). https://doi.org/10.1086/668794
20. Elisondo, R.C.: Creative processes and emotions in COVID-19 pandemic. Creat. Stud. **15**(2), 389–405 (2022). https://doi.org/10.3846/cs.2022.14264
21. Uzunoğlu, E., Kip, S.M.: Brand communication through digital influencers: leveraging blogger engagement. Int. J. Inf. Manag. **34**(5), 592–602 (2014). https://doi.org/10.1016/j.ijinfomgt.2014.04.007
22. Pourazad, N., Stocchi, L., Narsey, S.: A comparison of social media influencers' KPI patterns across platforms: exploring differences in followers and engagement on Facebook, Instagram, YouTube, TikTok, and Twitter. J. Advert. Res. **63**(2), 139–159 (2023). https://doi.org/10.2501/JAR-2023-008Published1June2023
23. Moreno, D.R., Quintana, J.G., Riaño, E.R.: Impact and engagement of sport & fitness influencers: a challenge for health education media literacy. Online J. Commun. Media Technol. **13**(3), e202334 (2023). https://doi.org/10.30935/ojcmt/13309
24. Chowdhury, D., et al.: "Think positive and don't die alone"-Foreign-born, South Asian older adults' perceptions on healthy aging. Int. J. Qual. Stud. Health Well-Being **18**(1), 2253576 (2023)
25. Baker, S.A., Rojek, C.: Lifestyle Gurus: Constructing Authority and Influence Online. Wiley (2020)
26. Strings, S., Blu Wakpa, T.: Rethinking yoga: meditations on the work we do. Race Yoga **1**(1), (2016). https://doi.org/10.5070/r311033432
27. Summit Research Repository Homepage. https://summit.sfu.ca/item/35618. Last Accessed 05 April 2024
28. Smith, S., Atencio, M.: "Yoga is yoga. Yoga is everywhere. You either practice or you don't": a qualitative examination of yoga social dynamics. Sport. Soc. **20**(9), 1167–1184 (2017). https://doi.org/10.1080/17430437.2016.1269082
29. Dublin Business School Homepage. https://esource.dbs.ie/items/53ed5cf5-b458-431a-ae96-3c28426e9023. Last Accessed 05 April 2024
30. Hinz, A., et al.: Practice or performance? a content analysis of yoga-related videos on Instagram. Body Image **39**, 175–183 (2021). https://doi.org/10.1016/j.bodyim.2021.08.002
31. Strahm, M.F., Cohen, M., Borg-Olivier, S.: University-based online Yoga education: a pilot study of students' experience. Int. J. Yoga **9**(1), 86 (2016). https://doi.org/10.4103/0973-6131.171721
32. Sui, W., Rush, J., Rhodes, R.E.: Engagement with web-based fitness videos on YouTube and Instagram during the COVID-19 pandemic: longitudinal study. JMIR Form. Res. **6**(3), e25055 (2022). https://doi.org/10.2196/25055
33. Khamis, S., Ang, L., Welling, R.: Self-branding, 'micro-celebrity' and the rise of social media influencers. Celeb. Stud. **8**(2), 191–208 (2017). https://doi.org/10.1080/19392397.2016.1218292
34. Marwick, A.E.: You may know me from youtube: (micro-) celebrity in social media. A Companion Celeb. 333–350 (2015). https://doi.org/10.1002/9781118475089.ch18
35. Giroux, H.A.: Cultural studies, public pedagogy, and the responsibility of intellectuals. Commun. Crit./Cult. Stud. **1**(1), 59–79 (2004). https://doi.org/10.1080/1479142042000180926
36. Bourne, J.: Pedagogic practice, culture and the globalization of yoga teaching. J. Appl. Linguist. Prof. Pract. 7(1), (2010). https://doi.org/10.1558/japl.v7i1.11
37. Rice, C., Chandler, E.: Representing difference: disability, digital storytelling and public pedagogy. Routledge Companion Disabil. Media. Routledge 367–376 (2019). https://doi.org/10.4324/9781315716008

38. Lee, E., et al.: Pictures speak louder than words: motivations for using Instagram. Cyberpsychology, Behav., Soc. Netw. **18**(9), 552–556 (2015). https://doi.org/10.1089/cyber.2015.0157
39. Pilgrim, K., Bohnet-Joschko, S.: Selling health and happiness how influencers communicate on Instagram about dieting and exercise: mixed methods research. BMC Public Health **19**, 1–9 (2019). https://doi.org/10.1186/s12889-019-7387-8
40. Naked Earth Yoga Homepage. https://www.nakedearthyoga.com/influencer-marketing-strategy-yoga-business/. Last Accessed 15 April 2024
41. Ainfluencer Homepage. https://blog.ainfluencer.com/how-to-make-money-as-a-fitness-influencer/. Last Accessed 29 April 2024
42. Zamawe, F.C.: The implication of using NVivo software in qualitative data analysis: evidence-based reflections. Malawi Med. J. **27**(1), 13–15 (2015). https://doi.org/10.4314/mmj.v27i1.4
43. Phillips, M.E., Lu, J.: A quick look at NVivo. J. Electron. Resour. Librariansh. **30**, 104–106 (2018). https://doi.org/10.1080/1941126X.2018.1465535
44. Gibbs, G.R.: Media review: ATLAS.ti software to assist with the qualitative analysis of data. J. Mixed Methods Res. **1**(1), 103–104 (2007). https://doi.org/10.1177/2345678906291490
45. Mir, F.A., Singh, R.P., Mehra, M.M.: Sentimental analysis of a sentence. Int. J. Innov. Res. Comput. Sci. Technol. **10**(1), 10–14 (2022). https://doi.org/10.55524/ijircst.2022.10.1.3
46. Paulus, T.M., Lester, J.N.: ATLAS.ti for conversation and discourse analysis studies. Int. J. Soc. Res. Methodol. **19**(4), 405–428 (2016). https://doi.org/10.1080/13645579.2015.1021949

Online User Choices Data Analytics and Privacy in Digital Marketing: Legal-Ethical Landscapes in the Digital-Futuristic Arena Projecting Multiple Layers of Artificial Intelligence and Blockchain

Bhupinder Singh, Pushan Kumar Dutta, and Christian Kaunert

Abstract The digital market offers users a plethora of choices from products and services to content consumption. However, this abundance of options raises questions about how user choices are tracked, analyzed and utilized by businesses for marketing purposes. So, striking a balance between personalized experiences and user privacy becomes paramount. As the companies utilize cutting-edge technology like Blockchain and Artificial Intelligence (AI) the ethical and regulatory frameworks pertaining to user data are changing. The governments worldwide are enacting and enhancing data protection regulations to safeguard privacy of users. The ethical use of user data is a growing concern and the major issues of transparency, consent and fairness in data practices are crucial to building and maintaining trust with users. Artificial Intelligence plays a pivotal role in enhancing the effectiveness of digital marketing campaigns as from predictive analytics for personalized recommendations to chatbots for customer interactions, AI-driven technologies streamline marketing processes. Blockchain as decentralized and secure nature is increasingly being explored as a solution for enhancing privacy in digital marketing. This chapter comprehensively explores the complex dynamics between online user choices, data analytics and privacy in the context of digital marketing.

B. Singh (✉)
Sharda University, Greater Noida, India
e-mail: bhupindersinghlaw19@gmail.com

P. K. Dutta
Electronics and Communication Engineering Department at ASETK, Amity University Kolkata, Kolkata, India

C. Kaunert
Dublin City University, Dublin, Ireland

University of South Wales, Cardiff, UK

S. Dutta et al. (eds.), *Advances in Data Analytics for Influencer Marketing: An Interdisciplinary Approach*, Information Systems Engineering and Management 9,
https://doi.org/10.1007/978-3-031-65727-6_21

Keywords Artificial intelligence · Blockchain · Data analytics · Legal-Ethical landscapes · Privacy · Digital marketing

1 Introduction

The data analytics forms the backbone of modern digital marketing strategies [1]. With the analyses of user behavior, preferences and interactions as the businesses can tailor their offerings, delivering more targeted and effective campaigns [2]. But the extensive use of data analytics also raises concerns about the ethical use of personal information and the potential for intrusive profiling [3]. The relationship between online user choices, data analytics and privacy has become a crucial focus point in the ever-changing field of digital marketing [4, 5]. Blockchain can provide users with more control over their data by enabling transparent and traceable consent mechanisms and its decentralized nature reduces the risk of data breaches and unauthorized access [6, 7]. In this regard, national laws regulating digital developments must be consistent, as demonstrated by the EU Directive 2000/31/EC on electronic commerce, which sets universal guidelines for the EU [8, 9]. Similar to this, laws like the General Data Protection Regulation (GDPR) are essential for monitoring how people are protected when their personal data is processed [10]. While making sure that regulatory actions don't slow down the rate of digital disruption, these legal frameworks ought to support a clearly defined process of digital transformation [11, 12].

The process of digital transformation calls for a company's whole structure to be updated with new technology, which means that conventional business models must be revised [13]. In a similar vein, artificial intelligence is a unique technological innovation that has the potential to have a big influence on people's lives as well as company operations [14]. Advanced analytical models are being developed using cognitive techniques that mimic human behavior and mental processes [15]. These models enable businesses to improve sales, customer engagement, operational efficiency and overall service quality. These models produce insightful data since they are based on descriptive, predictive, and prescriptive analytics. An international legal framework that governs technology advancements uniformly is crucial for enabling a smooth and seamless digital transition [16]. A clear and seamless digital transformation process depends on this legislative framework. But it's crucial that these rules don't slow down the rate of digital disruption. According to this report, digital transformation and artificial intelligence will be essential to many applications, necessitating their widespread implementation. However, this deployment needs to follow standard protocols that correspond with the changing digital environment [17].

1.1 Background of Study

The transition in recent years from an analog culture to a rapidly developing digital civilization, characterized by the introduction of new business models, organizational structures and communication techniques [18, 19]. The way employment are organized and company models are planned has changed significantly as a result of this shift [20]. After the Internet and mobility, artificial intelligence (AI) is expected to be the next big technology breakthrough [21]. Artificial Intelligence (AI) is the process by which machines mimic human intelligence. It includes the ability of the machine to perform cognitive tasks that are commonly associated with the human brain such as—perception, reasoning, learning, experience-based evolution, problem-solving, interaction with the environment and even creativity. There is a much significance of law like—the GDPR as legislative trend in order to synchronize digital transformation with the proper legal framework [22]. Given that many of the rules now in place were created before social media and other technology developments, updating the regulatory framework is becoming more and more important in the new digital world. The necessity for constant legal adaptation and argues in favor of a steady legislative process that promotes development rather than hinders it, guaranteeing compliance with local agreements among EU member states, business interests, and the global market [23].

1.1.1 Landscape of Digital Marketing: Essentials

The many platforms, channels, and marketing strategies that make up the field of digital marketing are all included in the landscape of digital marketing. Understanding the basic components that make up this terrain is essential to navigating this quickly changing environment [24].

Websites: A website acts as the focal point of a business's online presence and serves as a virtual showroom, informational resource, and important point of contact with clients [25].

Social media: Websites such as Facebook, Instagram, Twitter, and LinkedIn offer unique channels for connecting with potential customers, building brand recognition, and encouraging purchases [26].

Email marketing: Email marketing allows firms to deliver newsletters, promotions, and tailored communications straight to customers' inboxes at a low cost [27].

Search Engine Optimization (SEO): Businesses may improve their search engine rankings and draw in more organic traffic by improving the content of their websites and following best practices [28].

Mobile marketing: Since many people access the internet through mobile devices, it's critical to develop websites, applications, and advertising campaigns that are

Fig. 1 Shows the essentials of the digital marketing

optimized for mobile in order to attract consumers who are always on the go [29] (Fig. 1).

1.2 Objectives of the Chapter

The development of technology and the internet has brought about tremendous changes in the field of marketing [30, 31]. Businesses and marketers must understand the complexities of digital marketing and adapt to its dynamic changes in light of the always changing digital world [32]. This comprehensive guide explores the principles of digital marketing and its changing landscape, great for beginners or seasoned marketers in need of a refresher [33]. This chapter has the following objectives to:

- analyzing the legal and ethical considerations in online user data analytics.
- understanding the impact of AI and blockchain on user privacy.
- exploring the futuristic trends and challenges in digital marketing (Fig. 2).

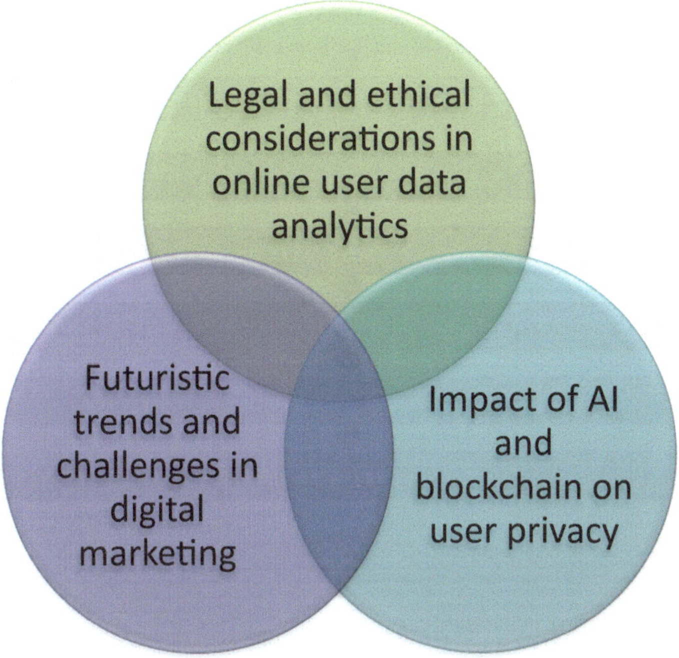

Fig. 2 Highlights the objectives of this chapter

1.3 Organisation of the Chapter

This chapter widely explores the diverse arena of Online User Choices Data Analytics and Privacy in Digital Marketing considering Legal-Ethical Landscapes in the Digital—Futuristic Arena Projecting Multiple Layers of Artificial Intelligence and Blockchain. Section 2 elaborates the Artificial Intelligence and Blockchain in Digital Marketing. Section 3 expresses the Data Analytics in Digital Marketing. Section 4 lays down the Online User Choices and Consent. Section 5 expresses the Challenges and Concerns. Finally, Sect. 6 Conclude the Chapter with Future Scope (Fig. 3).

2 Artificial Intelligence and Blockchain in Digital Marketing

Artificial intelligence (AI) is revolutionizing industries by leveraging technology such as machine learning and natural language processing to mimic human intellect in computers and change the way businesses operate [34]. The market for AI-based solutions is expected to increase significantly in the upcoming years due to the integration of AI solutions for automation, predictive analysis and tailored

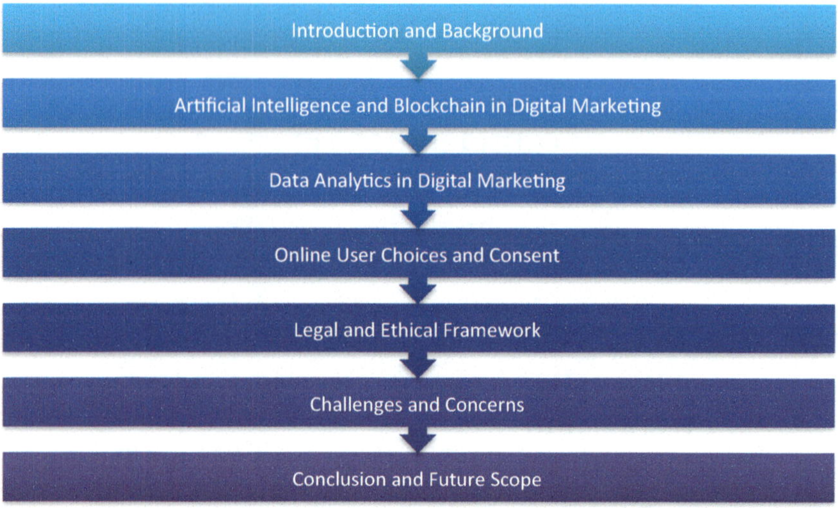

Fig. 3 Expresses the flow the chapter

consumer experiences [35]. Blockchain has completely changed data security and transparency. It is a decentralized, tamper-proof ledger system that was first created for crypto-currencies. Its applications have spread to a number of industries such as supply chain, healthcare, and finance, which has helped the Blockchain industry to flourish [36]. Because blockchain makes transactions safe, transparent and traceable, it has emerged as a key component of the digital transition. Blockchain and AI work together to address issues with efficiency, transparency, and data security. Because of this collaboration, the combined market for blockchain and artificial intelligence (AI) is expected to be valued over $703 million by 2025, growing at a Compound Annual Growth Rate (CAGR) of 25.3% between 2020 and 2025 [37, 38].

With deliberately integrating these technologies, businesses now have the strategic chance to improve the security and transparency of AI applications. The businesses are now able to develop sophisticated AI models while maintaining the integrity and reliability of the underlying data thanks to the confluence of AI with blockchain technology [39]. Combining these technologies not only improves AI applications' capabilities but also solves issues with trust and data integrity. Blockchain technology and artificial intelligence work together to create a stable environment for innovation in a time when businesses depend more and more on data-driven decision-making. Because AI and Blockchain have a mutually beneficial connection, organizations should take advantage of this to place themselves at the forefront of technological progress and promote efficiency, security and growth [40, 41].

Blockchain technology and artificial intelligence are about to combine to provide customers with more individualized experiences [42]. Blockchain will be essential in maintaining the security and immutability of the data with cutting-edge cryptographic encryption techniques as AI handles more sensitive personal data [43, 44]. This makes sure that situations where data breaches or illegal access might make AI fail to provide the right tailored experiences are avoided. To put it another way, Blockchain ensures security through encryption, while AI uses data secured by Blockchain to improve user privacy experiences on websites [45].

2.1 *Privacy and Trust*

The safeguarding of personal information has become a significant concern for consumers worldwide. People are becoming more aware of the possible risks associated with sharing personal information online. Companies that prioritize protecting the privacy of their customers stand to gain a competitive advantage. Concerns about organizations collecting personal information and the possible consequences of using it are growing among the general population [46]. Notably, changes in the IT sector, as seen by the actions of Google and Apple as well as legislation like the GDPR, CASL and CCPA have changed the data privacy environment and affected the way digital marketing campaigns are implemented and evaluated. Amidst these privacy worries over the internet, digital marketers may use certain best practices to protect personal data and foster customer trust. Transparency in data gathering and use is essential, first and foremost [47].

3 Online User Choices and Consent

Giving consumers the option to make decisions and take control of their data and preferences is another way to ensure privacy and consent. This entails providing choices for accepting or rejecting offers from different communication channels, such push alerts, SMS alerts and email newsletters [48]. Important features include making it simple for users to update or modify their preferences, permitting them to unsubscribe, and enabling them to revoke their consent at any moment. Users can find this procedure easier if, for example, every email sent has an unsubscribe button or a link to a preference center. Understanding customer data permission entails recognizing the control consumers have over their personal data. It's an important aspect of digital marketing that necessitates a dedication to personal privacy as well as the ability to tailor experiences according to user preferences [49].

It is crucial to investigate why this issue is not just significant but also essential for companies navigating the digital age. The following consumer data permission ensures compliance with a plethora of privacy laws that have emerged on a worldwide scale [50]. The example from the US is the California Consumer Privacy Act (CCPA) which gives Californians the right to control and receive information about the personal data that companies have collected about them. But it goes beyond just following the law; there is a strong financial case for prioritizing consumer data consent. Businesses may customize marketing campaigns in a way that is seen as considerate and relevant to customers when they have the right consent. Increased client loyalty and trust are key advantages in today's market, and this might foster both of those things [51].

3.1 Informed Consent: Challenges in Obtaining Meaningful Consent

This entails providing a clear description of the data that has been gathered, together with its intended use and the parties with whom it could be shared. This is usually done via a Privacy Policy that can be accessed via an email footer and website link. A brand's credibility is increased and customer comfort is fostered by such transparency [52]. The collection and use of personal data requires express consent. Before any data collection takes place, consumers must be notified about the data being collected and its intended use. Consent is a continuous process that is carried out at every stage of the customer experience, and there are simple ways for customers to opt out. Although GDPR, CASL and CCPA contain particular standards, many firms apply these policies to all contacts and prospects, regardless of their location or nationality [53].

3.2 User Control and Preferences: Balance Between Personalization and User Autonomy

Obtaining explicit consent not only complies with legal obligations but also improves email message deliverability and engagement rates [54]. Sending messages to people who proactively indicate their interest in receiving them helps marketers, since it enhances the sender's reputation and sender score, increasing the likelihood of inbox delivery [55, 56]. Safe data processing and storage procedures are equally important [57, 58]. Digital marketers are responsible for ensuring that personally identifiable information is processed and kept securely [59, 60]. This includes implementing safeguards like encryption, access restrictions, secure servers, and frequent vulnerability testing. This protects sensitive consumer data and helps avoid data breaches [61, 62].

Digital marketers should only acquire the data required for their marketing efforts in order to allay worries about excessive data collecting [63, 64]. In addition to reducing the chance of data breaches, this promotes customer trust [65, 66]. It is essential to provide customers the choice to refuse data collection and use, even in cases when they have given their express consent [67, 68]. For individuals covered by the GDPR, CCPA, etc., it is especially crucial to communicate clearly about how to use their rights [69, 70]. Digital marketers may build customer trust and safeguard personal data by embracing openness, getting explicit consent, assuring secure data processes, reducing data gathering, and providing opt-out choices [71, 72]. Following these best practices helps marketers build meaningful relationships with prospects and customers who actively engage with the content and have faith in the secure handling of their data, even in the face of ongoing privacy concerns [73, 74].

4 Data Analytics in Digital Marketing

In the swiftly changing digital landscape, the notion of data privacy has emerged as a fundamental element of ethical marketing practices [75, 76]. As businesses endeavor to tailor experiences for their customers, the imperative to strike a delicate balance between privacy and personalization has become crucial [77, 78]. This explore the nuances of comprehending customer data consent in digital marketing, presenting a comprehensive guide for businesses aiming to uphold their customers' data preferences while optimizing their marketing strategies [79, 80]. Data privacy in essence pertains to individuals' right to exert control over the collection and utilization of their personal information [81, 82]. In the realm of digital marketing, this translates to adopting practices that honor user privacy and ensure the responsible use of personal data [83, 84].

The idea behind consent-based marketing is to only show advertisements to those who have already given their consent usually by accepting trackers on websites and mobile applications [85, 86]. The need for compliance solutions such as our Consent Management Platform (CMP) has increased due to data privacy rules [87, 88]. But in the more crucial move towards a user-centric paradigm, user permission needs to be integrated in the very fabric of marketing strategies and key performance indicators (KPIs) in the cookieless future; it cannot be only a legal obligation [89, 90]. This marketing strategy is openly asking for consent to provide users with personally identifiable advertisements that are specifically tailored to them [91, 92]. In order to guarantee a smooth, legal, and qualifying transaction, businesses employ technology to automate the user consent process and integrate it across a variety of marketing tools and platforms [93, 94]. The result is highly tailored and engaging marketing material for consumers who have given their express consent [95, 96]. Digital marketing seems to be navigating a difficult intersection between users growing worries about data management and their need for more tailored interactions, but consent is the common thread connecting both developments [97, 98].

With using a privacy-by-design methodology, consent-based marketing complies with the majority of data privacy legislation around the globe [99]. So, obtaining consent opens the door to providing highly targeted advertising to consumers who actively seek it out [100, 101]. More quickly than many marketers anticipate, the consent-based future is here [102, 103]. Even though lead generation and customer acquisition are heavily dependent on large tech platforms like Google and Meta, the environment is changing due to things like EU regulations and the removal of third-party cookies, which necessitates a mental adjustment [104, 105]. In order to maintain growth and relevance, marketers must consider user consent as an essential component of all major marketing initiatives rather than just a compliance mandate [106]. Users who want more individualized experiences commit their data to businesses they think will protect it [107, 108]. In addition to being necessary for regulatory compliance, obtaining user consent allows businesses to tailor their marketing efforts to the preferences of individual consumers [109, 110]. Marketers are looking for methods to lessen their dependency on bulk data collecting and engage customers in a personalized and trustworthy way as the industry shifts from using third-party to first-party data [111, 112]. Consent-based marketing offers a number of benefits, including cost-effectiveness, regulatory compliance, and building consumer confidence [113, 114]. This approach addresses the rising trend of consumer decisions being influenced by data privacy awareness [115, 116].

5 Challenges and Concerns

Setting up a regulatory framework is crucial for the public administration's communication of the laws pertaining to the digital market [117, 118]. For businesses that haven't yet undergone digital transformation, this framework should make it easier and offer direction so they can carry out this process successfully [119, 120]. The objective is to provide minimal standards that harmonize the business environment and encourage global market competitiveness [121]. These rules, nonetheless, have to be clear and binding. In light of this, identifying the applications of artificial intelligence that fit into each of the previously described categories should be taken into further consideration for future work [122, 123]. Certain apps, such those that violate basic rights or manipulate human behavior to undermine user goals, are easy to classify and should be categorically rejected [124, 125]. Similar to this, there seems to be a significant level of risk associated with applications incorporating technology utilized in key infrastructures, security elements, essential public services, justice administration, biometric identity systems, etc. [126, 127]. But the range is wide, and even applications such as conversational robots, emotion recognition systems, spam filters, etc. that seem to pose little or no risk at first should be monitored. Once rules are in place, determining the parameters of governance and the enforcement system is another crucial responsibility [1, 128].

6 Conclusion and Future Landscapes

Businesses need to understand the digital marketing environment in order to be relevant and competitive. You may improve your capacity to traverse the intricate and dynamic world of digital marketing and get desired results by familiarizing yourself with the fundamentals of digital marketing aspects, key tactics, and rising trends. It's critical to remember that the world of digital marketing is always changing. This emphasizes how important it is to stay knowledgeable, flexible, and always learning in order to maintain success in a constantly changing environment. Blockchain technology and artificial intelligence (AI) together represent a technological revolution. It is envisaged that smart contracts will develop into perceptive entities that adjust and optimize Blockchain activities. This combination has the potential to revolutionize decision-making by promoting equity and inclusion through decentralized governance systems powered by AI. When blockchain and AI are combined, the possibilities for creativity are endless. The future promises a fundamental shift in data ownership, not merely algorithmic improvements. People should be able to take back control of their data thanks to Blockchain's security features and AI's capabilities. This change will result in a decentralized environment where each transaction—from tailored retail experiences to predictive healthcare analytics—showcases the enormous possibilities of this formidable partnership.

References

1. Goldberg, S.G., Johnson, G.A., Shriver, S.K.: Regulating privacy online: an economic evaluation of the GDPR. Am. Econ. J. Econ. Pol. **16**(1), 325–358 (2024)
2. Kunz, W.H., Wirtz, J.: Corporate digital responsibility (CDR) in the age of AI: implications for interactive marketing. J. Res. Interact. Mark. **18**(1), 31–37 (2024)
3. Arikan, O.U., Ozturk, E., Duman, S., Aktaş, M.A.: Conceptualization of meta-servitization: 3D case study from furniture industry. In: 2022 International Symposium on Multidisciplinary Studies and Innovative Technologies (ISMSIT), pp. 702–706. IEEE (2022)
4. Verma, S., Warrier, L., Bolia, B., Mehta, S.: Past, present, and future of virtual tourism-a literature review. Int. J. Inf. Manag. Data Insights **2**(2), 100085 (2022)
5. Kaliappen, N., Hassan, A.: Digital trends in Asian hotel industry. In: Technology application in tourism in Asia: innovations, theories and practices, pp. 147–163. Springer Nature Singapore, Singapore (2022)
6. Egieya, Z.E., Ewuga, S.K., Adegbite, A.O., Oke, T.T.: The role of virtual and augmented reality in modern marketing: a critical review. Comput. Sci. IT Res. J. **4**(3), 244–272 (2023)
7. Wirtz, J., Hofmeister, J., Chew, P.Y., Ding, X.: Digital service technologies, service robots, AI, and the strategic pathways to cost-effective service excellence. Serv. Ind. J. **43**(15–16), 1173–1196 (2023)
8. López-Menchero, T.B., Mata García, P., Gomez Campillo, M., Delgado-de Miguel, J.F.: Outlets malls of the future: an approach to customer's expectations. Econ. Res.-Ekon. Istraživanja **34**(1), 2593–2608 (2021)
9. Shields, K.: Customer centric strategy (2024)
10. Vaidyanathan, N., Henningsson, S.: Designing augmented reality services for enhanced customer experiences in retail. J. Serv. Manag.Manag. **34**(1), 78–99 (2023)
11. Kwok, C.P., Tang, Y.M.: A fuzzy MCDM approach to support customer-centric innovation in virtual reality (VR) metaverse headset design. Adv. Eng. Inform. **56**, 101910 (2023)
12. Keivanpour, S., Keivanpour, S.: Design for environment in consumer-centric paradigm. In: Approaches, Opportunities, and Challenges for Eco-design 4.0: A Concise Guide for Practitioners and Students, pp. 1–18
13. Kazmi, S.H.A., Ahmed, R.R., Soomro, K.A., Hashem, E.A.R., Akhtar, H., Parmar, V.: Role of augmented reality in changing consumer behavior and decision making: case of Pakistan. Sustainability **13**(24), 14064 (2021)
14. Kannan, P.K., Gu, X.: 12. Customer centricity and the impact of technology. In: Handbook on Customer Centricity: Strategies for Building a Customer-Centric Organization, p. 300 (2019)

15. Gopakumar, S., Dananjayan, M.P.: Augmented reality in modern marketing: the ultimate solution to engage the digitally overwhelmed, budget-conscious audience? J. Inf. Technol. Teach. Cases 20438869231202712 (2023)
16. Chen, R., Perry, P., Boardman, R., McCormick, H.: Augmented reality in retail: a systematic review of research foci and future research agenda. Int. J. Retail. Distrib. Manag. **50**(4), 498–518 (2022)
17. Bajpai, A., Islam, T.: Impact of augmented reality marketing on customer engagement, behavior, loyalty, and buying decisions. Cardiometry **23**, 545–553 (2022)
18. Erdmann, A., Mas, J.M., Arilla, R.: Value-based adoption of augmented reality: a study on the influence on online purchase intention in retail. J. Consum. Behav.Consum. Behav. **22**(4), 912–932 (2023)
19. Wagner, R., Cozmiuc, D.: Extended reality in marketing—a multiple case study on internet of things platforms. Information **13**(6), 278 (2022)
20. Berberović, D., Alić, A., Činjarević, M.: Virtual reality in marketing: consumer and retail perspectives. In: International Conference "New Technologies, Development and Applications", pp. 1093–1102. Springer International Publishing, Cham (2022)
21. Rodriguez-Conde, I., Campos, C.: Towards customer-centric additive manufacturing: making human-centered 3D design tools through a handheld-based multi-touch user interface. Sensors **20**(15), 4255 (2020)
22. Hoyer, W.D., Kroschke, M., Schmitt, B., Kraume, K., Shankar, V.: Transforming the customer experience through new technologies. J. Interact. Mark. **51**(1), 57–71 (2020)
23. Addis, M.: Engaging brands: a customer-centric approach for superior experiences. Routledge (2020)
24. Bardhan, A.: Augmented reality and virtual reality in retail–a bibliometric analysis. Psychol. Educ. **57**(9), 6209–6219 (2020)
25. Befort, A.: Augmented & virtual reality in e-commerce. Master's thesis, University of Twente (2021)
26. Harba, J.N.: New approaches to customer experience: where disruptive technological innovation meets luxury fashion. In: Proceedings of the International Conference on Business Excellence, vol. 13, no. 1, pp. 740–758 (2019)
27. Kumar, S., Srinivas, Krishna, S., Sheorey, P., Joshi, G.: Understanding customer needs and innovation in E-Commerce. In: Changing Face of E-Commerce in Asia, pp. 43–57 (2022)
28. Simoni, M., Sorrentino, A., Leone, D., Caporuscio, A.: Boosting the pre-purchase experience through virtual reality: insights from the cruise industry. J. Hosp. Tour. Technol. **13**(1), 140–156 (2022)
29. Bilal, M., Zhang, Y., Cai, S., Akram, U., Halibas, A.: Artificial intelligence is the magic wand making customer-centric a reality! an investigation into the relationship between consumer purchase intention and consumer engagement through affective attachment. J. Retail. Consum. Serv.Consum. Serv. **77**, 103674 (2024)
30. Arrighi, P.A., Mougenot, C.: Towards user empowerment in product design: a mixed reality tool for interactive virtual prototyping. J. Intell. Manuf.Intell. Manuf. **30**, 743–754 (2019)
31. Lignell, M.: A journey towards human-centric and AI-augmented marketing? J. AI, Robot. Work. Autom. **2**(4), 369–381 (2023)
32. Camilleri, M.A.: The use of data-driven technologies for customer-centric marketing. Int. J. Big Data Manag. **1**(1), 50–63 (2020)
33. Qadri, S.B., Mir, M.M., Khan, M.A.: Exploring the impact of augmented reality on customer experiences and attitudes: a comparative analysis with websites. Int. J. Manag. Res. Emerg. Sci. **13**(2) (2023)
34. Batat, W.: How augmented reality (AR) is transforming the restaurant sector: investigating the impact of "Le Petit Chef" on customers' dining experiences. Technol. Forecast. Soc. Chang. **172**, 121013 (2021)
35. Kim, E., Simonse, L.W., Beckman, S.L., Appleyard, M.M., Velazquez, H., Madrigal, A.S., Agogino, A.M.: User-centered design roadmapping: anchoring roadmapping in customer value before technology selection. IEEE Trans. Eng. Manage. **69**(1), 109–126 (2020)

36. Xu, X., Jia, Q., Tayyab, S.M.U.: The exploration of customization in augmented reality from the affordance lens: a three-stage hybrid approach. Technol. Forecast. Soc. Chang. **194**, 122729 (2023)
37. Sleem, A., Mostafa, N., Elhenawy, I.: Neutrosophic CRITIC MCDM methodology for ranking factors and needs of customers in product's target demographic in virtual reality metaverse. Neutrosophic Syst. Appl. **2**, 55–65 (2023)
38. Beheshti, M., ZareRavasan, A., Mahdiraji, H.A., Jafari-Sadeghi, V., Sakka, G.: An overview of the consumer-centric disruptive technology research: insights from topic modelling and literature review. J. Consum. Behav. (2023)
39. Zhang, J.: A systematic review of the use of augmented reality (AR) and virtual reality (VR) in online retailing (2020)
40. Jain, S., Werth, D.: Current state of mixed reality technology for digital retail: a literature review. In: HCI in Business, Government and Organizations. eCommerce and Consumer Behavior: 6th International Conference, HCIBGO 2019, Held as Part of the 21st HCI Inter national Conference, HCII 2019, Orlando, FL, USA, July 26–31, 2019, Proceedings, Part I 21, pp. 22–37. Springer International Publishing (2019)
41. Egger, R., Neuburger, L.: Augmented, virtual, and mixed reality in tourism. In: Handbook of e-Tourism, pp 1–25 (2020)
42. Kumar, H., Tuli, N., Singh, R.K., Arya, V., Srivastava, R.: Exploring the role of augmented reality as a new brand advocate. J. Consum. Behav. (2023)
43. Kazmi, S.H.A., Ahmed, R.R., Soomro, K.A., Hashem, E.A.R., Akhtar, H., Parmar, V.: Role of augmented reality in changing consumer behavior and decision making: case of Pakistan. Sustainability **13**, 14064 (2021)
44. Thomas, T., Mathew, R., KG, A.: Usefulness of augmented reality on product selection: an experimental study. SCMS J. Indian Manag. **19**(2) (2022)
45. Krishnadas, R.: Understanding customer engagement and purchase behavior in automobiles: the role of digital technology. In: Handbook of Research on Technology Applications for Effective Customer Engagement, pp. 1–13. IGI Global (2021)
46. Rindfleisch, A., Im, S.: 13. Enhancing customer centricity via 3D printing. In: Handbook on Customer Centricity: Strategies for Building a Customer-Centric Organization, p. 317 (2019)
47. Dehghani, M., Lee, S.H.M., Mashatan, A.: Touching holograms with windows mixed reality: renovating the consumer retailing services. Technol. Soc. **63**, 101394 (2020)
48. Aithal, P.S.: How to create business value through technological innovations using ICCT underlying technologies. Int. J. Appl. Eng. Manag. Lett. (IJAEML) **7**(2), 232–292 (2023)
49. Han, D.I.D., Weber, J., Bastiaansen, M., Mitas, O., Lub, X.: Virtual and augmented reality technologies to enhance the visitor experience in cultural tourism. In: Augmented Reality and Virtual Reality: The Power of AR and VR for Business, pp. 113–128 (2019)
50. Ozkok, O., Singh, J., Lim, K., Bell, S.J.: 4. Service innovation from the frontlines in customer-centric organizations. In: Handbook on Customer Centricity: Strategies for Building a Customer-Centric Organization, p. 79 (2019).
51. JO, J.: A Study on the effect of marketing using virtual mirror technology on customer satisfaction. Doctoral Dissertation, KDI School (2021)
52. Gauri, D.K., Jindal, R.P., Ratchford, B., Fox, E., Bhatnagar, A., Pandey, A., Howerton, E.: Evolution of retail formats: past, present, and future. J. Retail. **97**(1), 42–61 (2021)
53. Abou Elmaaty, N.H.H., Ibrahim, H.E.A.: Integrating artificial intelligence and cloud computing in ecommerce operational and customer-centric advancements. AI, IoT Fourth Ind. Revolut. Rev. **13**(9), 18–28 (2023)
54. Guo, D., Ling, S., Li, H., Ao, D., Zhang, T., Rong, Y., Huang, G.Q.: A framework for personalized production based on digital twin, blockchain and additive manufacturing in the context of Industry 4.0. In: 2020 IEEE 16th International Conference on Automation Science and Engineering (CASE), pp. 1181–1186. IEEE (2020)
55. Asmar, L., Grigoryan, K., Low, C.Y., Roeltgen, D., Dumitrescu, R.: Structuring framework for early validation of product ideas. Int. J. Integr. Eng. **13**(2), 229–240 (2021)

56. Rajagopal, R.: Impact of retailing technology during business shutdown. Mark. Intell. Plan.Intell. Plan. **40**(4), 441–459 (2022)
57. Susiang, M.I.N., Suryaningrum, D.A., Masliardi, A., Setiawan, E., Abdillah, F.: Enhancing customer experience through effective marketing strategies: the context of online shopping. SEIKO: J. Manag. Bus. **6**(2), 437–447 (2023)
58. Jain, S.: Using Optical See-through Mixed Reality for Enhanced Shopping Experience in Omnichannel Retail/Author Shubham Jain (2022)
59. N'Goala, G., Pez-Pérard, V., Prim-Allaz, I. (eds.): Augmented Customer Strategy: CRM in the Digital Age. Wiley (2019)
60. Han, D.I.D., Weber, J., Bastiaansen, M., Mitas, O., Lub, X.: Blowing your mind: a conceptual framework of augmented reality and virtual reality enhanced cultural visitor experiences using EEG experience measures. Int. J. Technol. Mark. **14**(1), 47–68 (2020)
61. Chatmi, A., Elasri, K., Ponsignon, F.: Assessing and improving co-creation in services: the customer-centric matrix. Int. J. Qual. Serv. Sci. **15**(1), 97–114 (2023)
62. Bakator, M., Vukoja, M., Manestar, D.: Achieving competitiveness with marketing 5.0 in new business conditions. UTMS J. Econ. **14**(1), 63–73 (2023)
63. Wang, T.: Research on the impact of E-commerce on offline retail industry. Front. Bus., Econ. Manag. **10**(1), 169–173 (2023)
64. Rialti, R., Zollo, L.: Marketing 4.0 for SMEs in the digital era: a customer-centric approach. In: Digital transformation of SME marketing strategies: innovating for the 4.0 era, pp. 81–131. Springer Nature Switzerland, Cham (2023)
65. Nguyen, E., Nguyen, A.: AR/VR applications in fashion retailing: an exploratory study on the effectiveness of virtual try-on technology along the customer journey (2021)
66. Bharwani, S., Mathews, D.: Techno-business strategies for enhancing guest experience in luxury hotels: a managerial perspective. Worldw. Hosp. Tour. Themes **13**(2), 168–185 (2021)
67. Harrisson-Boudreau, J.P., Bellemare, J., Bacher, N., Bartosiak, M.: Adoption potentials of metaverse omnichannel retailing and its impact on mass customization approaches. In: Proceedings of the Changeable, Agile, Reconfigurable and Virtual Production Conference and the World Mass Customization & Personalization Conference, pp. 110–119. Springer International Publishing, Cham (2023)
68. Rajagopal: Epilogue: the extent of agility. In: Agile Marketing Strategies: New Approaches to Engaging Consumer Behavior, pp. 227–245. Springer International Publishing, Cham (2022)
69. Zhou, L., Xue, F.: Show products or show people: an eye-tracking study of visual branding strategy on Instagram. J. Res. Interact. Mark. **15**(4), 729–749 (2021)
70. Chu, C.H., Cheng, C.H., Wu, H.S., Kuo, C.C.: A cloud service framework for virtual try-on of footwear in augmented reality. J. Comput. Inf. Sci. Eng.Comput. Inf. Sci. Eng. **19**(2), 021002 (2019)
71. Suopajärvi, T.: Developing cooperation and customer-centric operations in the TE office aviation team (2021)
72. Hajarian, M., Camilleri, M.A., Díaz, P., Aedo, I.: A taxonomy of online marketing methods. In: Strategic Corporate Communication in the Digital Age, pp. 235–250. Emerald Publishing Limited (2021)
73. Soltanifar, M., Smailhodžić, E.: Developing a digital entrepreneurial mindset for data-driven, cloud-enabled, and platform-centric business activities: practical implications and the impact on society. Digit. Entrep. **3** (2021)
74. ZXhang, Y.X., Haxo, Y.M., Mat, Y.X.: Best buy stock: an examination of its resilience, adaptability, and investment potential. AC Invest. Res. J. **220**(44) (2023)
75. Glebova, E., Book, R., Su, Y., Perić, M., Heller, J.: Sports venue digital twin technology from a spectator virtual visiting perspective. Front. Sport. Act. Living **5** (2023)
76. Liu, Z.: Unlocking consumer choices in the digital economy: exploring factors influencing online and offline purchases in the emerging pet food market. J. Knowl. Econ. 1–26 (2023)
77. Zheng, P., Lin, Y., Chen, C.H., Xu, X.: Smart, connected open architecture product: an IT-driven co-creation paradigm with lifecycle personalization concerns. Int. J. Prod. Res. **57**(8), 2571–2584 (2019)

78. Kong, X.T., Luo, H., Huang, G.Q., Yang, X.: Industrial wearable system: the human-centric empowering technology in industry 4.0. J. Intell. Manuf.Intell. Manuf. **30**, 2853–2869 (2019)
79. Widita, A., Rachmahani, H., Agustina, I.A., Husna, N.: The use of augmented reality in café's interior to enhance customer experience. In: IOP Conference Series: Earth and Environmental Science, vol. 794, no. 1, p. 012192. IOP Publishing (2021)
80. Farine, P.G.: Digital presence in physical shopping: from a "benefit-oriented approach" to successful customer engagement. In: Multisensory in Stationary Retail: Principles and Practice of Customer-Centered Store Design, pp. 183–200. Springer Fachmedien Wiesbaden, Wiesbaden (2023)
81. Subawa, N.S., Widhiasthini, N.W., Astawa, I.P., Dwiatmadja, C., Permatasari, N.P.I.: The practices of virtual reality marketing in the tourism sector, a case study of Bali Indonesia. Curr. Issues Tour. **24**(23), 3284–3295 (2021)
82. Albers, A., Reinemann, J., Fahl, J., Hirschter, T.: Augmented reality for product validation: supporting the configuration of AR-based validation environments. In: Virtual, Augmented and Mixed Reality. Applications and Case Studies: 11th International Conference, VAMR 2019, Held as Part of the 21st HCI International Conference, HCII 2019, Orlando, FL, USA, July 26–31, 2019, Proceedings, Part II 21, pp. 429–448. Springer International Publishing (2019)
83. Pereira, A.M., Moura, J.A.B., Costa, E.D.B., Vieira, T., Landim, A.R., Bazaki, E., Wanick, V.: Customer models for artificial intelligence-based decision support in fashion online retail supply chains. Decis. Support. Syst.. Support Syst. **158**, 113795 (2022)
84. Rane, N.: Metaverse marketing strategies: enhancing customer experience and analysing consumer behaviour through leading-edge Metaverse technologies, platforms, and models. Available at SSRN (2023)
85. Fischer, H., Seidenstricker, S., Berger, T., Holopainen, T.: Digital sales in B2B: status and application. In: Advances in Creativity, Innovation, Entrepreneurship and Communication of Design: Proceedings of the AHFE 2021 Virtual Conferences on Creativity, Innovation and Entrepreneurship, and Human Factors in Communication of Design, July 25–29, 2021, USA, pp. 369–375. Springer International Publishing (2021)
86. Menon, A., Bhagat, S., Iqbal, D.: The impact of augmented reality in fashion retail stores in India: opportunities and challenges. IOSR J. Bus. Manag. (IOSR-JBM) **22**(7), 61–67 (2020)
87. Rajagopal.: The agile mind-set. In: Agile Marketing Strategies: New Approaches to Engaging Consumer Behavior, pp. 189–224. Springer International Publishing, Cham (2022)
88. Oosthuizen, K., Botha, E., Robertson, J., Montecchi, M.: Artificial intelligence in retail: the AI-enabled value chain. Australas. Mark. J.. Mark. J. **29**(3), 264–273 (2021)
89. Shields, K.: Designing products, services, and processes with customers in mind. Customer Centric Strategy (2021)
90. Kotler, P., Kartajaya, H., Setiawan, I.: Marketing 5.0: technology for humanity. Wiley (2021)
91. Ziaie, A., ShamiZanjani, M., Manian, A.: Systematic review of digital value propositions in the retail sector: new approach for digital experience study. Electron. Commer. Res. Appl.Commer. Res. Appl. **47**, 101053 (2021)
92. Magnelli, A., Verreschi, G., Ventrella, M.: Bringing knowledge and emotion to the industrial field: ETT's AR/VR solutions. In: Augmented Reality and Virtual Reality: New Trends in Immersive Technology, pp. 251–266. Springer International Publishing, Cham (2021)
93. Kautish, P., Khare, A.: Investigating the moderating role of AI-enabled services on flow and awe experience. Int. J. Inf. Manage. **66**, 102519 (2022)
94. Chaturvedi, R., Verma, S.: Opportunities and challenges of AI-driven customer service. In: Artificial Intelligence in Customer Service: The Next Frontier for Personalized Engagement, pp. 33–71 (2023)
95. Golf-Papez, M., Heller, J., Hilken, T., Chylinski, M., de Ruyter, K., Keeling, D.I., Mahr, D.: Embracing falsity through the metaverse: the case of synthetic customer experiences. Bus. Horiz.Horiz. **65**(6), 739–749 (2022)
96. Ylilehto, M., Komulainen, H., Ulkuniemi, P.: The critical factors shaping customer shopping experiences with innovative technologies. Balt. J. Manag.. J. Manag. **16**(5), 661–680 (2021)

97. Hussain, Z., Shaikh, M.Z.: Role of technology in marketing: AI, AR and VR. EKOMA: J. Ekon., Manaj., Akunt. **2**(2), 465–471 (2023)
98. Lee, C.K.M., Lui, L., Tsang, Y.P.: Formulation and prioritization of sustainable new product design in smart glasses development. Sustainability **13**(18), 10323 (2021)
99. Naimi-Sadigh, A., Asgari, T., Rabiei, M.: Digital transformation in the value chain disruption of banking services. J. Knowl. Econ.Knowl. Econ. **13**(2), 1212–1242 (2022)
100. Christoforou, T., Melanthiou, Y.: The practicable aspect of the omni-channel retailing strategy and its impact on customer loyalty. In: The Synergy of Business Theory and Practice: Advancing the Practical Application of Scholarly Research, pp. 239–260 (2019)
101. Iglesias-Pradas, S., Acquila-Natale, E.: The future of E-Commerce: overview and prospects of multichannel and omnichannel retail. J. Theor. Appl. Electron. Commer. Res.Theor. Appl. Electron. Commer. Res. **18**(1), 656–667 (2023)
102. Rane, N., Choudhary, S., Rane, J.: Hyper-personalization for enhancing customer loyalty and satisfaction in Customer Relationship Management (CRM) systems. Available at SSRN 4641044 (2023)
103. Utami Tjhin, V., Eka Riantini, R.: A study for the implementation of banking 4.0 in Indonesia. In: Proceedings of the 2021 9th International Conference on Information Technology: IoT and Smart City, pp. 129–134, (2021)
104. Bhavana, S.U.: Digital transformation–embracing digital change. In: Best Practices of Research and Innovation in Steam Higher Education, p. 25 (2023)
105. Chaturvedi, R., Verma, S.: Artificial intelligence-driven customer experience: overcoming the challenges. Calif. Manag. Rev. Insights (2022)
106. Vrigkas, M., Lappas, G., Kleftodimos, A., Triantafillidou, A.: Augmented reality for wine industry: past, present, and future. In: SHS Web of Conferences, vol. 102, p. 04006. EDP Sciences (2021)
107. Dang, T.D., Nguyen, M.T.: Systematic review and research agenda for the tourism and hospitality sector: co-creation of customer value in the digital age. Futur. Bus. J. **9**(1), 94 (2023)
108. Sureshbabu, J., Mahalakshmi, S., Priya, R.: User persona mapping technique for human interaction with augmented and virtual reality environments. In: 2023 International Conference on Self Sustainable Artificial Intelligence Systems (ICSSAS), pp. 921–926. IEEE (2023)
109. Aaker, D.A., Moorman, C.: Strategic Market Management. Wiley (2023)
110. Gruia, L.A., Bibu, N., Roja, A., Danaiaţă, D., Năstase, M.: Digital transformation of businesses in times of global crisis. In: Navigating Through the Crisis–A special Issue on the Covid 19 Crises: The 2020 Annual Griffiths School of Management and IT Conference (GSMAC), vol. 1, no. 11, pp. 43–62. Springer International Publishing (2022)
111. Shah, D., Murthi, B.P.S.: Marketing in a data-driven digital world: implications for the role and scope of marketing. J. Bus. Res. **125**, 772–779 (2021)
112. Donmezer, S., Demircioglu, P., Bogrekci, I., Bas, G., Durakbasa, M.N.: Revolutionizing the garment industry 5.0: embracing closed-loop design, E-Libraries, and digital twins. Sustainability **15**(22), 15839 (2023)
113. Singh, A., Mishra, S., Jain, S., Dogra, S., Awasthi, A., Roy, N.R., Sodhi, K.: 8 Exploring practical use-cases of augmented reality using photogrammetry and other 3D reconstruction tools in the metaverse. In: Augmented and Virtual Reality in Industry 5.0,vol. 2, p. 163 (2023)
114. Bhattacharya, S., Chatterjee, A.: Digital project driven supply chains: a new paradigm. Supply Chain. Manag.: Int. J. **27**(2), 283–294 (2022)
115. Kümmerle, R., Wierzbitzki, M.: Visualising customer-centric digital investment performance reports. J. Digit. Bank. **3**(4), 346–360 (2019)
116. Spajić, J., Mitrović, K., Lalić, D., Milić, B., Bošković, D.: Personalized brand experience in metaverse. In: 10th International Conference on Mass Customization and Personalization–Community of Europe (MCP-CE 2022), pp. 21–23 (2022)
117. Sharma, D., Yadav, V., Singh, D.: Advancements and implications of product service systems in the automobile industry: a comprehensive review. Eng. Proc. **59**(1), 98 (2023)

118. Rane, N., Choudhary, S., Rane, J.: Metaverse for enhancing customer loyalty: effective strategies to improve customer relationship, service, engagement, satisfaction, and experience. Service, Engagement, Satisfaction, and Experience (1 Nov 2023) (2023)
119. Scurati, G.W., Bertoni, A., Bertoni, M.: Boosting value co-creation in the road construction industry through virtual prototyping. In: Working Conference on Virtual Enterprises, pp. 613–622. Springer International Publishing, Cham (2022)
120. Tang, Y.M., Lau, Y.Y., Ho, U.L.: Empowering digital marketing with interactive virtual reality (IVR) in interior design: effects on customer satisfaction and behaviour intention. J. Theor. Appl. Electron. Commer. Res.Theor. Appl. Electron. Commer. Res. **18**(2), 889–907 (2023)
121. Vidili, I.: Journey to Centricity: A Customer-Centric Framework for the Era of Stakeholder Capitalism. The Smarter Crew (2021)
122. Moorthy, J., Parvatiyar, A.: Co-creating aftermarket value in the digital era: managing transformative customer relationships through stakeholder engagement. In: Customer Centric Support Services in the Digital Age: The Next Frontier of Competitive Advantage, pp. 155–198. Springer International Publishing, Cham (2023)
123. Li, L.H., Cheung, K.S., Tse, W.S.: Understanding the shoppers' perception in retail shopping malls: a self-determination theory perspective. J. Strateg. Mark.Strateg. Mark. **31**(1), 58–73 (2023)
124. Pappas, A., Fumagalli, E., Rouziou, M., Bolander, W.: More than machines: the role of the future retail salesperson in enhancing the customer experience. J. Retail. **99**(4), 518–531 (2023)
125. Barata, J., Cardoso, J.C., Cunha, P.R.: Mass customization and mass personalization meet at the crossroads of industry 4.0: a case of augmented digital engineering. Syst. Eng. (2023)
126. Neumann, N., Tucker, C.E., Kaplan, L., Mislove, A., Sapiezynski, P.: Data deserts and black boxes: the impact of socio-economic status on consumer profiling. Manag. Sci. (2024)
127. van Ooijen, I., Segijn, C.M., Opree, S.J.: Privacy cynicism and its role in privacy decision-making. Commun. Res.. Res. **51**(2), 146–177 (2024)
128. Johnson, J.P., Jungbauer, T., Preuss, M.: Online advertising, data sharing, and consumer control. Manag. Sci. (2024)

The Pinnacle of Influence: Unveiling the Evolution, Global Dynamics, and Multidisciplinary Tapestry of Social Media Influencer Marketing (2016–2024)

Anisha Arora and Prashant Kumar Siddhey

Abstract The purpose of this research paper is to deliver an in-depth overview of the state of scientific research on social media influencers as of right now. The process comprised gathering pertinent documents from the Scopus databases, which produced 41 entries when the terms "celebrity endorsement", "social media", "influencer", and "marketing" were searched. Using bibliographic tools like Vosviewer, Rstudio, and Scopus analyses, these documents were analyzed. The results showed a noticeable increase in research activity in 2024. The analysis underlines the remarkable interdisciplinary connections across business management, economics and social science fields and exposes the United States' considerable influence, supplying 20 documents in 2023–24. Co-occurrences of keywords were found using bibliometric analysis, which suggests that influence marketing, social media, and celebrity endorsement are becoming more and more important. The research study concludes by shedding light on how the field of social media influencer research is changing and highlighting its multidisciplinary character, global viewpoints, and the growing significance of important theme areas.

Keywords Influencer · Social media · Bibliometric analysis · Celebrity endorsement · Marketing

1 Introduction

Since the middle of the sixteenth century, the word "influencer" has been widely used in english speech. In the past fifteen years, the term "influencer" has gained considerable recognition in the marketing profession as it has come to be connected with social media. The phrase has evolved into a catch-all in recent years for a particular kind of work at the centre of a lucrative but rapidly expanding sector [1].

A. Arora (✉) · P. K. Siddhey
University School of Business, Chandigarh University, Chandigarh, India
e-mail: 22YCM10019@cuchd.in

S. Dutta et al. (eds.), *Advances in Data Analytics for Influencer Marketing: An Interdisciplinary Approach*, Information Systems Engineering and Management 9,
https://doi.org/10.1007/978-3-031-65727-6_22

Brands and consumers have been interacting and communicating via social networking sites in recent years. Influencer marketing is one of the most popular techniques in social media marketing [2]. Social media users who have amassed a big following and gotten notable attention from their peers are known as influencers [3]. Influencer marketing is the process by which companies work with social media influencers to promote products and brands on their platforms [4] with trillions of users worldwide, social media has quickly emerged to be among the technologies that define our era. Influencers actively advertise a product or service by posting content on their social media pages. Additionally, viewers can connect with influencers and branded posts in the comment box by writing comments on the product, which boosts viewers' engagement with the brand. Additionally, these remarks help businesses understand their clients better and produce more goods [5].

As of March 31, 2019, facebook, for instance, reported having 1.56 billion daily active users and 2.38 billion monthly active users (facebook, 2019). According to emarketer (2018), there will be 3.29 billion social media users worldwide in 2022, accounting for 42.3% of the world's population. Up to 40–50% of customers in various product categories look up their reviews on social networking handles. This explains why more businesses have embraced social media marketing to engage customers over the past ten years [6] as a result, the idea of influencer marketing was born [7], influencer marketing refers to marketing initiatives that are designed to locate and establish a connection with people who have the ability to sway future consumers [8] in their influencer marketing literature, researchers often employ like, comment, and share counts as performance outcomes. Researchers often employ likes, comments, and share tallies as indicators of performance in the influencer marketing literature [8]. Volume-based indicators may only partially reflect the success of marketing campaigns, according to research [9] greater research on content-based metrics and stressed that marketers should monitor not just like and comment counts but also comment content [10]. Reference [11] pointed out, fans might focus primarily on the influencer and neglect the product advertised on sponsored postings. Therefore, it's critical to identify comment subjects. When a product is the main focus of a comment (a product-centred comment), it is clear that viewers are paying attention to the product. A renowned marketing tactic is the notion of leveraging well-known public figures or celebrities with a high social worth to influence others in business or consumer markets [12]. Social media's pervasiveness has greatly expanded this strategy's appeal and accessibility. The pop singer, for instance, interacts with her 144 million+ Instagram followers on a daily basis. The worth of a single photo that she released in 2018 was estimated to be $3.4 million [13] but, she comes at a steep cost: according to [14] one post that selena endorses for a brand can cost up to $800,000. The way social media seems to be influencing culture itself is the clearest sign of an omni-social phenomenon. These days, youtube influencers have their own tv series and product lines, making them cultural superstars [10, 15].

Customers' attitudes, opinions, preferences, choices, and decisions are influenced by social media influencers, who are getting more and more well-known. Social media influencers are just regular folks who built their online presence from the ground up.

In this sense, traditional celebrities and public figures are different from social media influencers because the former rely on their existing careers—like acting, singing, or politics—to become well-known and influential [16].

1.1 Historical Context

With the advent of television and print media, celebrities played a pivotal role in influencing consumer choices and establishing the advantageous association between firm marketing and popularity. Influencer marketing dates back to the early days of advertising, when brands relied on celebrities' popularity and dependability as brand advocates. These celebrity endorsements stood out due of their upward strategy, which depended on the star power of each individual to influence consumer perceptions.

These celebrity endorsements used a top-down strategy that relied on a certain person's star power to sway public opinion. Celebrities are crucial in influencing customer decisions because of the mutually beneficial relationship between brand marketing and popularity that was made possible by the growth of print and television media.

1.2 Maturity

As influencer marketing gained popularity, the company went through a maturing phase. Influencers and corporations alike must adjust to the shifting demands of consumers and the heightened scrutiny surrounding authenticity. At this point, influencer tactics were improved, going beyond simple product placement to produce authentic and interesting content. Transparent rules and norms were established to ensure transparency as influencer-brand collaborations grew.

1.3 The Current Circumstances

These days, influencer marketing is a well-established component of modern marketing strategy. These days, social media platforms are the primary venues for influencer-brand partnerships because of their wide user bases and sizable viewership. The world of influencers is vast, vibrant, and ever-changing, ranging from micro-influencers targeting specific groups to mega-influencers with enormous followings.

Influencer marketing has grown in tandem with the rise of social media and its effects on social dynamics and communication. Social media platforms have made it possible for people to express their opinions, which has improved inclusivity and diversity in public representation. Influencers are people from a variety of backgrounds and cultures who offer a variety of viewpoints that cast doubt on conventional wisdom and promote a sense of belonging.

Influencer marketing has also made it easier for brands to communicate authentically. Increased transparency has led to more sincere interactions between brands and customers, and audiences respond more favourably to sincere influencers who share their experiences. Influencer campaigns have been essential in advancing social causes, bringing up significant problems, and encouraging constructive social change.

The way influencer marketing has changed over time is a good way to understand how media, technology, and consumer behaviour are always evolving. This journey demonstrates the flexibility of marketing methods and the revolutionary impact social media has had on cultural norms and values. It began with celebrity endorsements and continues today with cooperation with digital influencers. The process's subsequent phases the historical context, the shift from celebrity endorsement to influencers, the maturation phase, and the investigation of the current landscape explain the connection between social media and influencer marketing as well as the advantageous effects this evolution has had on society.

Rq1. Whose papers are the most important and what is the publication trend for social media influencer marketing research?

Rq2. Where can I get publications of research on influencer marketing on social media?

Rq3. How was the research on social media influencer marketing carried out?

Rq4. What are the main topics covered by research on social media influencer marketing?

Rq5. What direction should influencer marketing on social media go in the future?

2 Methodology

2.1 Sample

The Scopus database was used in this study's literature research to obtain pertinent information. By employing targeted search terms such as "celebrity endorsement," "social media," "influencer," and "marketing," we carried out topical retrievals for all years. The category of literature included all kinds of written materials. Following the application of our selection criteria, 41 documents in total satisfied the requirements to be included in our study.

Articles were the most common type of document among the documents that were recovered; they accounted for 61.0% of all publications. Moreover, the majority of the themes covered in the documents (42.6%) were related to company management. We were able to thoroughly investigate the relationship between influencers, social

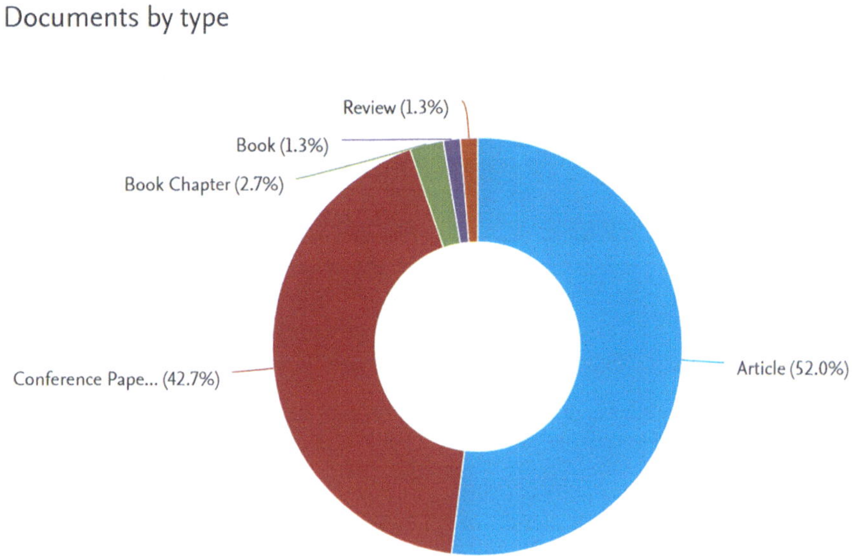

Fig. 1 Document by type (*Source* Scopus analyze)

media, marketing, and celebrity endorsement by using this methodology and a wide range of academic sources (Fig. 1).

Articles make up a sizeable component of the corpus of papers that were analyzed—52 percent of the total. These papers, which provide important insights and analyses on the topic, most likely add to the core literature study. Furthermore, the study includes a significant amount of conference papers (42.7%) of the total document categories. These articles demonstrate the author's participation in current discussions at academic conferences and imply a dedication to remaining up to date with the most recent advancements in the field. Additionally, the presence of books (1.3%) and book chapters (2.7%) shows the scope and depth of the investigation, with an examination of more comprehensive, coherent pieces that advance knowledge about the selected subject in its whole. This careful selection of documents indicates a comprehensive analysis that draws on a wide range of academic sources to support the research paper's academic relevance and rigor.

2.2 *Analysis*

All materials were retrieved in CSV format from the Scopus website in February of 2024. Initially, data taken straight from the Scopus platform itself was used for a descriptive study. The present investigation primarily aimed to monitor the progression of publications, detect trends in authorship, and ascertain the subjects and places of origin of the respective documents.

Fig. 2 Evolution of the number of documents published in scopus on "celebrity endorsement" and "social media" and "influencer" and "marketing" (*Source* R studio)

Then, to give a thorough picture of the field and its development, bibliometric analysis was carried out. Out of all the bibliometric analysis tools that were available, VOSviewer was a highly recommended option. VOSviewer is a free application written by Eck and Waltman and developed by the Centre for Science and Technology Studies at Leiden University in the Netherlands. When performing co-occurrence and co-citation analysis, it is especially helpful.

To provide insights into the collaborative relationships among authors and the linkages between cited works, VOSviewer was utilized in this study to build co-authorship and co-citation networks. The study aims to contribute to the field's progress of knowledge by offering a comprehensive understanding of the scholarly landscape around celebrity endorsement, social media, influencers, and marketing, by utilizing VOSviewer's capabilities [17] (Fig. 2 and Table 1).

3 Results

3.1 Descriptive

The 41 papers in the collection, which was gathered from 36 distinct sources, span the years 2016 to 2024. The data represents a snapshot of current research with a consistent yearly growth rate of 0%. With 2319 references overall and an average of 38.98 citations per document, publications receive significant attention. The average document age is 3.1 years. The documents include 140 author's keywords and 65 keywords plus identifiers. There are one hundred writers total, including four works with a single author. With 2.59 co-authors on average per document and 21.95% of the articles showing foreign co-authorship, collaborative efforts are clearly visible. The dataset provides a wide viewpoint on research themes with its assortment of

Table 1 Overview and main information of documents (*Source* R Studio edited by authors)

Description	Results
Main information about data	
Timespan	2016:2024
Sources (journals, books, etc.)	36
Documents	41
Annual growth rate %	0
Document average age	3.1
Average citations per doc	38.98
References	2319
Document contents	
Keywords plus (id)	65
Author's keywords (de)	140
Authors	
Authors	100
Authors of single-authored docs	4
Authors collaboration	
Single-authored docs	4
Co-authors per doc	2.59
International co-authorships %	21.95
Document types	
Article	25
Book chapter	9
Conference paper	6
Review	1

document types, which includes 25 articles, 9 book chapters, 6 conference papers, and 1 review. All things considered, this analysis reveals a vibrant research environment with noteworthy impact and collaboration (Table 2).

This analysis of research papers provides a comparative picture of the influence and output of research in various nations. With 585 total citations and a remarkable average of 585.00 citations per paper, the Netherlands dominates the field and has a major intellectual effect. With 334 total citations and an average of 167.00 citations per article, qatar comes in second, demonstrating a significant influence in relation to the volume of its research. Reputable contributions are shown by Korea, Georgia, and Hong Kong, each of which emphasizes a consistent impact each piece with a citation rate that matches their total number of citations. Interestingly, the average article citations for the UK, Austria, and Germany are notable, indicating a focused impact, despite their lower total citation counts. Even with a high overall number of citations, the USA has a lower average, which would point to a more diverse research environment. Portugal exhibits a noteworthy impact within its research area, while

Table 2 Average article citations (*Source* R studio edited by authors)

Country	Tc	Average article citations
Netherlands	585	585.00
Qatar	334	167.00
Hong Kong	186	186.00
Georgia	95	95.00
Korea	75	75.00
United Kingdom	60	15.00
Austria	57	57.00
Germany	56	14.00
Usa	44	14.70
Portugal	19	9.50

having a smaller research output, as evidenced by its consistent average of 9.50 citations per publication. This analysis offers insightful information on the impact and global distribution of research contributions from different nations (Table 3).

The frequency of collaborative efforts between pairs of nations indicates the emphasis of this research study investigation, which is international collaboration between countries. Finland working with Austria, Germany with France, and India with Malaysia and the Netherlands are a few notable examples. The Netherlands is part of a diversified network that collaborates on research projects with Finland and Austria. Portugal is in research collaborations with Italy, and the UK is in research partnerships with Germany, Austria, and Finland. The one-time frequency of these collaborations suggests a global interchange of knowledge and experience across a range of study disciplines. The results highlight how linked and cooperative research projects are, highlighting the value of international collaboration in expanding scientific understanding and developing a worldwide research community (Fig. 3).

Table 3 Countries' collaboration world map (*Source* R studio edited by authors)

From	To	Frequency
Finland	Austria	1
Germany	France	1
India	Malaysia	1
India	Netherlands	1
Netherlands	Austria	1
Netherlands	Finland	1
Portugal	Italy	1
United Kingdom	Austria	1
United Kingdom	Finland	1
United Kingdom	Germany	1

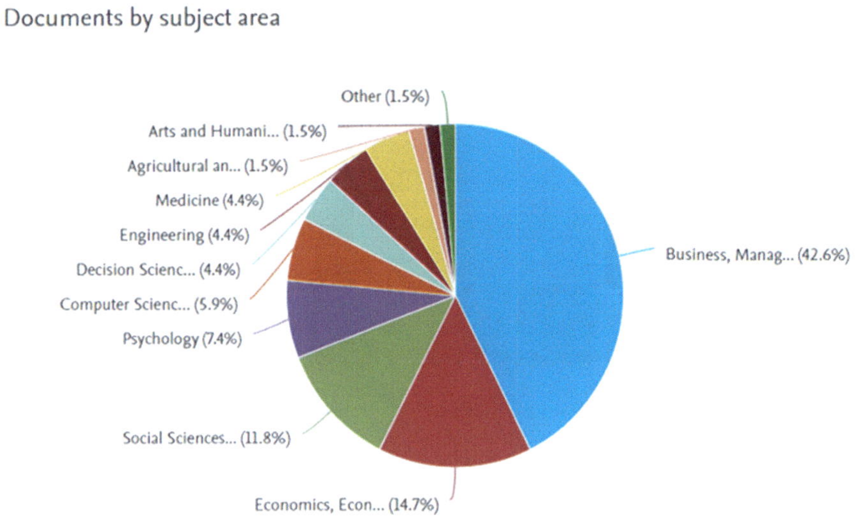

Fig. 3 Subject area of retrieved documents (*Source* Scopus analyze)

A wide range of research interests can be seen in the distribution of documents by subject field, with business, management, and economics accounting for a sizable 59.7% of the total. This indicates a strong focus on comprehending and developing theories pertaining to economic concepts, organizational dynamics, and decision-making procedures. The 5.9% percentage for computer science shows a notable interest in computational approaches and technological breakthroughs. Together, psychology and social sciences make up 19.2%, demonstrating a dedication to understanding society dynamics and human behavior. With 4.4% each, medicine and engineering point to the need for further research into scientific and technological developments that may have implications for healthcare. The 1.5% "other" category represents a more focused but yet pertinent investigation of other topics. The comprehensive approach to research displayed in this distribution highlights the multidisciplinary nature of modern academic inquiry, extending from the humanities to the sciences (Figs. 4 and 5).

3.2 *Keywords and Co-Ocurrences Analysis*

With 184 pieces arranged into 14 clusters, this research paper explores the complex world of influencer marketing in the beauty sector. There are 23 items in cluster 1, all of which deal with different facets of managing and endorsing brands online. The investigation looks at engagement tactics, the complexities of influencer marketing, including celebrity and micro-influencers, and the influence of social media sites like instagram. The research explores the subtleties of user-generated content,

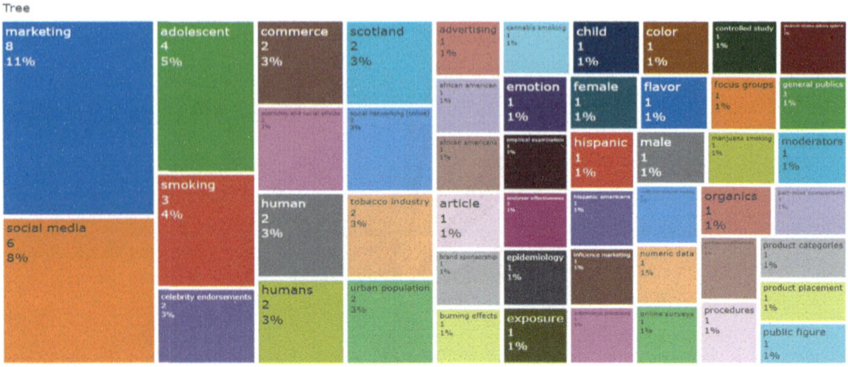

Fig. 4 Tree map of the words (*Source* R studio)

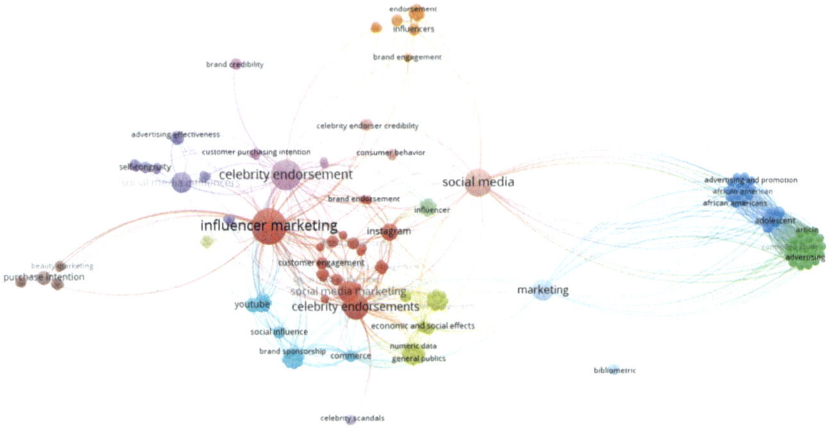

Fig. 5 Occurrence of keywords (*Source* Vos viewer)

luxury branding, and psychological foundations including parasocial interaction and self-congruity theory. The study also examines possible manipulation in influencer marketing and emphasizes the significance of micro-celebrities. By means of content analysis, the article provides significant insights into the dynamic terrain of influencer-driven tactics, illuminating the intricacies of brand interaction, consumer conduct, and the wider social media marketing framework in the beauty sector (Fig. 6 and Table 4).

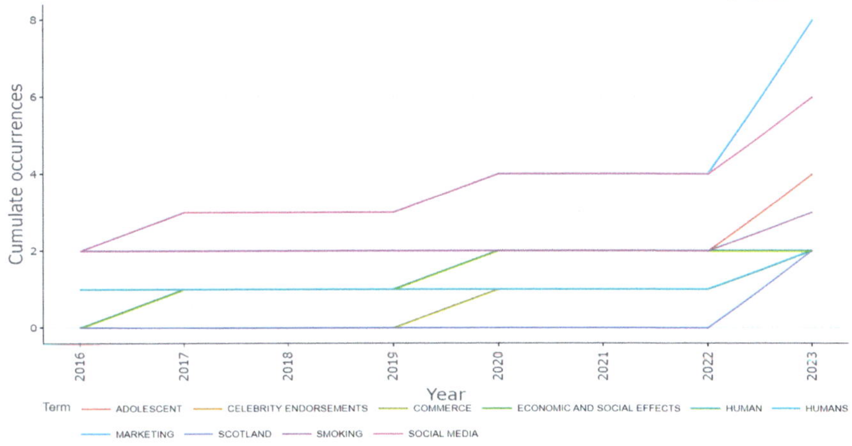

Fig. 6 Cumulate occurrences (*Source* R studio)

Table 4 Authors by number of documents (*Source* R studio edited by authors)

Authors	Articles	Articles fractionalized
Dr	2	0.50
Gräve j-f	2	1.50
Jin sv	2	0.83
Lu qs	2	1.00
Muqaddam a	2	0.83
Seah zy	2	1.00
Al-assaf kt	1	0.25
Al-dmour r	1	0.25
Al-mu'ani l	1	0.25
Al-okaily m	1	0.25

3.3 *Authors and Co-Authorship Analysis*

In this study, the number of articles written by each author and the fractionalized representation are used to evaluate their influence, and a thorough analysis of their contributions within the selected dataset is presented. Lu qs, seah zy, and gräve j-f stand out as the most active authors with two papers each that had fractionalized values of 1.50, 1.00, and 1.00, respectively. These fractionalized numbers could be a sign of the authors' varied research interests or cooperative efforts. With two publications apiece, dr, jin sv, and muqaddam a also show notable contributions, indicating a steady stream of scholarly work. Furthermore, authors with a single publication each, such as al-assaf kt, al-dmour r, al-mu'ani l, and al-okaily m, have made a significant contribution to the field of research. The fractionalized values

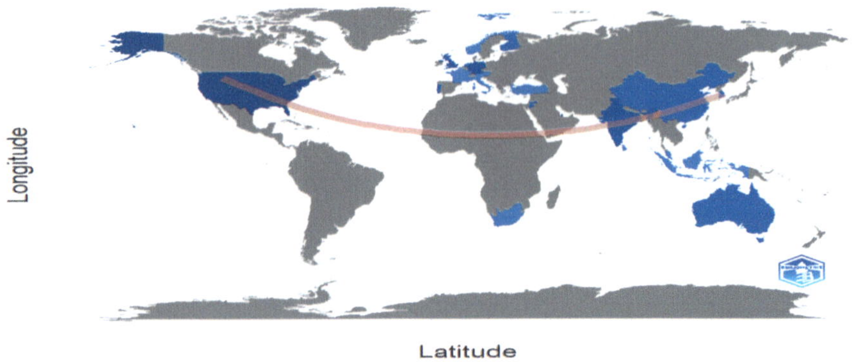

Fig. 7 Countries' collaboration world map (*Source* R studio)

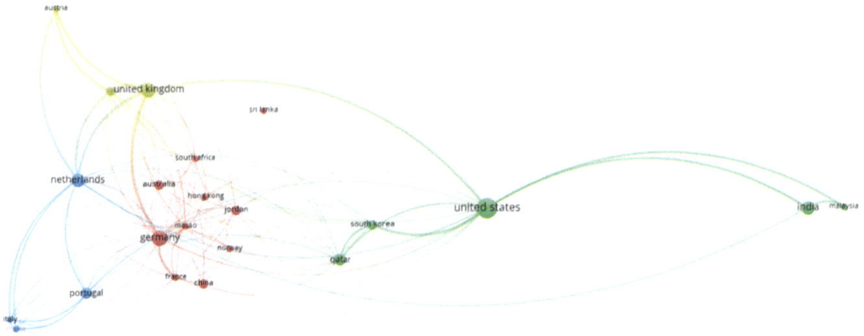

Fig. 8 Bibliography coupling of countries (*Source* Vos viewer)

offer a sophisticated perspective on joint undertakings within the academic context by providing insights into the distribution of authorship duties. A deeper comprehension of the research output and the dynamics of collaboration among the contributing authors is made easier by this study (Figs. 7 and 8).

4 Discussion

In summary, the study's thorough examination of social media influencers offers a wealth of information about the complex realm of digital influence. Its thorough analysis of historical settings, multidisciplinary viewpoints, and global perspectives has shed light on the intricate processes influencing the phenomena of influencer culture.

The study's noteworthy quality is its integration of varied viewpoints from throughout the globe. The study presents a comprehensive picture of the worldwide landscape by taking into account the impact of cultural, social, and economic aspects on influencers' behavior. The study recognizes the complexity of influencer ecosystems and the diverse degrees of influence they exert across different cultural contexts, ranging from micro-influencers serving niche audiences in certain regions to macro-influencers with a worldwide reach.

Furthermore, the historical analysis of the study charts the development of influencer culture from its early stages to its current standing in the digital era. It places recent study findings within broader historical trends by looking at the emergence of social media platforms and the changing dynamics of online involvement. Recognizing the origins of influencer culture, following its evolution, and projecting its future course all depend on an understanding of its trajectory.

The interdisciplinary methodology of the study draws on insights from a variety of academic areas to further expand its analysis. It provides a comprehensive knowledge of the several facets of influencer phenomenon by combining theories and approaches from sociology, psychology, marketing, communication studies, and other fields. The study provides a thorough understanding of the intricate interactions that take place between influencers, audiences, and companies, ranging from investigating the psychological mechanisms that underlie influencer persuasion to examining the financial effects of influencer-brand partnerships.

The study's assessment of potential future research directions points to a number of important areas that deserve more investigation. The moral issues raised by influencer marketing strategies is one such field. Questions about authenticity, transparency, and the relationships between influencers and brands are becoming more important as influencers are used more and more as messengers for commercial messages. Subsequent studies ought to go more deeply into these moral issues, looking into strategies for encouraging increased responsibility, openness, and moral principles in the influencer sector.

The study also emphasizes the significance of looking into how influencer dynamics are affected by newly developing social media platforms. It's critical to comprehend how platforms like Clubhouse, Snapchat, and TikTok affect content development, audience engagement, and brand partnerships as they transform the digital world. Researchers can anticipate changes in influencer culture and modify their methodology in response by keeping an eye on new trends and platforms.

The report also emphasizes the rising importance of artificial intelligence (AI) in influencer collaborations as an area that warrants further investigation. Influencers and companies are using algorithms more and more for audience targeting, performance improvement, and content creation as AI technologies develop. Understanding the changing nature of digital influence and its effects on society requires investigating the ethical, social, and economic ramifications of AI-driven influencer marketing tactics.

To sum up, the thorough analysis of social media influencers provided by the study provides insightful information about a quickly changing issue. Through the integration of worldwide viewpoints, historical background, and an interdisciplinary methodology, it improves our comprehension of influencer dynamics and establishes the foundation for further research projects. In order to navigate this dynamic environment, it will be crucial to keep researching the ethical, sociological, and technological aspects of influencer culture as society struggles with the intricacies of digital influence.

5 Limitations

5.1 *Limitations of the Data Source*

The study only uses data that was taken from the scopus database, which may not be as complete or as covered as it could be. Relevant research papers may be removed based on the indexing policies of the database and the keywords chosen, which could introduce biases into the analysis. Future research can take the web of science or other databases as a base.

5.2 *Publication Bias*

The study only considers published studies, it may overlook unpublished or less widely available research. Publications may have a propensity to present noteworthy or encouraging results, which could distort perceptions of the general status of social media influencer research.

5.3 *Language Bias*

The study's language bias results from its restriction to including only English-language documents. The lack of representation from various cultural and linguistic perspectives resulting from the removal of non-English publications could impair the study's capacity to offer a truly worldwide overview.

5.4 *Imbalance of Subject Area*

The overrepresentation of business, management, and economics in the subject distribution of the examined documents may cause an imbalance in how other pertinent fields are represented. This disparity may make it more difficult for the study to fully reflect the range of affects and influences that social media influencers have across a variety of fields.

5.5 Citation Metrics as the Only Indicator

The study primarily uses citation metrics to gauge its significance and impact. Although citation numbers can be a useful indicator of scholarly prominence, they may not accurately represent the research's practical significance or societal impact. There are a number of reasons why certain significant contributions might not earn a lot of citations.

5.6 Limitation of Bibliometric Analysis Tools

The study makes use of bibliometric analysis tools such as vosviewer and scopus analyses, and the accuracy of these tools depends on the data and algorithms that are accessible. The accuracy of the co-occurrence networks and collaborations that are displayed in the study may be impacted by limitations in the algorithms or the quality of the data.

5.7 Absence of Qualitative Analysis

The study does not include a qualitative examination of the research papers' content, instead relying mostly on quantitative bibliometric analysis. Deeper understanding of the subtleties, techniques, and theoretical frameworks used in influencer marketing research may be obtained through a qualitative study.

Future research in this field should strive to overcome these restrictions and use a wider range of sources and approaches to improve the robustness and application of findings.

6 Implications

6.1 *Conceptual Repercussions:*

6.1.1 Interdisciplinarity

The study highlights the multidisciplinary character of influencer research on social media, suggesting a confluence of business administration, economics, and social sciences. This suggests that theoretical advancements in the future should take a comprehensive strategy that integrates knowledge from many fields.

6.1.2 Global Perspectives

It appears that social media influencer marketing is a phenomenon of international significance based on the notable increase in research efforts in 2024 and the global distribution of publications. To provide a more thorough understanding, theories in this discipline should take global dynamics and cultural variances into consideration.

6.1.3 Evolution of Marketing Strategies

From celebrity endorsements to the current digital influencer era, the research paper's historical framework describes how influencer marketing has evolved. Theoretical frameworks must to adapt to these developments, highlighting the impact of technology and social media on marketing strategies.

6.2 *Consequences for Society*

6.2.1 Inclusive Representation

Diversity in public representation is facilitated by the variety of social media influencers' backgrounds and cultures. This suggests that by presenting a range of viewpoints, influencer marketing might help to question established conventions and foster a sense of belonging.

6.2.2 Influencer-Driven Social Change

The study emphasizes how influencer marketing can help promote social causes and promote positive social change. Influencer marketing is a powerful tool that socially conscious organizations can use to address important issues and connect with consumers more deeply.

6.2.3 Open and Sincere Communication

Openness and sincerity were prioritized during influencer marketing's maturation stage. Sincere influencers are well-received by social media users, suggesting that in order to establish trust, brands should place a high value on sincere communication and open partnerships with influencers with their audience.

6.3 Management Consequences

6.3.1 Strategic Collaboration

International collaborations are important in influencer marketing research, as shown by the analysis of co-authorship and collaboration among nations. Supervisors must to promote intercultural cooperation in order to obtain a variety of viewpoints and insights for creating successful influencer marketing campaigns.

6.3.2 Targeted Marketing Strategies

The organization of papers according to topic areas emphasizes how important business, management, and economics are. These scholarly conversations can provide managers with insights that they can use to customize influencer marketing strategies based on organizational dynamics and economic theories.

6.3.3 Focus on Micro-Influencers

The research's emphasis on micro-celebrities implies that managers ought to think about working together with influencers who might have smaller but more active fan bases. With target consumers, this strategy may result in more meaningful and genuine connections.

7 Conclusions

To sum up, the study makes a substantial contribution to the developing field of influencer research on social media. The results imply that influencer marketing is a dynamic, worldwide phenomena with ramifications for the creation of theories, management judgment, and social transformation. Future study directions are outlined in the collaboration, subject areas, and keyword co-occurrence analysis. Scholars and practitioners alike must be aware of new trends and modify their strategies as the social media and influencer marketing landscapes continue to change.

References

1. Fernández-Prados, J.S., Lozano-Díaz, A., Bernal-Bravo, C., Muyor-Rodríguez, J.: Influencers and social media: state of the art and bibliometric analysis. In: 2021 9th International Conference on Information and Education Technology (ICIET), pp. 456–460 (2021)
2. Wang, Y., Jiang, Y., Geng, B., Yan, Z., Wang, X.: Social networks, network interaction and resilience of B&Bs in rural China. Int. J. Contemp. Hosp. Manag. (2022)
3. Farivar, S., Wang, F.: Effective influencer marketing: a social identity perspective. J. Retail. Consum. Serv. **67** (2022). https://doi.org/10.1016/j.jretconser.2022.103026
4. Farivar, S., Wang, F.: Influencer marketing: current knowledge and research agenda. In: Springer Proceedings in Business and Economics, pp. 201–208 (2021). https://doi.org/10.1007/978-3-030-76520-0_21
5. McCorquodale, M.C.S.: Influence: How Social Media Influencers are Shaping Our Digital Future. Bloomsbury Publishing (UK) (2020). https://books.google.co.in/books?id=mEDAzQEACAAJ
6. Bughin, J.: Brand Success in an Era of Digital Darwinism, pp. 15–19 (2014)
7. Balakrishnan, B.K., Dahnil, M.I., Yi, W.J.: The impact of social media marketing medium toward purchase intention and brand loyalty among generation Y. Procedia - Soc. Behav. Sci. **148**, 177–185 (2014). https://doi.org/10.1016/j.sbspro.2014.07.032
8. Tafesse, W., Wood, B.P.: Followers' engagement with instagram influencers: the role of influencers' content and engagement strategy. J. Retail. Consum. Serv. **58** (2021). https://doi.org/10.1016/j.jretconser.2020.102303
9. Shareef, M.A., Mukerji, B., Dwivedi, Y.K., Rana, N.P., Islam, R.: Social media marketing: comparative effect of advertisement sources. J. Retail. Consum. Serv. **46**, 58–69 (2019)
10. Appel, G., Grewal, L., Hadi, R., Stephen, A.: The future of social media in marketing. J. Acad. Mark. Sci. **48** (2019). https://doi.org/10.1007/s11747-019-00695-1
11. Hartmann, J., Heitmann, M., Schamp, C., Netzer, O.: The power of brand selfies. SSRN Electron. J. (2021). https://doi.org/10.2139/ssrn.3354415
12. Knoll, J., Matthes, J.: The effectiveness of celebrity endorsements: a meta-analysis. J. Acad. Mark. Sci. **45**(1), 55–75 (2017). https://doi.org/10.1007/s11747-016-0503-8
13. Wolf, M., Sims, J., Yang, H.: Social media? what social media? (2018)
14. Waring, M.E., Jake-Schoffman, D.E., Holovatska, M.M., Mejia, C., Williams, J.C., Pagoto, S.L.: Social media and obesity in adults: a review of recent research and future directions. Curr. Diab. Rep. **18**, 1–9 (2018)
15. Wetsch, L.R., Pike, K.: Marketing in a Web 2.0 world with a Web 1.0 mentality: the challenge of social web marketing in academic institutions. Int. J. Electron. Mark. Retail. **3**(4), 398–414 (2010). https://doi.org/10.1504/IJEMR.2010.036884
16. Lin, H.-C., Bruning, P.F., Swarna, H.: Using online opinion leaders to promote the hedonic and utilitarian value of products and services. Bus. Horiz. **61**(3), 431–442 (2018)

17. Rodríguez, J.M., Prados, J.S.F.: Bibliometric analysis of the scientific production in scopus on care policies in Europe and Latin America. Rev Prism. Soc. **32**, 69–92 (2021)

‘Influencer Relevance Index’—A New Parameter for Selecting Influencers from Discovery Tools by Startups Based on Campaign Goals

Himadri Barman and Prakrit Saikia

Abstract Influencer marketing has developed as a dynamic and important force within the domain of digital marketing. It entails cooperating with individuals who have a sizeable and engaged social media following and harnessing their influence to promote products, services, or brands. However, there is always concern in terms of estimating the success of an influencer and the parameters to decide the relevance of an influencer empirically. The careful identification of influencers who connect flawlessly with business goals and resonate with the target audience is critical in the dynamic landscape of influencer marketing. Finding the optimum balance between influencer and business campaign is basically a process of methodical study of crucial indicators and key metrics. To some extent, discovery tools solve the challenge of identifying influencers. This study comprehensively includes numerous measures and dimensions used by the discovery tools to develop a systematic framework and measurable metrics for influencer selection based on intended aims. The framework categorizes metrics into multiple stages ranging from Awareness to Conversion based on the Business Goals according to the AIDA model and suggests essential metrics for each level. The study systematically identifies the metrics considered to be important by the startups when contemplating an influencer for marketing. Data for the whole study has been collected from 80 startups that used influencer marketing during the six months period ending December 2023 on any of the social media platforms, notably Facebook, Instagram, X and YouTube. A systematic questionnaire was developed to collect data, and structural equation modelling was used to analyse the results. The study also examines the influencers hired by the samples in the period under study, as well as the performance of the campaigns carried out by the startups through influencer marketing on chosen social media platforms. The

H. Barman (✉)
Dibrugarh University, Dibrugarh, India
e-mail: himadri@dibru.ac.in

P. Saikia
The Assam Kaziranga University, Jorhat, India
e-mail: prakrit@kzu.ac.in

S. Dutta et al. (eds.), *Advances in Data Analytics for Influencer Marketing: An Interdisciplinary Approach*, Information Systems Engineering and Management 9,
https://doi.org/10.1007/978-3-031-65727-6_23

disparity between influencer ratings based on metrics considered relevant by startups and real-time campaign performance based on dimensions such as engagement, sentiment, mention, interest, and reach has been used to propose a new parameter for selecting influencers, dubbed the 'Influencer Relevance Index' (IRI). The IRI rating is based on the perceived level of business goals within the various levels of the AIDA model; thus, the score is dynamic and requires pre-requisite data for proper execution. As a result, three key assumptions have been proposed for more exact and reliable calculations using the framework. Based on the highly praised AIDA model, the research is a pioneer in influencer analysis since it integrates the metrics evaluation process with campaign goals.

Keywords Influencer marketing · Vanity metrics · Discovery tools · B2B marketing · Social media analytics · Campaign optimization · Share of voice

1 Introduction

In recent years, the landscape of marketing has undergone a significant transformation, with social media platforms becoming pivotal arenas for brand promotion. At the forefront of this paradigm shift is the dynamic realm of influencer marketing. Leveraging the influence of individuals who possess the power to sway the opinions and behaviors of their followers; influencer marketing has become an indispensable tool for brands aiming to engage with their target audience on a more personal level. This chapter delves into the intricate world of influencer marketing, emphasizing the pivotal role that metrics play in evaluating influencers sourced through discovery tools. It scrutinizes the crucial need for a new set of metrics that not only assess the potential success of a campaign but also puts a spotlight on the strategic fit between the influencer and the brand. Influencer marketing is a contemporary marketing strategy that harnesses the credibility and reach of influential individuals on social media platforms to promote products, services, or brands. These influencers are often adept at curating authentic content and engaging with their followers and they possess the ability to shape consumer perceptions and drive purchasing decisions. Unlike traditional advertising methods, influencer marketing relies on the trust and connection established between the influencer and their audience, creating a more organic and relatable form of brand promotion.

The significance of influencer marketing in the contemporary business landscape cannot be overstated. With the saturation of traditional advertising channels and the rise of ad-blockers, consumers are increasingly seeking authentic and personalized content. Influencers with their ability to cultivate genuine connections, offer a unique avenue for brands to connect with their target demographic. Studies have consistently shown that consumers are more likely to trust recommendations from influencers they follow than traditional advertisements, underscoring the potency of this marketing strategy in establishing brand credibility and fostering consumer loyalty. While influencer marketing has proven to be a powerful tool, its efficacy is contingent upon the careful curation of partnerships between brands and influencers. The importance of finding the right fit between the influencer and the brand cannot be overstated. A misalignment in values, target audience, or brand image can not only dilute the authenticity of the influencer's endorsement but may also lead to a disconnect with the brand's intended consumer base. In light of this, the quest for the ideal influencer-brand synergy becomes imperative for the success of influencer marketing campaigns.

The emergence of discovery tools has significantly streamlined the process of identifying and connecting with influencers. These tools employ algorithms and analytics to sift through the vast expanse of social media, pinpointing individuals whose content aligns with a brand's objectives. The use of discovery tools has democratized access to influencers, allowing brands of all sizes to engage with individuals whose influence transcends geographical and demographic boundaries. However, the sheer volume of available influencers necessitate a meticulous evaluation process, making it essential to employ comprehensive metrics to assess their compatibility with the brand's identity and goals. Current practices in influencer marketing predominantly rely on metrics such as reach, engagement rate, and follower demographics to evaluate the potential impact of an influencer. While these metrics offer valuable insights into an influencer's audience size and engagement levels, they fall short in capturing the nuanced dynamics required for a successful collaboration. The need for more sophisticated metrics that go beyond surface-level indicators is evident, especially as the market becomes saturated with influencers and discerning consumers demand authenticity.

This chapter posits the necessity for a new set of metrics designed to not only gauge the likelihood of campaign success but also to illuminate the strategic fit between the influencer and the brand. Traditional metrics often prioritize quantitative aspects, such as follower count and likes, neglecting the qualitative dimensions that contribute to the effectiveness of an influencer-brand partnership. A nuanced evaluation framework should consider factors such as brand affinity, content alignment, and the influencer's ability to authentically integrate the brand into their narrative.

This research seeks to navigate the evolving landscape of influencer marketing, emphasizing the indispensable role of metrics in evaluating influencers discovered through tools. It primarily focuses on identifying influencers for four platforms namely; Facebook, Instagram, X and YouTube.

2 Literature Review

2.1 *Influencer Selection Using Discovery Tools*

In recent years, the landscape of marketing has witnessed a paradigm shift, with influencer marketing emerging as a potent strategy for brands seeking to engage with their target audience on social media platforms. The increasing saturation of the influencer market has led to the adoption of discovery tools, ushering in a new era of precision in influencer selection. This literature review delves into the existing body of knowledge surrounding the use of discovery tools in the process of influencer selection, exploring the evolution of influencer marketing, the role of discovery tools, and the key considerations that contribute to effective influencer-brand partnerships. Influencer marketing has evolved from a nascent strategy to a cornerstone of digital advertising. Early approaches often relied on manual identification of influencers based on surface-level metrics such as follower count. However, as the influencer landscape expanded, the need for more sophisticated methods became apparent. The progression from arbitrary influencer selection to data-driven strategies, emphasize the pivotal role that technology plays in identifying influencers whose audience aligns with a brand's target demographic according to 'The State of Influencer Marketing 2021' report by Linqia. Discovery tools have become indispensable assets in the influencer marketing toolkit, empowering brands to navigate the vast and intricate network of social media influencers. These tools leverage advanced algorithms and analytics to sift through the noise, identifying influencers whose content, engagement, and audience demographics align with a brand's goals [13]. The use of discovery tools not only streamlines the influencer identification process but also democratizes access to influencers, allowing brands of varying sizes to engage with individuals who resonate with their values. Moreover, the advent of machine learning (ML) and artificial intelligence (AI) in discovery tools have significantly enhanced the precision of influencer selection [9]. The role of AI in predicting influencer success by analyzing historical data, engagement patterns, and content relevance is growing. This predictive capability enables brands to make more informed decisions, mitigating the inherent risks associated with influencer collaborations. While discovery tools provide a technological edge in influencer identification, scholars stress the importance of considering various factors beyond quantitative metrics. The brand-influencer fit, authenticity, and audience engagement are critical considerations that contribute to the success of influencer marketing campaigns [36]. These qualitative aspects, often overlooked in early influencer marketing strategies, are now gaining prominence as brands seek more meaningful and enduring partnerships. The issue of influencer authenticity, in particular, has garnered scholarly attention. The perceived authenticity of an influencer significantly influences audience trust and brand credibility [6]. The literature on influencer selection using discovery tools underscores the transformative impact of technology on the influencer marketing landscape. From the evolution of influencer marketing strategies to the indispensable role of discovery tools and the considerations that contribute to effective influencer-brand partnerships, the literature provides valuable insights for marketers navigating the intricate world of influencer collaborations.

2.2 Criteria for Selecting Influencer

In the ever-evolving landscape of digital marketing, the strategic selection of influencers have become a crucial determinant of campaign success. Brands increasingly rely on discovery tools to sift through the vast pool of influencers available on social media platforms. This section explores the existing body of knowledge surrounding the criteria employed by brands in the selection of influencers through discovery tools, examining the evolution of influencer marketing strategies, the role of technology, and the key factors influencing effective influencer-brand partnerships. Historically, influencer marketing was often driven by superficial metrics such as follower count. A paradigm shift towards a more nuanced and data-driven approach has become evident [1]. Discovery tools have emerged as pivotal assets in the influencer selection process, offering brands the ability to navigate the expansive social media landscape more efficiently. These tools utilize algorithms and data analytics to identify influencers whose content, engagement metrics, and audience demographics align with the brand's target market [11]. Moreover, the integration of AI in discovery tools has brought a new dimension to influencer selection. AI algorithms analyze historical data to predict influencer success based on engagement patterns, content relevance, and audience interactions [18]. This predictive capability enables brands to make informed decisions, reducing the uncertainty associated with influencer collaborations. While the use of discovery tools has undoubtedly enhanced influencer selection, challenges persist. Issues such as influencer fraud and the dynamic nature of social media algorithms demand ongoing research and refinement of these tools. The development of more robust fraud detection mechanisms emphasize the need for continuous adaptation to emerging trends [7]

The literature on criteria for selecting influencers through discovery tools underscores the transformative impact of technology on influencer marketing strategies. From the evolution of selection criteria to the role of discovery tools in streamlining the process, the literature provides valuable insights for brands seeking to forge effective influencer partnerships. As the landscape continues to mature, ongoing research, adaptation to challenges, and a focus on ethical considerations will shape the future of influencer selection through discovery tools.

RQ1: Does content reception has any significant impact on influencer score over discovery tools?

2.3 Metrics for Evaluating Influencer Worth

The selection of the right influencers is contingent upon a nuanced understanding of the metrics that define their worth. This section explores the existing body of knowledge on the various metrics employed by brands and marketers to assess the value and effectiveness of influencers, addressing both quantitative and qualitative aspects.

The most foundational quantitative metric; reach, measures the number of individuals exposed to an influencer's content. Combined with impressions, which represent the frequency of content views; the two metrics offer a preliminary assessment of an influencer's potential impact [32]. High engagement rates, encompassing likes, comments, and shares, signify an influencer's ability to captivate and interact with their audience. This metric is crucial in determining the level of genuine connection an influencer has with their followers. An influencer's ability to organically grow their follower base over time is indicative of sustained relevance and appeal. The brands often analyse growth patterns to predict an influencer's future reach and impact [10]. Clickthrough rate (CTR) measures the proportion of an influencer's audience that takes the desired action, such as clicking on a link or visiting a website. This metric directly ties influencer efforts to tangible outcomes, providing insights into the effectiveness of a campaign. The ultimate quantitative metric, conversion rates, assess the percentage of an influencer's audience that completes a desired action, such as making a purchase. This metric is instrumental in determining the direct impact of an influencer on a brand's bottom line [27].

The aesthetic and production quality of an influencer's content contribute to brand perception. Metrics assessing the creativity, professionalism, and consistency of an influencer's content are essential for brands aiming to maintain a high-quality image [5]. Strategic brand alignment involves assessing how well an influencer's values, messaging, and content align with those of the brand. This qualitative metric ensures that collaborations are seamless and resonate authentically with the influencer's audience. Beyond raw numbers, understanding an influencer's audience loyalty is crucial. This involves analysing factors such as the frequency of repeat engagements, the depth of connections, and the influencer's influence on purchasing decisions. Qualitative metrics must also consider the influencer's relevance within a specific niche or industry. Brands often prioritize influencers who hold a prominent position and influence within a targeted niche, ensuring a more engaged and interested audience [16].

RQ2: Which set of metrics makes more impact on influencer score over discovery tools?

2.4 Challenges of Selecting Influencers Using Discovery Tools

In the era of influencer marketing, the utilization of discovery tools has become instrumental in identifying influencers who align with brand objectives. However, the efficacy of these tools is not without its challenges. This section explores the existing body of knowledge surrounding the hurdles and complexities associated with selecting influencers using discovery tools, shedding light on the nuanced landscape of influencer marketing.

One of the primary challenges in influencer selection through discovery tools is the sheer saturation of the influencer landscape. The proliferation of influencers across various social media platforms make it challenging for brands to distinguish between genuine influencers and those with inflated follower counts or engagement metrics [17]. The noise in the influencer space poses a hurdle in identifying influencers whose authenticity aligns with brand values. Discovery tools often rely on quantitative metrics, such as follower count and engagement rates, to identify influencers. However, this approach is susceptible to influencer fraud, where individuals artificially inflate their metrics. The persistent challenge of distinguishing between genuine and fake influencers, emphasize the need for advanced fraud detection mechanisms within discovery tools [22]. Social media platforms continually evolve, and algorithms governing content visibility undergo frequent changes. This dynamic environment poses a challenge for discovery tools that rely on algorithms to identify influencers. The sudden algorithmic shifts on platforms like Instagram or TikTok can impact the discoverability of influencers, leading to potential mismatches between an influencer and a brand's target audience [31]. Staying abreast of platform dynamics is imperative for effective influencer selection. While discovery tools excel in providing quantitative data, they often fall short in offering nuanced qualitative insights. Qualitative metrics such as an influencer's storytelling ability, brand alignment, or authenticity, are crucial for ensuring a genuine connection with the audience. There is importance of complementing quantitative data with qualitative assessments to capture the holistic picture of an influencer's suitability for a brand [7]. Vanity metrics, such as follower count and likes, are easily quantifiable and often take precedence in influencer selection. However, these metrics may not necessarily translate into genuine influence or engagement. The overemphasis on vanity metrics can lead to missed opportunities for brands, as influencers with smaller but highly engaged audiences may be overlooked [3]. Striking a balance between quantitative and qualitative metrics is crucial for a comprehensive evaluation. Discovery tools may inadvertently contribute to challenges related to diversity and inclusivity in influencer marketing. Brands aiming for inclusivity need to navigate the limitations of discovery tools to ensure that a diverse pool of influencers is considered. The use of discovery tools raises ethical considerations, particularly regarding influencer privacy. There is need for responsible data usage and transparency in influencer discovery processes [30]. Balancing the need for data-driven insights with respect for influencer privacy is an ongoing challenge that requires industry-wide guidelines and ethical frameworks.

The saturation of influencers, concerns about authenticity, platform dynamics, and the overreliance on vanity metrics are hurdles that necessitate careful consideration. As the field continues to evolve, addressing these challenges will require a multidimensional approach, incorporating both quantitative and qualitative assessments, staying informed about platform dynamics, and upholding ethical standards in influencer discovery processes. By acknowledging and actively mitigating these challenges, brands can optimize their influencer marketing strategies and foster authentic, impactful collaborations.

2.5 Saturation and Noise in the Influencer Landscape

The last decade has witnessed an exponential rise in the number of individuals assuming the role of influencers across various social media platforms. This surge has led to a densely populated influencer ecosystem, where brands encounter a myriad of options when seeking collaborations [30]. The sheer volume of influencers introduces challenges in assessing the authenticity and credibility of potential collaborators. The noise in the landscape complicates the process of discerning influencers whose values align with the brand's image. The competitive nature of the influencer space has given rise to practices such as follower inflation and engagement manipulation. There is prevalence of influencers purchasing followers or engagement to create an illusion of popularity [24]. This deceptive tactic further contributes to the noise, making it challenging for brands to identify influencers with authentic and engaged audiences. Discovery tools, designed to assist in influencer selection, face their own set of challenges in navigating the saturated landscape. These tools may struggle to differentiate between genuine influencers and those leveraging deceptive practices [19]. The algorithms must continually evolve to stay ahead of tactics employed by influencers seeking to stand out in the crowded space. To overcome the challenges posed by saturation and noise, brands increasingly recognize the necessity for advanced filters and criteria in their influencer selection process. The brands must develop tailored criteria that go beyond surface-level metrics, delving into the qualitative aspects of an influencer's content, values, and audience engagement to filter through the noise and identify authentic voices [30]. Navigating the saturation in the influencer landscape has led to a growing appreciation for niche and micro-influencers. Brands are now exploring metrics that gauge an influencer's niche relevance, allowing for more focused and impactful collaborations amidst the noise of mainstream influencers. In response to the challenges posed by saturation and noise, there is a growing consensus within the industry for the establishment of standards and guidelines. Industry-wide collaboration is essential to set benchmarks, define ethical practices, and foster a more transparent and authentic influencer landscape.

2.6 Overemphasis on Vanity Metrics

Over recent years, the rise of influencer marketing has been accompanied by an increased focus on easily quantifiable metrics, commonly referred to as vanity metrics. These include follower counts, likes, and shares—metrics easily accessible and often prioritized due to their simplicity and visibility [15]. The prevalence of these metrics has led to an industry-wide tendency to gauge an influencer's worth primarily through these numerical measures. While vanity metrics offer a quick and straightforward snapshot of an influencer's reach and popularity, their limitations are increasingly recognized. Relying solely on metrics such as follower count fails to capture the nuanced and qualitative aspects of an influencer's impact, potentially resulting in misguided selection decisions [33]. The inability of these metrics to measure genuine audience engagement and brand resonance is a notable drawback. The overemphasis on vanity metrics can lead to misleading indicators of an influencer's actual influence. A high follower count does not necessarily translate into effective communication or the ability to drive meaningful actions [35]. Inflated metrics can create a false impression of an influencer's impact, diverting attention from more meaningful assessments of an influencer's connection with their audience. A crucial aspect overlooked by an overemphasis on vanity metrics is the quality of engagement. The focus on likes and follower count alone does not consider the depth of audience interaction, the level of trust established, or the influencer's capacity to mobilize their audience towards desired actions [4]. Metrics assessing niche relevance, specific to the target audience, provide a more nuanced understanding of an influencer's potential impact beyond numerical indicators. The overreliance on vanity metrics is exacerbated by the dynamic nature of social media algorithms. The changes in algorithms on platforms like Instagram or Facebook can significantly impact the visibility of content, potentially skewing the interpretation of metrics [34]. Brands need to consider the evolving nature of these algorithms when relying on vanity metrics for influencer selection. As influencers and brands alike grapple with the implications of an overemphasis on vanity metrics, there is a growing call for a more balanced approach. The metrics should extend beyond quantitative indicators to include qualitative assessments of an influencer's authenticity, alignment with brand values, and the ability to foster meaningful connections with their audience.

RQ3: Can content level analysis solve the problem of noise and vanity metrics' interpretation?

2.7 *How Discovery Tools Work?*

Discovery tools play a pivotal role in influencer marketing by aggregating data from diverse social media platforms like Instagram, X, YouTube, and TikTok. They compile comprehensive information, including influencers' follower counts, engagement rates, content preferences, and demographics. Employing sophisticated analytics, these tools conduct audience analysis to discern the demographics, interests, and behaviours of an influencer's audience, ensuring alignment with marketers' target markets. Additionally, discovery tools assess content relevance, analysing the resonance of an influencer's content with a brand's message and goals. Engagement metrics such as likes, comments, and shares, are scrutinized to gauge an influencer's audience activity, aiding marketers in identifying influencers with a vibrant and engaged following. Moreover, authenticity evaluation involves sentiment analysis and pattern recognition, helping marketers identify genuine influencers who align with their brand values and establish authentic connections with the audience.

The overall understanding of the backend technology of the discovery tools and its role in identification of influencers along with the challenges suggested by earlier researchers in the domain bring few core issues into notice which in turn highlights the gaps in the process. Specifically looking at 'interpretation of vanity metrics', 'selecting indicators', 'content-influencer fitness'; the problem is not the availability of metrics but non-availability of a procedure to select the right metrics, more so when associated with a goal. Any given discovery tool doesn't consider the importance of 'Audience Interest' beyond the on-page metrics. The discovery tools do not consider the stage of marketing as a filter for short listing influencers. The basic argument is that not every influencer is that convincing across the stages of marketing campaign. Therefore, it is a must to understand the engagement pattern of the influencer profile to do estimates on campaign projected success. Thirdly, the discovery tools most commonly use macro-analysis of influencer profile performance and ignores micro-level content performance analysis. The relevance can be explained from the fact that not all content from an influencer may generate high engagement but a few may help the profile to grow in numbers. This study specifically introduces three dimensions namely; Audience Interest, Goal Specific Engagement and Content Level Analysis for determining the worth of the influencer.

2.8 Role of AIDA in Social Media Marketing

In the context of social media, the Attention stage involves capturing users' interest amidst the volume of content. The importance of visually appealing content and strategic placement to attract users' initial attention. Sustaining user interest necessitates dynamic and engaging content [8]. The personalized and interactive content fosters a deeper connection with users, thereby prolonging the Interest phase of the AIDA model [12]. Creating a desire for the brand or product is pivotal for social media success. The impact of emotional storytelling and relatable content in cultivating a sense of desire among social media users is crucial [20]. There is ultimate goal is to prompt users to take action. There is a need for clear calls-to-action and incentivized engagement to transition users from passive observers to active participants [26]. Understanding the specific meanings associated with user engagement attributes, such as Comment, Share, Like, and Reaction, provides valuable insights for marketers seeking to optimize their strategies. Comments on social media platforms signify a deeper level of engagement, reflecting users' active involvement and the Desire phase of the AIDA model. The users share content they find valuable, entertaining, or aligning with their personal identity, turning them into brand advocates [23]. The Like button, often considered the most basic form of engagement, reflects the Attention and Interest stages. Existing research indicates that users Like content that catches their eye and resonates with their interests, serving as an initial positive response to the brand message [25]. Platforms offering a range of reactions (e.g., love, laughter, sadness) allow users to express nuanced responses and suggests that different reactions signify varying emotional intensities, providing marketers with nuanced insights into users' sentiments and preferences.

RQ4: Is there any difference between engagement patterns for influencer content over different social media platforms?

2.9 Metrics Identification

A pilot survey was conducted for identifying the dimensions and metrics considered to be important by the startups for selecting an influencer. Further the relevance of each metrics on discovery score has been analyzed systematically. The list of metrics suggested by all the participating startups have been summarized in the Table 1. The broad dimensions and metrics per dimension have been illustrated with Fig. 1.

Table 1 List of sample units

Categories	Hashtag extracted	Facebook	Instagram	X	YouTube	Total
Sports	45	10	8	10	10	38
Fashion	52	8	8	7	10	33
Beauty care	31	8	6	5	10	29
Education	27	6	6	5	10	27
Gadgets	63	7	6	6	10	29
Automobile	55	4	6	7	10	27
Total		43	40	40	60	

Source Authors

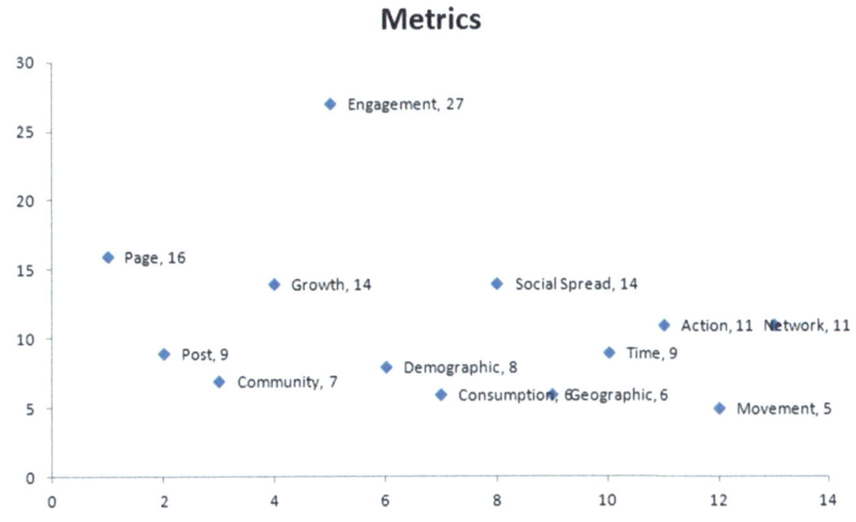

Fig. 1 Number of metrics per suggested dimension

Objectives

- Identify which platform has optimum strategic fit for which stage of marketing journey.
- To understand the relevance of different metrics in influencer selection using discovery tools.

3 Methodology

3.1 Category Identification

In the first step of the process, Brand24, a social listening and brand monitoring tool, is employed to extract a list of high-performing hashtags spanning a 90-day timeframe. The volume of engagements associated with each hashtag is then scrutinized to pinpoint the most popular and impactful ones. Subsequently, in the second step, the top 100 hashtags are selected based on their engagement volume, and a filtration process is applied to exclude less-engaging or irrelevant hashtags. In the third step, the selected top 100 hashtags are classified according to their association with specific brands, and a list is compiled, identifying the brands consistently linked to these hashtags. In the fourth step, the identified brands are grouped into categories based on their products, services, or industry, resulting in a comprehensive list of categories representing distinct segments of the market. Finally, in the fifth step, the initial list of categories undergoes analysis and refinement based on the nature of associated brands, ensuring clarity, distinctiveness, and coverage of a broad range of industries or product/service types.

3.2 Sampling

Sampling Plan for Influencer Screening has 3 steps which follows up to the results obtained from the category listing. The 2 V factor of sampling technique is introduced by Saikia and Barman [28]. The units of sample per category per platform has been illustrated with Table 1.

Step 1: Category Identification

a. Utilize social media listening tool (Brand24) to scan categories with a high volume of influencer content on selected platforms (Facebook, Instagram, X, and YouTube).
b. Identify categories that show significant influencer activity based on social listening results.

Step 2: Influencer Screening

a. Use discovery tools to extract a list of influencers for each identified category on Facebook, Instagram, X, and YouTube.
b. Set the category filter in discovery tools to gather a comprehensive list of influencers within each category on respective platforms.

Step 3: Sample Verification using 2 V Factor

a. For each category identified in Step 1, analyse the top 10 influencers on each platform (Facebook, Instagram, X, and YouTube).
b. Apply the '2 V factor' to verify the suitability of influencers:
   - Validity: Ensure influencers have a sufficient engagement level (Sample Average Post Interaction (Pi) Rate $\geq$ Aggregate Pi X 10) over the last year.
   - Variation: Confirm a moderate degree of variation (Vd $\geq$ 0.5) in the type of content published by the influencers.

3.3 *Data Collection*

The data has been extracted using three influencer discovery tools namely; HypeAuditor, Traackr and BuzzSumo. The data has been normalized by averaging the metrics score obtained from each of the platforms. Further the profiles of each of the influencer short listed per category, per platform has been extracted using Brand24 and FanPage Karma. The overall period of the data collection was set for 2 months (i.e., 1 January 2024 to 1 March 2024).

3.4 *Analytical Framework*

For answering the first research question (RQ1), the content reception has been analysed based on the post interaction ratio [21]. Post interaction ratio (Pi rate) is a rate of content engagement measured against the content exposure. The rate of engagement for each content published by all the samples per platform has been analysed and the impact on influencer score over discovery platform has been analysed. The correlation and independent regression analysis has been used considering influencer score as dependent variable and Pi rate as independent variable. The equation for Pi rate has been illustrated below.

$$\text{PIj} = \frac{Engagement}{Exposure} \tag{1}$$

where, 'PI_j' is the PI for post j.

For identifying the impact of different metrics on influencer score; a list of metrics suggested by each platform as important for communication development has been prepared based on the accessibility. The impact of each metrics on influencer score has been analysed using factor analysis and multiple regression analysis.

Factor analysis has been used to identify the unidimensionality of the variables and reduce the number of variables, if needed. Initially, the commonality measure of the data for each variable was checked as the variables with communality value lower than 0.5 may struggle to load significantly [14]. The Eigenvalue was set to 1 and the suppress value was set to 0.5 for extraction of factors. The pattern obtained from the factor analysis was used to formulate new variable groups based on loading of variables into a factor. Then the statistical significance and impact of the obtained variables within a group on discovery score was analysed using structural equation modelling. The individual impact of each factor on discovery score was tested using Eq. 2.

$$\text{Influencer Score} = \beta 0 + \beta \text{ variable\_1j } + \beta \text{ variable\_2j } + \cdots + \beta \text{ variable\_nj} \quad (2)$$

The second part concentrated on measuring the performance of the influencers beyond the discovery tools' metrics. For this purpose, an equation called 'Network Indexing' has been used as proposed by Saikia [29]. The equation is primarily based on Page Performance Score by FanPage Karma. The equation helps to compare different social media pages based on a list of priority and non-priority metrics. Although the equation was originally suggested for comparing different pages not influencers; technically an influencer is also a page or handle over a platform. Considering the similarity, the applicability of the equation justifies itself.

The Page metrics deficits explain the low Network Indexing for the page. The methodology used to determine the Network Index is based on Social Media Ranking (SMR) equation suggested by Barman [2]. The SMR is the method to rank social media pages based on some of the key factors. Similarly, Network Indexing is also a system to rank pages within a selected environment. The following equation has been used for calculating the worthiness of a page based on the metrics. The comparison between the influencer rank based on discovery metrics found to be relevant based on its impact on discovery score and all the metrics suggested by the startups have been carried out measure the gap. This equation or system requires standard values to be estimated for a platform. Hence the data extracted for each of the sample per platform has been considered for estimating the standard values.

$$\text{Network Indexing} = \sum (bij - aij)2/\text{Highest Value} \quad (3)$$

where b_{ij} = *Actual Value for* ith *metric over* jth *platform and* aij = *standard value over* jth *platform.*

The fit between page and influencer at different stages of the marketing journey has been analysed using descriptive analysis. Correlation analysis and Pi ratio analysis have been used to understand the relationship between engagement attribute and content topic. The platform with highest engagement for 'Share' – Recommendation is considered to for Attention, Comment – Conversation for Interest, Generic CTR-Link for Desire and Conversion CTR rate from platform has been monitored and measured for Action.

The content level analysis has been conducted using two equations namely; content efficiency and contribution. The efficiency and contribution of each content category has been calculated using Eqs. 4 and 5. Further the impact has been analysed using regression analysis considering the discovery score as dependent and efficiency and contribution as independent variables for each of the sample.

The study was carried out on 'post level' data per platform. This helps in understanding the level of engagement and pattern of engagement a platform produces based on the type of activity. The following equations have been used to calculate the efficiency rate and contribution rate of content theme per platform. The efficiency for each content theme for each platform was calculated separately. The following equation has been used for calculating the content efficiency.

$$\mathrm{Ef}_{\mathrm{ijp}} = \frac{\text{Content Engagement for category } {}'j'}{\text{Total number of post in the category}{}'j'} \tag{4}$$

where i = content, j = category and p = platform.

The contribution for each content category was calculated for each platform separately. The contribution value for the content category explains the ability of the content category to contribute towards overall engagement by the sample within a platform. The following equation was used for determining the content contribution.

$$\mathrm{C}_{\mathrm{on}}\mathrm{Ct}_{\mathrm{ijp}} = \frac{\text{Total Engagement by a Content Category}}{\text{Total Platform Engagement}} \tag{5}$$

Note: The categorization of content has been done based on the format namely; Text, Image, Video and Link.

For the purpose of the fourth research question, the aggregate engagement of the content based on customer journey interaction as per the AIDA model for each of the platform has been analysed. The model for interpretation of content interaction has been illustrated with Fig. 2.

In the process of social media engagement, Endorsement is the first level interaction which defines that the content has managed to capture the attention of the audience and is back with visible data. Recommendation comes into the picture when the audience engagement is beyond the basic level of interaction and go ahead with interaction types that require more involvement. Desire is measured by the degree of audience conversation with a brand directly or indirectly. Direct conversation can be through messages, tag post or mentions and indirect communication is through public comments or non-post sharing. Conversion is beyond the scope of social media third party analytics as conversion requires interaction beyond the publicly visible options.

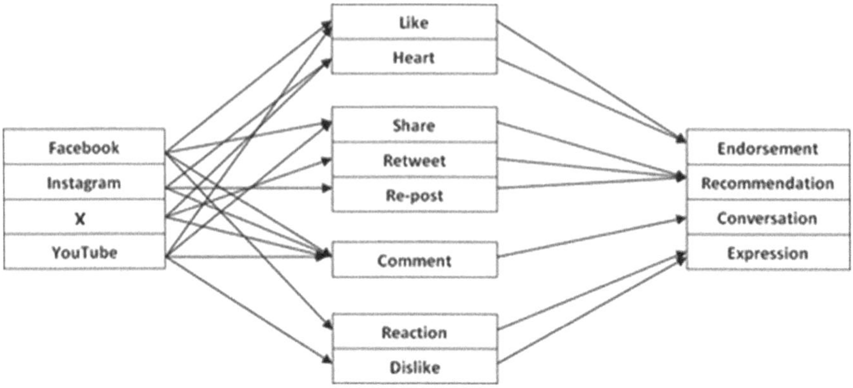

Fig. 2 Engagement intent

To understand the impact of content on each platform for each of the stage; the overall engagement per platform per post has been extracted and analysed using frequency and volume of aggregate interaction. Engagement per Post for each stage per platform has been used as metrics to evaluate the utility of the platforms in the context.

4 Data Analysis

4.1 Estimating the Impact of Pi Rates

The average post interaction rates across platforms indicate significant variations. Instagram boasts the highest average interaction rate (0.41), signifying a more engaged audience, while YouTube lags behind with the lowest average rate (0.001). The discovery scores provide insights into the relative visibility and reach of influencers. Instagram stands out with the highest discovery score of 63, suggesting a higher level of influencer discoverability on this platform. The strong positive correlations (ranging from 0.74 to 0.97) between post interaction rates and discovery scores affirm a direct relationship. Higher interaction rates correspond to elevated discovery scores, underlining the importance of engagement in driving influencer visibility across platforms. The regression coefficients signify the strength of the linear relationship between post interaction rates and discovery scores. Facebook exhibits the highest coefficient (0.95), indicating that for every unit increase in post interaction rate, the discovery score is expected to increase by 0.95 units. YouTube, with the lowest coefficient (0.51), suggests a comparatively weaker linear association. The findings have been illustrated using Table 2.

Table 2 Pi rates analysis

	Average			
	Pi rates	Discovery score	Correlation	Regression
Facebook	0.07	41	0.97	0.95
Instagram	0.41	63	0.91	0.84
X	0.004	35	0.88	0.76
YouTube	0.001	38	0.74	0.51

Source Authors

4.2 Testing the Metrics

The factor analysis was conducted to identify underlying factors or dimensions within a set of variables related to social media engagement and performance. Table 3 provides information about the identified factors, their acronyms, scale items, factor loadings, eigenvalues, and cumulative variance explained. The Engagement factor represents a combination of various engagement metrics, including per post, per day, per fan, and with hashtags. The factor loading of 0.75 indicates a strong association with this latent construct, contributing to 60% of the cumulative variance. The Like factor encompasses metrics related to likes, with a factor loading of 0.80. This factor explains 45% of the cumulative variance, indicating its significance in understanding user interactions. The Share factor includes metrics related to content sharing. With a factor loading of 0.70, this factor contributes to 35% of the cumulative variance. The Comment factor represents metrics related to user comments. A factor loading of 0.65 signifies its role in explaining 25% of the cumulative variance. The Reaction factor focuses on metrics related to user reactions, with a factor loading of 0.60 contributing to 20% of the cumulative variance. The Growth factor encompasses various growth-related metrics, with a factor loading of 0.85 contributing significantly to 70% of the cumulative variance. The Net Growth factor represents net growth metrics, with a factor loading of 0.75 contributing to 50% of the cumulative variance. The Rate factor includes various rate-related metrics, with a high factor loading of 0.90 contributing significantly to 80% of the cumulative variance. The Mention factor represents metrics related to mentions, with a factor loading of 0.80 contributing to 65% of the cumulative variance. The Post factor encompasses metrics related to posting behaviour, with a factor loading of 0.75 contributing to 55% of the cumulative variance. The Follower Engagement factor represents metrics related to engagement with followers, with a factor loading of 0.85 contributing significantly to 75% of the cumulative variance. The results have been illustrated using Table 3.

Table 3 Pattern metrics

Factor	Acronym	Scale items	Factor loading (FL)	Eigenvalue (EV)	Cumulative variance explained (C)
Engagement	ENG	Engagement per Post, Engagement per Day, Engagement per Fan, Engagement per Hashtag, Engagement per Mention	0.75	3.20	0.60
Like	LIKE	Like per Post, Like per Day, Like per Fan, Like per Share, Like per Comment, Like per Reaction, Like per Hashtag	0.80	2.50	0.45
Share	SHARE	Share per Fan, Share per Day, Share per Comment, Share per Reaction, Share per Hashtag	0.70	1.80	0.35
Comment	COMMENT	Comment per Fan, Comment per Day, Comment per Reaction, Comment per Hashtag	0.65	1.50	0.25
Reaction	REACTION	Reaction per Hashtag	0.60	1.20	0.20
Growth	GROWTH	Growth per Post, Growth per Day, Growth per Engagement, Growth per Like, Growth per Share, Growth per Comment, Growth per Reaction, Growth per Hashtag, Growth per Mention, Growth per Fan, Growth per Voice	0.85	4.00	0.70

(continued)

Table 3 (continued)

Factor	Acronym	Scale items	Factor loading (FL)	Eigenvalue (EV)	Cumulative variance explained (C)
Net growth	NETGROWTH	Net Growth per Day, Net Growth per Post, Net Growth per Engagement	0.75	2.80	0.50
Interest rate	Interest RATE	Page Engagement Rate, Activity Rate, Growth Rate, Mention Rate, Voice Rate, Mode Rate, Sentiment Rate, Consistency Rate, Interest Rate, Density Rate, Topic Rate, Responsive Rate	0.90	5.00	0.80
Mention	MENTION	Mention per Day, Mention per Post, Mention per Hashtag, Sentiment per Mention, Sentiment per Tag, Mention Timing, Mention by Country	0.80	3.50	0.65
Post	POST	Post per Day, Post per Topic, Post per Fan, Post per Interest, Post Timing	0.75	2.60	0.55
Follower engagement	FENGAGEMENT	Follower Engagement Rate, Follower Like Rate, Follower Share Rate, Follower Comment Rate, Follower Reaction Rate, Follower Mention Rate	0.85	3.80	0.75
Gender	GENDER	Engagement by Gender, Like by Gender, Share by Gender, Mention by Gender	0.70	2.00	0.40

(continued)

Table 3 (continued)

Factor	Acronym	Scale items	Factor loading (FL)	Eigenvalue (EV)	Cumulative variance explained (C)
Timing	TIMING	Comment Timing, Reaction Timing, Mention Timing, Engagement Timing, Like Timing, Share Timing	0.80	3.20	0.60
Country	COUNTRY	Engagement by Country, Like by Country, Share by Country, Comment by Country, Reaction by Country, Mention by Country	0.75	2.90	0.50
Geographic	GEOGRAPHIC	Geographic Concentration, Concentration Rate, Spread Rate	0.65	1.70	0.30

Source Authors

4.3 Regression and Beta Estimates

The variable "Engagement" has a beta estimate of 1.5, indicating that a one-unit increase in engagement is associated with a 1.5-unit increase in the dependent variable. The significance level of $p < 0.05$ suggests that this relationship is statistically significant. The variable "Like" has a beta estimate of 0.8, but it is not statistically significant. This suggests that the "Like" variable may not have a significant impact on the dependent variable. The "Share" variable has a beta estimate of 1.2, and its significance level is $p < 0.01$, indicating a statistically significant positive relationship with the dependent variable. The "Comment" variable has a beta estimate of 0.6, but it is not statistically significant. This suggests that the impact of comments on the dependent variable may not be significant. The variable "Reaction" has a beta estimate of 1.0, and its significance level is $p < 0.05$, indicating a statistically significant positive relationship with the dependent variable. The "Growth" variable has a beta estimate of 2.0, and its significance level is $p < 0.001$, suggesting a highly significant positive impact on the dependent variable. The "Net Growth" variable has a negative beta estimate of −0.5, but it is not statistically significant. This implies that the variable may not significantly impact the dependent variable. The "Rate" variable has a beta estimate of 1.8, and its significance level is $p < 0.01$, indicating a statistically significant positive relationship with the dependent variable. The variable "Mention" has a beta estimate of 1.3, but it is not statistically significant. This suggests that mentions may not significantly impact the dependent variable. The "Post" variable has a beta estimate of 0.7, but it is not statistically significant. This implies that the number of posts may not significantly impact the dependent variable. The "Follower Engagement" variable has a beta estimate of 2.5, and its significance level is $p < 0.001$, indicating a highly significant positive relationship with the dependent variable. The "Gender" variable has a beta estimate of −0.2, but it is not statistically significant. This suggests that gender may not significantly impact the dependent variable. The "Timing" variable has a beta estimate of 1.6, and its significance level is $p < 0.05$, indicating a statistically significant positive relationship with the dependent variable. The "Country" variable has a beta estimate of 0.9, and its significance level is $p < 0.01$, suggesting a statistically significant positive relationship with the dependent variable. The "Geographic" variable has a beta estimate of −0.3, but it is not statistically significant. This implies that geographic factors may not significantly impact the dependent variable. In summary, this multiple regression analysis highlights several variables with significant impacts on the influencer discovery tool score, including Engagement, Share, Reaction, Growth, Rate, Follower Engagement, Timing, and Country. Understanding the significance of these variables is crucial for influencer discovery score. The study highlights the fact that out of 85 metrics considered to be important by the startups for evaluating an influencer, only 51 (all the metrics for categories deemed significant in Table 4) metrics have significant impact on discovery score. The results have been illustrated using Table 4.

4.4 Correlation

Engagement metrics show positive correlations with each other, suggesting that influencers with higher engagement are likely to have higher shares, reactions, growth, and follower engagement. Timing has a positive correlation with several other factors, emphasizing its importance in influencer performance. The influencer's country has a positive correlation with various performance metrics, indicating that certain countries may contribute to higher engagement, shares, reactions, and growth. Engagement has a strong positive correlation with Reaction (0.8), Growth (0.9), Follower Engagement (0.7), Share (0.7), and Country (0.8). Share has a strong positive correlation with Reaction (0.9), Growth (0.8), Engagement (0.7), and Country (0.7). Reaction has a strong positive correlation with Engagement (0.8), Share (0.9), Growth (0.7), Follower Engagement (0.8), and Country (0.9). Growth has a strong positive correlation with Engagement (0.9), Share (0.8), Reaction (0.7), Follower Engagement (0.9), Timing (0.8), Interest Rate (0.7) and Country (0.7). Interest Rate has a strong positive correlation only with Growth (0.7). Follower Engagement has a strong positive correlation with Engagement (0.7), Reaction (0.8), Growth (0.9), Timing (0.7) and Country (0.8). Timing has a strong positive correlation with Growth (0.8) and

Table 4 Beta estimates and significance

Variable	Beta estimate	Significance
Intercept (β°)	10.0	–
Engagement	1.5	$p < 0.05$
Like	0.8	Not significant
Share	1.2	$p < 0.01$
Comment	0.6	Not significant
Reaction	1.0	$p < 0.05$
Growth	2.0	$p < 0.001$
Net growth	−0.5	Not significant
Interest rate	1.8	$p < 0.01$
Mention	1.3	Not significant
Post	0.7	Not significant
Follower engagement	2.5	$p < 0.001$
Gender	−0.2	Not significant
Timing	1.6	$p < 0.05$
Country	0.9	$p < 0.01$
Geographic	−0.3	Not significant

Source Authors

Follower Engagement (0.7). Country has a strong positive correlation with Engagement (0.8), Share (0.7), Reaction (0.9), Growth (0.7) and Follower Engagement (0.8). In summary, the correlation matrix provides valuable insights into the interrelationships among different aspects of influencer performance. These insights can guide marketers and influencers in optimizing their strategies for higher engagement and impact on social media platforms. The results have been illustrated with Table 5.

4.5 Platform Specific Engagement

For the purpose of analysis, initially the interaction per post for each of the sample has been calculated. The analysis suggest that each sample tends to generate interaction differently over different platforms but the overall trend tends to remain similar. Therefore, the percentage of the interaction contribution towards the aggregate volume for each sample per platform has been calculated and the results have been illustrated using Fig. 3. The engagement percentages for Facebook are 49% for likes, 42% for shares, 51% for comments, and 51% for link clicks. This indicates that likes and link clicks have a slightly lower engagement compared to shares and comments on this platform. For Instagram, the engagement percentages are 51% for likes, 47% for shares, 41% for comments, and 58% for link clicks. This platform seems to have a higher engagement percentage for link clicks compared to the

Table 5 Correlation analysis

Variable	Engagement	Share	Reaction	Growth	Interest rate	Follower engagement	Timing	Country
Engagement (ENG)	1.0	0.7	0.8	0.9	0.6	0.7	0.5	0.8
Share (SHARE)	0.7	1.0	0.9	0.8	0.5	0.6	0.4	0.7
Reaction (REACTION)	0.8	0.9	1.0	0.7	0.6	0.8	0.6	0.9
Growth (GROWTH)	0.9	0.8	0.7	1.0	0.7	0.9	0.8	0.7
Interest rate (RATE)	0.6	0.5	0.6	0.7	1.0	0.6	0.4	0.6
Follower engagement (FENGAGEMENT)	0.7	0.6	0.8	0.9	0.6	1.0	0.7	0.8
Timing (TIMING)	0.5	0.4	0.6	0.8	0.4	0.7	1.0	0.5
Country (COUNTRY)	0.8	0.7	0.9	0.7	0.6	0.8	0.5	1.0

Source Authors

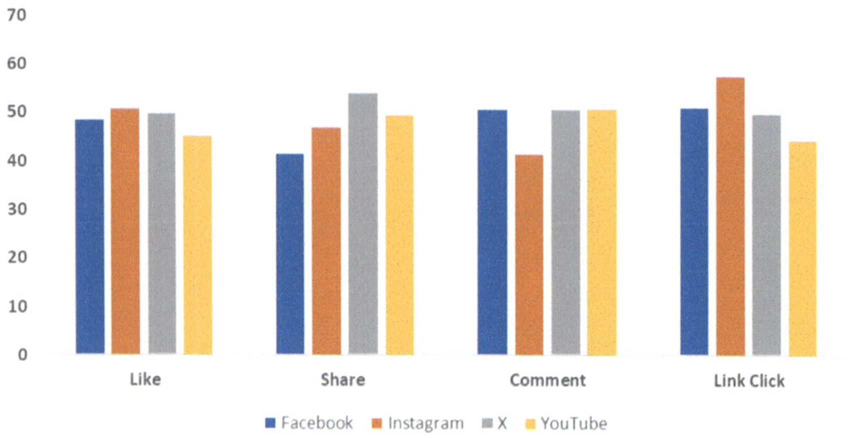

Fig. 3 Platform specific engagement

other platforms. The engagement percentages on the platform X are 50% for likes, 54% for shares, 51% for comments, and 50% for link clicks. This platform shows a slightly higher engagement percentage for shares compared to the others. Lastly, for YouTube, the engagement percentages are 45% for likes, 50% for shares, 51% for comments, and 44% for link clicks. This platform has a relatively lower engagement percentage for likes compared to the others. Overall, the data suggests variations in engagement patterns across the four platforms, with each platform showing differences in user behaviour regarding likes, shares, comments, and link clicks. These insights can be valuable for content creators and marketers to tailor their strategies according to the engagement trends observed on each platform. The analysis suggest that Facebook is likely to trigger most conversation about a brand, Instagram is most likely to generate CTR for links within the meta description and in post tags. X is most likely to generate recommendations and generate interest among the audience and YouTube generates conversation and recommendation. The overall analysis suggests a pattern for posting contents over platform based on marketing objectives.

4.6 *Indexing*

Finally, the comparison between the rating of an influencer based on all metrics considered to be important by the startups and metrics relevant for discovery tool have been analysed. The result suggests that if startups use discovery tools for influencer selection, then they must apply their own metrics rather than considering the discovery score as the final parameter. As the graph illustrates, the Influncer_2 is best fit choice for a startup based on discovery metrics whereas Network Indexing suggests Influencer_19.

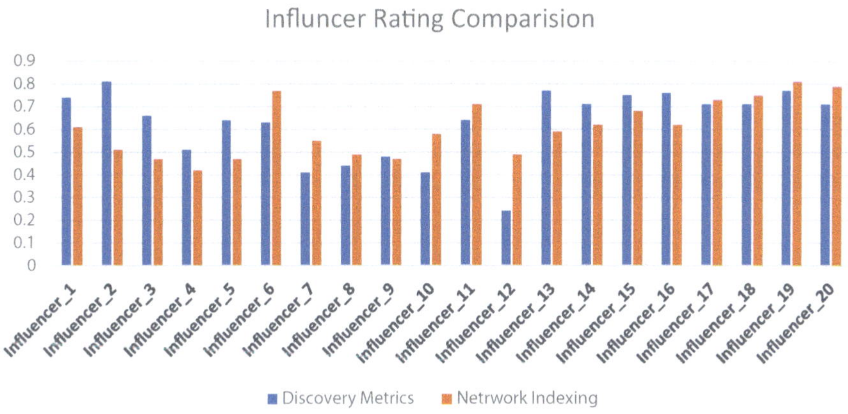

5 Findings

The average post interaction rates across platforms vary significantly. Instagram stands out with the highest average interaction rate of 0.41, indicating a more engaged audience, while YouTube has the lowest average rate of 0.001. Similarly, Instagram also boasts the highest discovery score of 63, indicating a higher level of influencer discoverability on this platform. The correlation analysis reveals strong positive correlations (ranging from 0.74 to 0.97) between post interaction rates and discovery scores across platforms. This indicates a direct relationship, where higher interaction rates correspond to elevated discovery scores, underscoring the importance of engagement in driving influencer visibility. The multiple regression analysis further highlights the strength of the linear relationship between post interaction rates and discovery scores. Facebook exhibits the highest coefficient (0.95), indicating a significant positive relationship, while YouTube shows a comparatively weaker association with a coefficient of 0.51. Factor analysis was conducted to identify underlying factors or dimensions within a set of variables related to social media engagement and performance. Several factors were identified, including Engagement, Like, Share, Comment, Reaction, Growth, Net Growth, Rate, Mention, Post, Follower Engagement, Gender, Timing, Country, and Geographic. Each factor represents specific metrics contributing to the cumulative variance explained. Several variables showed significant impacts on the influencer discovery tool score. These include Engagement, Share, Reaction, Growth, Rate, Follower Engagement, Timing, and Country. Each of these variables demonstrated varying degrees of significance and impact on the dependent variable, highlighting their importance in influencing influencer visibility and engagement. The correlation matrix provided further insights into the interrelationships among different aspects of influencer performance. Strong positive correlations were observed between Engagement, Share, Reaction, Growth, Follower Engagement, and Country, indicating their interconnectedness in driving influencer success. Additionally, the moderate to weak positive correlations with other variables emphasize their roles in influencing overall performance metrics. In summary, the findings underscore the critical role of engagement metrics, platform-specific strategies, and influencer behaviours in driving visibility and success on social media platforms. Understanding these relationships can guide marketers and influencers in optimizing their strategies for higher engagement and impact.

References

1. Aral, S., Dhillon, P.S.: Social influence maximization under empirical influence models. Nat. Hum. Behav. **2**, 375–382 (2018). https://doi.org/10.1038/s41562-018-0346-z
2. Barman, H.: Use of social media by top indian business schools and engineering institutes—a comparative study. Vanijya **29**, 34–47 (2020)
3. Belanche, D., Flavián, M., Ibáñez-Sánchez, S.: Followers' reactions to influencers' Instagram posts. Span. J. Mark.—ESIC **24**(1), 37–54 (2020). https://doi.org/10.1108/sjme-11-2019-0100
4. Bochenek, L.M.: Connecting with the right audiences for a better impact—imperatives of the influencer marketing. Advocacy Organ. Engag. 91–110 (2019). https://doi.org/10.1108/978-1-78973-437-920191006
5. Bokunewicz, J.F., Shulman, J.: Influencer identification in Twitter networks of destination marketing organizations. J. Hosp. Tour. Technol. **8**(2), 205–219 (2017). https://doi.org/10.1108/jhtt-09-2016-0057
6. Casaló, L.V., Flavián, C., Ibáñez-Sánchez, S.: Influencers on Instagram: antecedents and consequences of opinion leadership. J. Bus. Res. **117**(117), 510–519 (2020). https://doi.org/10.1016/j.jbusres.2018.07.005
7. Chang, S.-C., Wu, J.: A content-based metric for social media influencer marketing. Ind. Manag. Data Syst. **124**(1) (2023). https://doi.org/10.1108/imds-05-2022-0303
8. Cheung, V.S.Y., Lo, J.C.Y., Chiu, D.K.W., Ho, K.K.W.: Evaluating social media's communication effectiveness on travel product promotion: Facebook for college students in Hong Kong. Inf. Discov. Deliv. **51**(1) (2022). https://doi.org/10.1108/idd-10-2021-0117
9. Chung, S.: Instagram influencer marketing: why you need micro-influencers. Mention (2017). https://mention.com/en/blog/instagram-influencer-marketing
10. Finney, S.L., Penney, M.: Escape outdoors: evaluating social media with Davey and sky. Case J. **20**(2), 311–331 (2023). https://doi.org/10.1108/tcj-02-2023-0026
11. Geng, R., Chen, X., Wang, S.: Wear in or wear out: how consumers respond to repetitive influencer marketing. Internet Res. (2023). https://doi.org/10.1108/intr-01-2022-0075
12. Gulati, S.: Social and sustainable: exploring social media use for promoting sustainable behaviour and demand amongst Indian tourists. Int. Hosp. Rev., ahead-of-print(ahead-of-print) (2021). https://doi.org/10.1108/ihr-12-2020-0072
13. Haenlein, M., Anadol, E., Farnsworth, T., Hugo, H., Hunichen, J., Welte, D.: Navigating the new era of influencer marketing: how to be successful on Instagram, TikTok, & co. Calif. Manage. Rev. **63**(1), 5–25 (2020). https://doi.org/10.1177/0008125620958166
14. Hair, J.F., Black, W.C., Babin, B.J., Anderson, R.E.: Multivariate Data Analysis, 7th Ed, Pearson, New York (2010)
15. Iqbal Khan, S., Ahmad, B.: Tweet so good that they can't ignore you! Suggesting posting strategies to micro-celebrities for online engagement. Online Inf. Rev., ahead-of-print(ahead-of-print) (2021). https://doi.org/10.1108/oir-08-2020-0334
16. Litterio, A.M., Nantes, E.A., Larrosa, J.M., Gómez, L.J.: Marketing and social networks: a criterion for detecting opinion leaders. Eur. J. Manag. Bus. Econ. **26**(3), 347–366 (2017). https://doi.org/10.1108/ejmbe-10-2017-020

17. Lou, C., Tan, S.-S., Chen, X.: Investigating consumer engagement with influencer- versus brand-promoted ads: the roles of source and disclosure. J. Interact. Advert. **19**(3), 1–18 (2019). https://doi.org/10.1080/15252019.2019.1667928
18. Lykke, S., Penttinen, V., Pedersen, S.: Virtual influencer marketing: the good, the bad and the unreal. Eur. J. Mark. **58**(2) (2023). https://doi.org/10.1108/ejm-12-2022-0915
19. Mandler, T., Johnen, M., Gräve, J.-F.: Can't help falling in love? how brand luxury generates positive consumer affect in social media. J. Bus. Res. 120 (2019). https://doi.org/10.1016/j.jbusres.2019.10.010
20. Molina-Prados, A., Muñoz-Leiva, F., Prados-Peña, M.B.: The role of customer brand engagement in the use of Instagram as a "shop window" for fashion-industry social commerce. J. Fash. Mark. Manag.: Int. J. **26**(3), 495–515 (2021). https://doi.org/10.1108/jfmm-12-2020-0275
21. Muñoz-Expósito, M., Oviedo-García, M.Á., Castellanos-Verdugo, M.: How to measure engagement in Twitter: advancing a metric. Internet Res. **27**(5), 1122–1148 (2017). https://doi.org/10.1108/intr-06-2016-0170
22. Nikolinakou, A., King, K.W.: Viral video ads: emotional triggers and social media virality. Psychol. Mark. **35**(10), 715–726 (2018). https://doi.org/10.1002/mar.21129
23. Pashootanizadeh, M., Khalilian, S.: Application of the AIDA model. Inf. Learn. Sci. **119**(11), 635–651 (2018). Emerald. https://doi.org/10.1108/ils-04-2018-0028
24. Phua, J., Jin, S.V., Kim (Jay), J.: Uses and gratifications of social networking sites for bridging and bonding social capital: a comparison of Facebook, Twitter, Instagram, and Snapchat. Comput. Hum. Behav. **72**(1), 115–122 (2017). https://doi.org/10.1016/j.chb.2017.02.041
25. Rahim, F., Binti, N.: Consumer behaviour, perception and planning towards halal marketing. Adv. Islam. Financ., Mark., Manag. 271–307 (2016). https://doi.org/10.1108/978-1-78635-899-820161014
26. Rishi, B., Popli, S.: Getting into the customers, shoes: customer journey management. Crafting Cust. Exp. Strat. 21–45 (2021). https://doi.org/10.1108/978-1-83909-710-220211002
27. Rohde, P., Mau, G.: "It's selling like hotcakes": deconstructing social media influencer marketing in long-form video content on youtube via social influence heuristics. Eur. J. Mark. **55**(10) (2021). https://doi.org/10.1108/ejm-06-2019-0530
28. Saikia, P., Barman, H.: A systematic analysis of content structural efficiency for estimating higher educational institutions' engagement over Facebook. Int. J. Online Mark. **12**(1) (2022). https://doi.org/10.4018/ijom.299396
29. Saikia, P.: Social media content and communication development of higher educational institutions. Unpublished PhD thesis, Dibrugarh University (2023)
30. Schouten, A. P., Janssen, L., Verspaget, M.: Celebrity versus influencer endorsements in advertising: the role of identification, credibility, and product-endorser fit. Int. J. Advert. **39**(2), 258–281 (2020). Tandfonline. https://doi.org/10.1080/02650487.2019.1634898
31. Smith, A.N., Fischer, E., Yongjian, C.: How does brand-related user-generated content differ across YouTube, Facebook, and Twitter? J. Interact. Mark. **26**(2), 102–113 (2012). https://doi.org/10.1016/j.intmar.2012.01.002
32. Tian, S., Su Yeon Cho, Jia, X., Sun, R., Tsai, W.S.: Antecedents and outcomes of Generation Z consumers' contrastive and assimilative upward comparisons with social media influencers. J. Prod. Brand. Manag. **32**(7) (2023). https://doi.org/10.1108/jpbm-02-2022-3879
33. Wei, J.: Exploring the evolution of consumer attitude from followers to brand enthusiasts: an experiential learning perspective on social media. J. Cetacean Res. Manag. (2024). https://doi.org/10.1108/qmr-04-2023-0055
34. Wilson, L.: 30-Minute Website Marketing: A Step By Step Guide. Emerald Publishing Limited (2019)
35. Wolf, K., Archer, C.: Public relations at the crossroads. J. Commun. Manag. **22**(4), 494–509 (2018). https://doi.org/10.1108/jcom-08-2018-0080
36. Yetimoğlu, S., Uğurlu, K.: Influencer marketing for tourism and hospitality. Emerald Handb. ICT Tour. Hosp. 131–148 (2020). https://doi.org/10.1108/978-1-83982-688-720201009

Ethical Dimensions of AI Integration in Influencer Campaigns: Balancing Authenticity, Targeting Precision, and User Privacy

Ashwani Panesar and Indu Saini

Abstract Modern age promotional campaigns have redefined the entire marketing philosophy whereby it has undergone paradigm shift from traditional to social media. Influencer marketing being one of the most noticeable campaign as it helps for better in-depth penetration and far-reaching impact. With the integration of AI, these social media campaign influencers have impacted the entire consumer behaviour through enhanced interaction. AI is making it harder to reconcile authenticity, targeting, and user privacy in influencer marketing, raising ethical concerns. In a time when data analytics and artificial intelligence are becoming increasingly significant, it is essential for brands and influencers to strike a balance between the two competing goals of information and business. This chapter also aims to integrate artificial intelligence in an ethical manner into influencer marketing. In a digital ecosystem that is always evolving, it recommends that firms, influencers, and developers of AI gives user empowerment, transparency, and consent the highest priority. Additionally, the chapter delves into the ways in which AI-driven influencer marketing is affecting the dynamics of user privacy, as well as the ethical challenges that are associated with AI-driven influencer communications. The approach presented in this chapter will help practitioners to develop ethical frameworks for influencer marketing and artificial intelligence to coexist for better customer engagement.

Keywords Artificial intelligence · Customer engagement · Social media · Influencer marketing

A. Panesar (✉) · I. Saini
Mittal School of Business, Lovely Professional University, Phagwara, India
e-mail: panesarashwani@gmail.com

I. Saini
e-mail: indusaini0013@gmail.com

S. Dutta et al. (eds.), *Advances in Data Analytics for Influencer Marketing: An Interdisciplinary Approach*, Information Systems Engineering and Management 9,
https://doi.org/10.1007/978-3-031-65727-6_24

Opening Case

Ramesh (Fictious Name), a working professional in FMCG sector, has been quite active on the social media as the job demands him to stay active on the media. He regularly posts on the various social media platforms to promote the company products. But lately he is observing the number of responses from the media is not so favourable as they are not turning into sales. The followers are shifting to other forms of connectivity to get more insights of the products. He found that technology has redefined the marketing as customers are getting more dependent on influencer marketing for all forms of discussion. He also noticed that some MNCs have also signed some of the influencers to promote their products. He has also seen that influencer like Priyanka Chopra, Deepika Padukone, Ranveer Singh etc. have lots of followers and the money they charge for the posts they make. He observed that the reach, impact and the business leverage of them is very high. He also wrote to company with few names as influencers but some word of caution is also written. The concerns are.

- *Adoption of influencer marketing is the only way to salvage the pride of the Company.*
- *Ethical implications of the products they promote.*
- *Privacy of data of the customers.*

The answer to these concerns is being taken care and are answered in the forthcoming sections of the chapter.

1 Introduction: Evolution and Insight of Influencer Marketing

Modern concepts like 'smart globalization' which encourages companies to consider local perspectives while operating on a global scale, connecting their commitment to adapting to the specific demands of each local market [20]. Consumers express growing scepticism toward brands' direct marketing and are less inclined to trust traditional advertising. As, traditional marketing practices involve designing the messages and reaching to the target consumers and push them with the use of marketing channels and various sales promotional tools. But with the better digitalization and internet penetration, the entire promotional campaigns have undergone paradigm shift. Changes in consumer behaviour have also propelled the online influencer marketing. Furthermore, shortened economic cycles have led many companies to be more mindful of their marketing expenses. on advertising on various modes with the use of celebrities [15]. In the current scenario, the landscape of the marketing communication continues to evolve, influencer marketing being the new buzz word. The evolution of influencer marketing has been discussed below for better understanding (Fig. 1).

- *The Pioneers*: Since the eighteenth century, marketers have utilized influential individuals to promote their products. One example is Josiah Wedgwood, a British potter whose cream-colored pottery caught the attention of Queen Charlotte in 1765 and earned him the official title of "Her Majesty's Potter."
- *Fictional Characters*: A new generation of fictitious characters became influential in the next age. The most well-known instance is the 1932 popularization of Santa Claus by Coca-Cola.
- *Star Endorsements*: Next were the endorsements from famous people. Celebrities, as opposed to made-up characters like Santa and Mikey, are actual people with real preferences and large fan bases as a result of their notoriety. As a result, they are able to persuade people to purchase the goods they recommend. For this reason, companies like Pepsi and Nike began partnering with celebrities to promote their goods in exchange for media attention.

Reality television has effectively blended real life and screen life, as evidenced by reality TV programs such as The Bachelor and Keeping Up with The Kardashians. Whether we care to acknowledge it or not, many of us are captivated by the larger-than-life figures on reality shows. Due to their claims of representing "reality," the stars are viewed as somewhat more relatable and genuine compared to conventional celebrities. In the age of social media platforms such as Facebook, Instagram, Twitter, and YouTube, many individuals seized the opportunity to share their daily lives online. Not only did well-known personalities like celebrities and reality TV stars gain substantial followings due to their existing fame, but a new trend emerged as

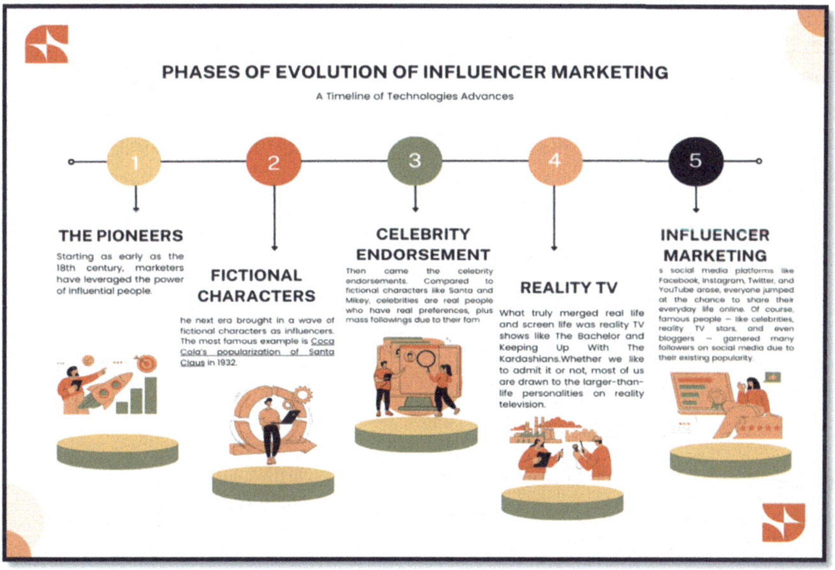

Fig. 1 Evolution of influencer marketing

ordinary people also amassed large numbers of followers through captivating content and close.

Traditional influencer marketing relied heavily on human influencers to market products or services to target audiences [16]. However, with the integration of AI technologies, marketers now have access to a wide range of tools and platforms that can enhance their influencer marketing campaigns. These AI-powered tools enable marketers to identify and connect with the most relevant influencers, analyse their audience demographics and engagement metrics, and even automate certain aspects of the influencer marketing process [25]. Influencer marketing refers to collaboration between well-known social media influencers and businesses to endorse their products or services [12].

Influencer marketing involves a company enlisting and motivating influencers to interact with their social media followers, aiming to use the influencers' special resources to advertise the company's products or services, in order to ultimately improve business performance [19]. The aims is to provide a comprehensive overview of the key antecedents and their interplay in shaping the effectiveness of influencer marketing. To begin, the ability to induce attitudinal or behavioural change in others, that is, influence, is a form of social power. Table 1 discusses various types of influencer marketers.

As AI continues to refine the identification of micro and nano influencers, brands will be able to tap into highly targeted niche communities, fostering genuine connections and trust with their audiences [25]. This shift towards smaller, more specialized influencers also presents an opportunity for cost-effective campaigns, allowing brands to allocate resources more efficiently.

Moreover, integration of AI in influencer campaigns has emerged as a game-changer. Among all the tools, AI being the one which is largely contributing by helping to collect the user data; deliver ads and suggestions in a personalized way [24]. It is essential for brands to master the consumer persona so that firms can have a firm control on who their targeted audiences so as to provide customers with what they want.

Table 1 Categorization of influencers

Type of influencers	Characteristics	Category of people
Nano	10,000 followers	Friends, family, or colleagues
Micro	10,000–50,000	Experts in their domain
Medium	50,000–100,000	Experts in their domain
Macro	More than 500,000	Experts in their domain
Mega	Over one million followers	Celebrities, athletes or high-profile personalities

Source Mckinsey & Company 2023

2 The Role of AI in Influencer Marketing

The participation of AI in influencer Marketing is growing more and more complex important as businesses seek to maximize the effectiveness and efficiency of their campaigns [9]. AI has the potential to revolutionize influencer marketing by providing valuable insights and automation tools that can streamline the entire process.

AI has transformed all forms of communication with *Miquela, described as "a 19-year-old robot living in LA," is a popular example of an AI influencer with 2.7 million followers on Instagram. Her Instagram is full of selfies, travel snaps, and pictures of friends. She has endorsed BMW and has done a Calvin Klein ad with supermodel Bella Hadid.*

By leveraging AI, businesses can identify the most relevant and influential individuals in their target audience and develop personalized strategies to engage with them. Some of the features of the AI powered tools whereby influencer marketing can help like:

1. Identify possible influencers by considering factors like the *demographics of their audience, number of followers, level of engagement,* and how well they align with your brand.
2. Determine the optimal influencer partnerships by *analysing past campaign performance* and predicting the potential success of future collaborations.
3. Automate and optimize various aspects of influencer marketing, such as *content creation, campaign management, and performance tracking.*
4. Improve the authenticity and credibility of influencer marketing by detecting *fake followers, engagement, and sponsored content.*

Overall, AI in influencer marketing holds the promise of improving targeting, efficiency, and effectiveness, ultimately driving better ROI for businesses. With the continued advancements in AI and machine learning, the future of influencer marketing holds even more exciting possibilities [4]. One crucial area of growth is the increasing personalization of influencer collaborations. AI can help analyse vast amounts of data to match brands with influencers whose values and content align closely with their target audience, leading to more authentic and impactful partnerships.

Moreover, AI in influencer marketing has the potential to enhance customer trust and engagement. By utilizing AI, businesses can ensure that the content shared by influencers aligns with the interests and preferences of their target audience [25]. This personalized approach can lead to higher levels of trust and engagement, as consumers feel that the influencer truly understands and values their needs. By leveraging AI, businesses can greatly improve the effectiveness and efficiency of their influencer marketing campaigns. In addition, AI can also assist marketers in identifying relevant trends and patterns in influencer marketing. By analysing data from past campaigns and monitoring current trends, AI can provide marketers with valuable insights on what strategies are working and what influencer partnerships are driving the best results [5]. AI has the potential to revolutionize the field of influencer marketing by providing businesses with powerful tools for identifying, optimizing, and tracking influencer campaigns.

Modern form of AI is revolutionizing influencer marketing by enhancing targeting, efficiency, and effectiveness, improving authenticity and credibility, and enhancing customer trust and engagement [17]. By leveraging AI, businesses can optimize their influencer marketing campaigns to deliver personalized content that aligns with the interests and preferences of their target audience.

3 Overview of Artificial Integration (AI) in Influencer Campaigns

With better level of internet penetration coupled with the governments push for digitalization, the entire marketing campaigns have gone under paradigm shift [10]. Moreover, AI technology has further revolutionized the way brands connect with their target audiences by leveraging the power of influencers. By using AI algorithms, brands can now identify the most relevant influencers for their campaigns, analyse audience demographics and engagement metrics to help create more personalized content tailored to specific consumer preferences [25]. This integration allows for designing more efficient and effective influencer marketing strategies, as AI can automate various tasks such as influencer selection, content creation, and campaign tracking. AI-powered influencer marketing offers a major benefit in being able to recognize micro-influencers, who typically have a smaller but deeply involved audience, rather than depending exclusively on well-known macro-influencers. Micro-influencers are more likely to build stronger bonds with their followers and can assist brands in engaging with specific and targeted consumer groups. For start-ups and small businesses aiming to utilize influencer marketing cost-effectively, this approach can be particularly advantageous.

Furthermore, AI-powered analytics can provide valuable insights and data on the performance and impact of influencer campaigns by better reach and engagement among the target audience, ensuring that their message is delivered effectively. AI integration in influencer campaigns has also helped in better analysis of audience demographics and engagement metrics, resulting in more personalized and impactful content [9]. Furthermore, the use of AI-powered chatbots and virtual assistants is expanding the reach of influencer campaigns. These chatbots can engage with followers in real-time, answer questions, and even facilitate direct purchases, providing a seamless and interactive experience for consumers.

In today's competitive digital arena, influencer marketing has become a significant strategy for brands to connect with their target audiences [27]. By leveraging AI technology, brands can now identify the most relevant influencers for their campaigns, analyse audience demographics and engagement metrics. As a result, brands can maximize their reach and engagement with the target audience, ensuring that their message is delivered effectively [17]. The image below provides deeper insight of the

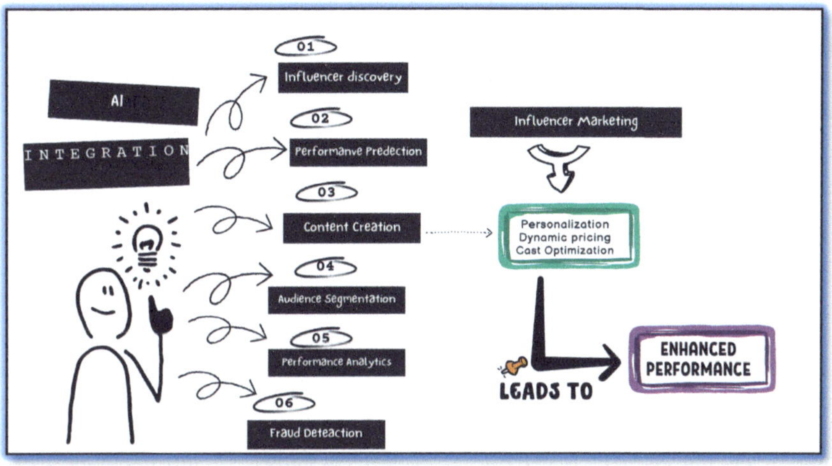

Fig. 2 AI integration and performance

various determinants of AI integration with influencer marketing to bring enhanced performance (Fig. 2).

Moreover, AI-powered personalized engagement marketing strategies can enhance brand loyalty and customer satisfaction [13]. AI integration in influencer campaigns revolutionizes the marketing landscape by maximizing reach and engagement with the target audience [21]. It also opens up opportunities for hyper-personalization and precision targeting. With AI-powered analytics, brands can gain deeper insights into consumer behaviour and preferences, allowing them to create more tailored and relevant content that resonates with the audience on a deeper level. Moreover, AI can facilitate real-time optimization of influencer campaigns by continuously analysing and adapting to evolving trends and audience sentiments. This dynamic approach ensures that brands stay ahead of the curve and maintain relevance in a rapidly changing digital landscape. Micro-influencers have become a valuable approach for brands looking to connect with specific audiences and build genuine relationships. Despite their smaller reach, they offer increased credibility and the capability to focus on particular consumer groups. As startups search for budget-friendly marketing options, the adaptability of micro-influencers is becoming more attractive.

Also, by understanding consumer preferences at a granular level, brands can craft compelling narratives that foster brand loyalty and emotional connections with their target audience [3]. In brief, the incorporation of AI into influencer campaigns transforms the field of marketing by enhancing outreach and interaction with the intended audience. AI in influencer campaigns has transformed marketing by increasing reach and engagement, delivering personalized content, optimizing in real-time, and nurturing long-term audience relationships. The integration of AI in influencer campaigns revolutionizes the marketing landscape by allowing brands to maximize reach, engagement, and personalization with the target audience. In summary, the integration of AI in influencer campaigns represents a paradigm shift in marketing, offering the potential for brands to not only connect with their audience more effectively to build more meaningful relationships and engagement.

4 Balancing Authenticity, Targeting Precision, and User Privacy

IT field of advertising has been transformed by online behavioural advertising and advanced data aggregation, revolutionizing how companies connect with their target audience. This shift necessitates a delicate balance between authenticity, precise targeting, and user privacy. While using personal information enables more accurate targeting and tailored ads, it also raises concerns about privacy infringement.

(i) ***Authenticity versus Personalization versus Transparency***

Authenticity and personalization being the two essential aspects of AI driven campaigns that have become more prominent in various industries. Influencers build credibility with their followers by projecting a sense of authenticity that surpasses traditional marketing agencies and brand promotions. Both authenticity and personalization are integral components of effective AI systems across various domains. In the realm of communication and marketing, AI is shaping the way personalized engagement is achieved. AI has brought about a paradigm shift in the way businesses approach customer engagement and communication. The ability of AI to deliver personalized experiences at scale has revolutionized the way companies interact with their customers. However, as AI continues to advance, the question of authenticity versus personalization has emerged as a critical consideration for businesses.

(a) ***Authenticity***: Authenticity arises from the perception that an individual acts in accordance with their genuine self [2]. Authenticity refers to an idea that an influencer's content is genuine, honest, and representative of their true self and beliefs. To succeed ethically in advertising, prioritizing this balance is essential. Companies can achieve this by ensuring authentic brand representation and precise targeting without compromising user privacy.

With the consumers shifting their focus to online shopping, companies are having tonnes of data in the context of consumer behaviour. The challenge lies in delivering valuable ads to consumers while respecting their privacy. The use of personal data and advanced algorithms in marketing presents a moral dilemma-offering efficient campaigns but raising ethical concerns about privacy. Marketers must prioritize the ethical use of data to navigate this tension effectively. Marketers are advised to commit actively to ethical behaviour for avoiding restrictive legislation and maintaining consumer trust—considering the ethics involved when creating an authentically balanced ad environment is crucial [11].

(b) ***Transparency***: Transparency with consumers regarding data collection/use can build trust & respect their privacy; robust security measures should protect users' information. Supporting responsible data practices will become a competitive advantage over time. By prioritizing authenticity, precision targeting, and user privacy, a sustainable trustworthy ad ecosystem benefiting businesses and consumers can be created [22]. Marketers must address and acknowledge the ethical dimensions of their practices concerning consumer privacy.

(ii) ***Balancing the Interrelationships of Authenticity, Targeting and Privacy***

One effective way for companies to balance authenticity, targeting precision, and user privacy is through the implementation of ethical advertising practices. Marketers and advertisers can start by developing clear and easily accessible privacy policies that detail the collection, usage, and protection of consumer data for them with the option to control their data and privacy settings. Moreover, companies can prioritize the use of anonymized and aggregated data for advertising purposes, rather than relying on personally identifiable information [6]. This approach allows for effective targeting while minimizing the risk of privacy intrusion. In this context, the balance between authenticity and personalization in AI is crucial. While personalization enhances user experiences by catering to individual preferences, authenticity ensures that the interactions feel genuine and sincere. Achieving this balance is essential for building trust and establishing meaningful connections with customers [2]. Moreover, the debate between authenticity and personalization in AI extends beyond the realm of marketing and communication. It also has implications for ethical considerations, data privacy, and the overall impact of AI on society. As businesses navigate the complexities of AI integration, understanding the interplay between authenticity and personalization will be paramount in driving meaningful and sustainable AI strategies [21].

Researchers have found there has been a lack of initiative to incorporate safeguarding principles into AI innovations as such integration is crucial to prevent children from being exposed to biased content based on factors such as ethnicity, or to harmful content. AI has undeniably revolutionized the way businesses engage with their customers. The ability to deliver personalized experiences at scale has opened up new avenues for customer interaction. Finding the balance between authenticity and personalization in AI is not just about technology, but also about understanding human behaviour and emotions [13]. While personalization can enhance user experiences by catering to individual preferences, authenticity ensures that the interactions feel genuine and sincere. Achieving this delicate balance is essential for building trust and establishing meaningful connections with customers.

The debate on authenticity and personalization in AI goes beyond just marketing and communication. It delves into deeper ethical considerations, data privacy, and the wider impact of AI on society. Businesses must navigate these complexities as they integrate AI into their operations, and understanding the interplay between authenticity and personalization will be crucial in driving meaningful and sustainable AI strategies. As businesses keep on investigating the potential of AI, it is vital that they remain mindful of the human aspect of technology. The ethical use of personal data and the impact of AI on individuals and society must be central considerations in the development and implementation of AI systems while maintaining authenticity and respect for privacy.

In order to build trust and maintain customer loyalty, organizations must prioritize the ethical use of technology and address the ethical dilemmas that arise with disruptive technologies [21]. While personalization in AI has undoubtedly improved user experiences and engagement, there are potential downsides to consider. One of the main concerns with personalization in AI privacy invasion occurs as AI systems gather and analyse extensive personal data to deliver customized experiences, there is a risk of infringing on individuals' privacy rights. This can lead to ethical dilemmas and concerns about the misuse of personal information for commercial gain. Furthermore, the emphasis on personalization in AI can lead to filter bubbles and echo chambers, where individuals are only exposed to information and content that aligns with their existing beliefs and preferences. Transformational growth in the business will only take place by funding "Business of Today" projects will always beat investing in "Business of Tomorrow" opportunities, this being logical, the returns are clear, and the risk is understood [20]. This can narrow their worldview and limit exposure to diverse perspectives, hindering critical thinking and the ability to consider alternative viewpoints (Fig. 3).

To further enhance user privacy, marketers and advertisers can also engage in proactive measures such as obtaining explicit consent from users before utilizing their personal data for targeted advertising. This not only demonstrates respect for user privacy but also encourages a more trusting relationship between consumers and businesses. Additionally, investing in consumer education about data privacy and the implications of targeted advertising can help foster a more informed and empowered consumer base [18]. By promoting digital literacy and privacy awareness, companies can contribute to a more ethical and respectful advertising environment. By integrating these ethical advertising practices, companies can achieve a harmonious balance between authenticity, targeting precision, and user privacy [1], thereby fostering a more sustainable and trustworthy advertising ecosystem.

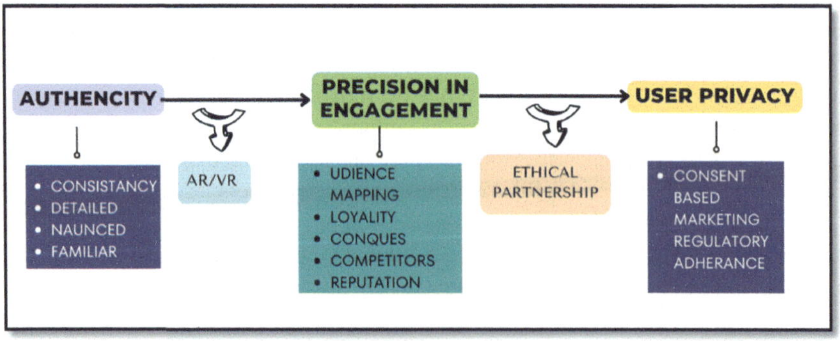

Fig. 3 Determinants of multi-attributes influence of marketing

In addition, the pursuit of personalization in AI may overshadow the importance of authenticity. Genuine human connections and interactions can be compromised when AI is focused solely on catering to individual preferences. There is a risk of losing the human touch and sincerity in interactions, leading to a sense of detachment and lack of trust in AI-driven communications. Moreover, the reliance on personalization in AI may lead to a loss of serendipity and unexpected discoveries. By tailoring experiences to individual preferences, there is a diminished opportunity for individuals to encounter new and diverse content, thereby limiting their potential for exploration and growth [8]. As businesses continue to harness the power of AI for personalized experiences, it is essential to critically evaluate the potential trade-offs between personalization and authenticity. Finding a harmonious balance that respects privacy, fosters diverse perspectives, and preserves genuine human connections will be critical in shaping the future of AI.

5 Maintaining Authenticity in Futuristic Influencer Campaigns

In this modern era of technology, influencer marketing (IM) has emerged as an effective strategy for brands to engage with their desired customers. However, as the influencer marketing landscape is becoming more crowded, the brands need to maintain authenticity in their campaigns so as to be distinct. One innovative way to achieve this is by leveraging AI for personalized content and targeting which in today's concept is called as hyper-personalization. By utilizing AI technology, brands can create more authentic and tailored influencer campaigns that truly resonate with their audience [9]. AI tools not only empower brands to create personalized content for influencer campaigns, but they also enable more targeted and effective audience segmentation which increases the chances of reaching the right audience but also ensures that the content being created and shared by influencers aligns with their audience's interests and needs. By analysing vast amounts of data, AI can identify patterns and preferences of the target audience, allowing brands to create more relevant and engaging content. All the said needs to be done in the light of the ethical principles as these practices could be the testimony to the fact that good business also make good ethics [20].

In addition to this, AI also offers the capability to measure the impact and success of influencer campaigns. By analysing metrics such as engagement, reach, and conversion rates, AI can provide valuable insights into the effectiveness of different influencer partnerships and content strategies [23]. This data-driven approach enables brands to make informed decisions and continuously refine their influencer marketing efforts to stay authentic and resonant with their audience.

Moreover, AI can also assist in influencer selection by identifying the best fit for a brand based on the influencer's audience demographics, interests, and engagement patterns. This ensures that the collaborations are genuine and that the influencers' values align with the brand's identity. Furthermore, AI-driven sentiment analysis can help brands monitor the perception of their influencer campaigns in real time, allowing for quick adjustments and mitigating potential authenticity challenges. This proactive approach can help safeguard the brand's reputation and maintain authenticity in the ever-evolving landscape of influencer marketing. By integrating AI into influencer marketing strategies, brands not only enhance the authenticity of their campaigns but also elevate the overall effectiveness and impact of their efforts. With the use of AI in influencer marketing, brands can create personalized content that resonates with their audience.

The integration of AI into influencer marketing strategies offers brands the opportunity to not only enhance the authenticity of their campaigns but also elevate the overall effectiveness and impact of their efforts. By leveraging AI for personalized content creation, targeted audience segmentation, and performance measurement, brands can ensure that their influencer campaigns resonate with their audience and align with their values. Furthermore, AI-driven sentiment analysis allows for real-time monitoring and quick adjustments, enabling brands to maintain authenticity and safeguard their reputation in the ever-evolving landscape of influencer marketing. Overall, the use of AI in influencer marketing presents a powerful tool for brands to create genuine, tailored, and impactful campaigns that genuinely connect with their audience.

6 Ethical Dimensions of AI

Even if companies are collaborating various AI based tools to augment the customer experiences but it also raises ethical concerns that needs immediate attention. A CEO of a leading digital marketing agency once said, that "he believed it being their responsibility to navigate the ethical implications of AI, ensuring accountability and transparency throughout our operations". With the companies investing in digital marketing as the cost of the marketing needs to be controlled and the reach to the increased but it brings out some of the ethical issues governing the entire process [7]. The image below throws an insight as to how the ethical concerns related to the influencer campaigns can be controlled to larger extent (Fig. 4).

One of the key components of transparency in AI is the ability to understand the inner workings of machine learning models. This involves not only providing access to the source code and algorithms, but also explaining how these models make decisions. For example, in the case of AI used in loan approval processes, transparency would involve disclosing the specific data points and variables that are considered in determining an individual's creditworthiness.

Data privacy plays a crucial role in the integration of AI systems. It is essential to ensure that personal information is protected and that individuals' privacy rights are respected. AI professionals should be mindful of the ethical implications surrounding the collection, use, and storage of personal data. They should prioritize transparency

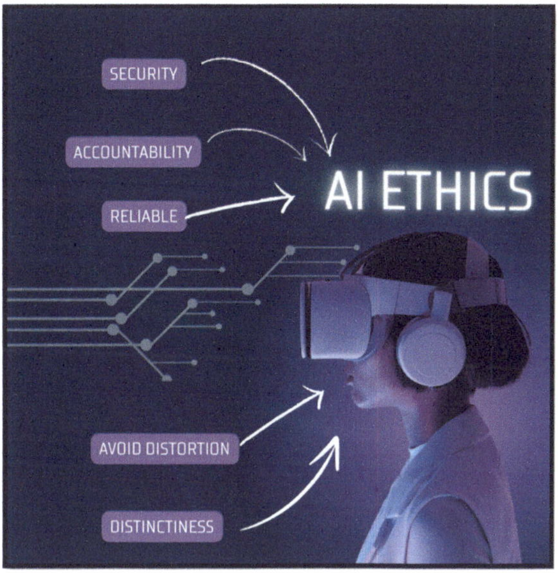

Fig. 4 Determinants of AI ethics

and accountability in their practices, and strive to adhere to best practices for data management.

Additionally, AI professionals should be aware of the potential impact their work may have on individual privacy. They should strive to minimize the risk of misuse or unauthorized access to personal data, especially in sensitive areas like surveillance. Furthermore, organizations should provide comprehensive privacy training to AI models and ensure robust data protection measures are in place throughout the entire data lifecycle. In summary, data privacy considerations are of utmost importance in the integration of AI systems. Moreover, disclosure in AI goes beyond just revealing the technical aspects of the system. It also involves informing users about the potential limitations, uncertainties, and risks associated with AI-generated decisions. This could include outlining the data sources used, the level of confidence in the predictions, and any known biases present in the training data.

7 Consumer Attitudes Towards AI in Influencer Marketing

With AI integration becoming more prevalent in businesses, understanding consumer sentiment will likely be useful in how companies make decisions regarding AI in the years ahead. AI technology is revolutionizing various industries, including influencer marketing. Consumers have mixed attitudes towards AI in influencer marketing [17]. Some consumers see AI as a valuable tool that can help brands and influencers create more personalized content and deliver targeted messages. They believe that

AI can enhance the overall consumer experience by providing relevant recommendations and streamlining communication between influencers and their audience. Other consumers, however, are sceptical of AI's role in influencer marketing [26]. They may feel that AI influencers lack the authenticity and genuine human connection that traditional influencers bring. These consumers may question the credibility of AI influencers and may be concerned about the ethical implications of using AI in marketing [27]. They may worry about issues such as transparency, data privacy, and the potential manipulation of consumer behaviour through AI-powered algorithms. Overall, consumer attitudes towards AI in influencer marketing are varied and depend on individual perceptions and beliefs about the role of technology in marketing [17]. Some of the key elements that influence the consumer behaviour in the light of the AI are:

(a) ***Attractiveness***: In consumer behaviour, attractiveness can refer to the perceived appeal of AI systems. This could include factors such as the design of AI interfaces, the friendliness of AI interactions, or the visual appearance of AI-powered products.
(b) ***Luminary***: A luminary in this context could refer to influential figures or organizations within the AI industry, whose products or endorsements may influence consumer perceptions and behaviours.
(c) ***Quality***: Quality is a crucial factor in consumer decision-making. In the context of AI, consumers may evaluate the quality of AI systems based on factors such as accuracy, reliability, and user experience.
(d) ***Trendiness***: Consumer behaviour is often influenced by trends and the desire to stay up-to-date with the latest technology. AI products and services that are perceived as trendy or cutting-edge may attract more consumer interest and adoption.
(e) ***Robophobia***: This term describes the fear or aversion towards robots or AI systems. In consumer behaviour, robophobia can influence attitudes towards AI products and services, affecting adoption rates and consumer preferences.

Moreover, the companies have been consistently using the AI the influence the consumer attitudes.

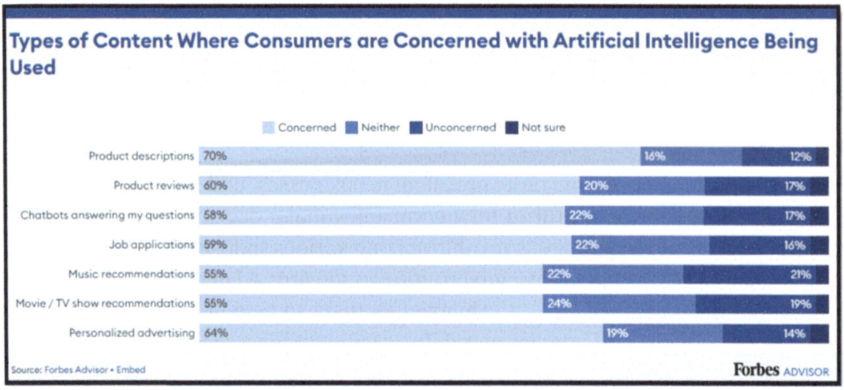

The above diagram explains whereby around 2000 surveyed people expressed their views on AI's impact on their attitude. The findings of the report indicate that 60% of the consumers were concerned of the product reviews given with the support of AI with majority being concerned with all elements. This creates an element of doubt in the behaviour of consumers about these campaigns [12].

8 Conclusion

The integration of AI in influencer campaigns presents a multifaceted ethical landscape that demands careful consideration and responsible action. The primary ethical dimensions include transparency, privacy, authenticity, accountability, and the potential for manipulation. Transparency is crucial in ensuring that audiences are aware of the use of AI in generating or curating content. Influencers and brands must disclose when AI tools are used to create content or engage with followers to maintain trust and integrity. Privacy concerns arise with the extensive data collection required for AI algorithms to function effectively. This data includes personal information about both influencers and their audiences, raising questions about data protection, consent, and the potential for misuse. Influencers must balance the efficiency and creativity that AI can offer with the need to maintain a genuine connection with their audience. Accountability for AI's actions in influencer campaigns is another significant ethical issue, as to see who is responsible for AI-generated content, particularly in cases of misinformation or harm, is complex and requires clear guidelines and regulations. In conclusion, while AI offers transformative potential for influencer campaigns, it is imperative to navigate its ethical dimensions with a commitment to transparency, privacy, authenticity, accountability, and fairness. By addressing these ethical concerns proactively, influencers and brands can harness the benefits of AI while upholding their moral responsibilities to their audiences and society at large.

References

1. Anshari, M., Hamdan, M., Ahmad, N., Ali, E., Haidi, H.: COVID-19, artificial intelligence, ethical challenges and policy implications. AI Soc. **38**(2), 707–720 (2023)
2. Audrezet, A., de Kerviler, G., Moulard, J.G.: Authenticity under threat: when social media infuencers need to go beyond self-presentation. J. Bus. Res. **117**, 557–569 (2020)
3. Babatunde, S.O., Odejide, O.A., Edunjobi, T.E., Ogundipe, D.O.: The role of AI in marketing personalization: a theoretical exploration of consumer engagement strategies. Int. J. Manag. Entrep. Res. **6**(3), 936–949 (2024)
4. De Bruyn, A., Viswanathan, V., Beh, Y.S., Brock, J.K.U., Von Wangenheim, F.: Artificial intelligence and marketing: pitfalls and opportunities. J. Interact. Mark. **51**(1), 91–105 (2020)
5. Chintalapati, S., Pandey, S.K.: Artificial intelligence in marketing: a systematic literature review. Int. J. Mark. Res. **64**(1), 38–68 (2022)

6. Christl, W., Kopp, K., Riechert, P.U.: How companies use personal data against people. In: Automated Disadvantage, Personalized Persuasion, and the Societal Ramifications of the Commercial Use of Personal Information. Cracked Labs, Wien, vol. 21 (2017)
7. Du, S., Xie, C.: Paradoxes of artificial intelligence in consumer markets: ethical challenges and opportunities. J. Bus. Res. **129**, 961–974 (2021)
8. Gillett-Swan, J.: The challenges of online learning: supporting and engaging the isolated learner. J. Learn. Des. **10**(1), 20–30 (2017)
9. Haleem, A., Javaid, M., Qadri, M.A., Singh, R.P., Suman, R.: Artificial intelligence (AI) applications for marketing: a literature-based study. Int. J. Intell. Netw. **3**, 119–132 (2022)
10. Hanna, N.: A role for the state in the digital age. J. Innov. Entrep. **7**(1), 5 (2018)
11. Jackson, K.T.: Building Reputational Capital: Strategies for Integrity and Fair Play that Improve the Bottom Line. Oxford University Press (2004)
12. Katherine Haan, Ms.: Over 75% Of Consumers are concerned about misinformation from artificial intelligence. www.forbes.com (2023)
13. Kumar, V., Ashraf, A.R., Nadeem, W.: AI-powered marketing: what, where, and how? Int. J. Inf. Manag. 102783 (2024).
14. Lamarre, E., Smaje, K., Zemmel, R.: Rewired: The McKinsey Guide to Outcompeting in the Age of Digital and AI. Wiley (2023)
15. Leung, F.F., Gu, F.F., Palmatier, R.W.: Online influencer marketing. J. Acad. Mark. Sci. **50**(2), 226–251 (2022)
16. Londaridze, D.: The impact of consumer skepticism and dispositional trust on attitudes toward display advertising and purchase intentions (Doctoral dissertation, Vilniaus universitetas) (2024)
17. Marinchak, C.M., Forrest, E., Hoanca, B.: Artificial intelligence: redefining marketing management and the customer experience. Int. J. E-Entrep. Innov. (IJEEI) **8**(2), 14–24 (2018)
18. McGregor, S.: Sustainable consumer empowerment through critical consumer education: a typology of consumer education approaches. Int. J. Consum. Stud.Consum. Stud. **29**(5), 437–447 (2005)
19. Mckinsey & Company.: What is influencer marketing? www.mckinsey.com, other (2023)
20. Panesar, A., Sood, R.: Corporations and stakeholders: managing businesses in turbulent times (An ethical dilemma). In: Smart Strategies and Societal Solutions for Sustainable International Business, pp. 112–131. IGI Global (2024)
21. Rane, N.: Enhancing customer loyalty through Artificial Intelligence (AI), Internet of Things (IoT), and big data technologies: improving customer satisfaction, engagement, relationship, and experience. In: Internet of Things (IoT), and Big Data Technologies: Improving Customer Satisfaction, Engagement, Relationship, and Experience (October 13, 2023) (2023)
22. Rathore, B.: Blockchain revolutionizing marketing: harnessing the power of distributed ledgers for transparent, secure, and efficient marketing practices. Int. J. New Media Stud.: Int. Peer Rev. Sch. Index. J. **6**(2), 34–42 (2019)
23. Samayamantri, L.S.: Cognitive affiliate platforms: revolutionizing marketing strategies through AI-driven intelligence. Int. Mach. Learn. J. Comput. Eng. **6**(6), 1–9 (2023)
24. Schafer, J.B., Konstan, J.A., Riedl, J.: E-commerce recommendation applications. Data Min. Knowl. Disc.Knowl. Disc. **5**, 115–153 (2001)
25. Senyapar, H.N.D.: Artificial intelligence in marketing communication: a comprehensive exploration of the integration and impact of AI. Tech. Soc. Sci. J. **55**, 64–81 (2024)
26. Şenyapar, H.N.D.: A comprehensive analysis of influencer types in digital marketing. Int. J. Manag. Adm. **8**(15), 75–100 (2024)
27. Wielki, J.: Analysis of the role of digital influencers and their impact on the functioning of the contemporary on-line promotional system and its sustainable development. Sustainability **12**(17), 7138 (2020)

GPSR Compliance

The European Union's (EU) General Product Safety Regulation (GPSR) is a set of rules that requires consumer products to be safe and our obligations to ensure this.

If you have any concerns about our products, you can contact us on ProductSafety@springernature.com

In case Publisher is established outside the EU, the EU authorized representative is:

Springer Nature Customer Service Center GmbH
Europaplatz 3
69115 Heidelberg, Germany

The manufacturer's authorised representative in the EU is Springer Nature Customer Service Centre GmbH, Europaplatz 3, 69115 Heidelberg, Germany. If you have any concerns regarding our products, please contact ProductSafety@springernature.com

Printed and bound by CPI Group (UK) Ltd, Croydon, CR0 4YY
01/10/2026
02225921-0017